Gut Remediation of Environmental Pollutants

Xiangkai Li • Pu Liu
Editors

Gut Remediation of Environmental Pollutants

Potential Roles of Probiotics and Gut Microbiota

Springer

Editors
Xiangkai Li
School of Life Sciences
Lanzhou University
Lanzhou, Gansu, China

Pu Liu
School of Life Sciences
Lanzhou University
Lanzhou, Gansu, China

ISBN 978-981-15-4761-4   ISBN 978-981-15-4759-1 (eBook)
https://doi.org/10.1007/978-981-15-4759-1

This Springer imprint is published by the registered company Springer Nature Singapore Pte Ltd.
The registered company address is: 152 Beach Road, #21-01/04 Gateway East, Singapore 189721, Singapore

Preface

> If the mountain will not come to Muhammad, then Muhammad must go to the mountain.
> (Francis Bacon, 1625)

As we are all aware that pollution is a global crisis that human race faces in the twenty-first century, there are overwhelming evidences that pollution has caused various health problems. For example, heavy metals are one of the major causes of increasing cancer rate, especially among the young population in China. Governments, societies, organizations, scientists, and individuals have paid very close attention to this issue. In the last two decades, a large number of human and financial resources have been devoted to environmental protection and remediation. As a research scientist in this field, these remedies have proven not be very effective. Our current strategy is focusing on the environments, including air, water, and soil. Physical, chemical, biological, and combined remediation approaches are able to reduce contaminants at small or medium scale. However, the scope of contamination is too huge. It is estimated that 20% of the agricultural soil is contaminated with heavy metals and the remediation capacity can only cover a very small portion of it. Organic contaminants, heavy metals, other emerging contaminants, etc. enter everyone's daily life through air, water, and food. And the situation is getting worse.

More than 100 years ago, Metchnikoff suggested manipulation of gut microbiota with probiotics can enhance human health, a fact that few people believed at that time but is proven over time. Most contaminants enter the human body through gut and lots of bacteria have the ability to remediate hazardous materials. Based on this, we proposed a novel approach against environmental contaminations, termed Gut Remediation. Compared to traditional remediation technology, gut remediation provides a new path for the protection of human health against pollutants. It is convenient *in vivo* because the functional gut microbiota can be enhanced by probiotic intervention, and probiotics colonize the gut in only a few weeks or a little longer. With the development of isolation of functional microbes, synthetic biology, and new microbial augmentation technology, gut remediation will further reduce pollutant accumulation and show a low-cost application in the future.

The multifaceted correlation of gut microbiota altered by environmental pollutants with human metabolism, nutrition, host physiology, and applications, have been discussed in this book, highlighting the importance for us to rationally manipulate the gut microbiota. Several reports indicated that coronavirus (COVID-19) can attack the human gut through ACE-2 protein and cause illness, which could be prevented by a healthy gut microbiota. We believe gut remediation can help us in lots of different ways and serve as an important adjunct to medical treatment in the near future.

Lanzhou, China
2020

Xiangkai Li

Acknowledgements

We sincerely thank all the individuals and companies for their support in attempting to theorize gut remediation and put it into practice. Thanks to every author who made this book possible. We also want to express our gratitude to Prof. Rajesh Sani of South Dakota School of Mines and Technology, Prof. Jidong Gu of Hong Kong University, Prof. Shuang-Jiang Liu of Chinese Academy of Sciences, Prof. Ning-Yi Zhou, Prof. Ping Xu of Shanghai Jiao Tong University, Prof. Gejiao Wang of Huazhong Agricultural University, and many other scientists from all parts of the world for their valuable suggestions and critics. This work is funded by Fundamental Research Funds for the Central Universities grant (No: lzujbky-2017-br01), Gansu province major science and technology projects (No: 17ZD2WA017), and National Natural Science Foundation Grant (No: 31870082). Lastly, we would like to thank Lanzhou University for the long-time support.

Contents

1 **Health Effects of Environmental Pollutants** 1
Toshiro Shigaki

2 **Gut Microbiota and Health** 31
Chunjiang Zhang, Amanpreet Kaur Virk, Israr Khan, and Haoran Qin

3 **Introduction to Probiotics and Their Potential Health Benefits** 81
Marwa M. El-Dalatony and Xiangkai Li

4 **Effects of the Bio-accumulative Environmental Pollutants on the Gut Microbiota** 109
Pengya Feng, Xingpeng Xiao, Tuoyu Zhou, and Xiangkai Li

5 **Environmental Pollutants That Can Be Metabolized by the Host (Gut Microbiota)** 145
Marwa M. El-Dalatony and Xiangkai Li

6 **Environmental Pollutants that Can Be Metabolized by the Host, but Would Be Harmful to Humans (e.g., Causing Cancers, etc.)** 169
Marwa M. El-Dalatony, Mostafa El-Sheekh, and Xiangkai Li

7 **Gut Remediation: Back to the Future** 199
Zhenmin Ling, Yiming Jiang, and Xiangkai Li

8 **Current Policies and Policy Implications for Environmental Pollution** 219
Huawen Han, Haiying Huang, and Xiangkai Li

Chapter 1
Health Effects of Environmental Pollutants

Toshiro Shigaki

1.1 Introduction

Environmental pollutants are everywhere surrounding us. They are in the food we eat every day, in the water we drink, and in the air we breathe in. They are also found in household cleaners, carpets, house dust, and mattresses. Such common occurrence of environmental pollutants will have harmful effects on our health. It is practically impossible to avoid each and every pollutant in our daily life. However, there are steps that you can follow to reduce the detrimental effects to live a healthier life. Oftentimes, there are also choices that keep the hazard to the minimum possible level.

Environmental pollutions can be both natural and anthropogenic. Natural causes include such events as forest fires and volcanic eruptions. Numerous foods contain natural toxins as part of the defense system in living organisms that we consume, such as animals, fish, plants, and mushrooms. Allergies can be caused by pollens from plants that are part of the ecosystem. Anthropogenic pollutants include pesticides, antibiotics, industrial products and by-products, and heavy metals from mining activities.

Human races, in their entire history, experimented on the foods they eat. Those who ate wrong types of food did not live long and left no offspring. In extreme cases, they simply died when the food they consume contained poisons. This process of selecting right foods that are available abundantly in their local area became the tribal knowledge and that guaranteed their survival. In a way, modern human beings are select elites who are the descendants of the people who ate the right foods for their survival and health.

T. Shigaki (✉)
Graduate School of Agricultural and Life Sciences, The University of Tokyo, Tokyo, Japan

Pacific Agriculture Alliance, Samukawa, Kanagawa, Japan
e-mail: tshigaki@ag-alliance.org

X. Li, P. Liu (eds.), *Gut Remediation of Environmental Pollutants*,
https://doi.org/10.1007/978-981-15-4759-1_1

In modern times, selecting right foods is becoming increasingly more difficult, because new synthetic chemical is made every day, and their short- and long-term effects are either hard to predict or unknown. Advancing technology enabled humans to develop and synthesize new materials and chemicals to make our life more convenient, healthier, and efficient. At the same time these same materials can be detrimental to us. Such negative effects of new materials and chemicals are often discovered long after they come into our life. By the time the effects are verified, the damage is already done. Smoking was once considered a healthy alternative to alcohol consumption as cigarette smoking did not cause hangover. It was even endorsed and recommended by medical doctors. The incidence taught us a valuable lesson, yet we may still be surrounded by many apparently benign materials with potentially harmful effects.

In this introduction, I would like to describe historical perspectives of environmental pollutants, suggest what can possibly happen, and what we should do to cope with the problems in the future.

1.1.1 Historical Perspectives

Environmental pollution is not new and has occurred throughout the human history, but especially since industrialization started. In the past, pollution was mainly of heavy metals and from biological sources. Some notable incidences in history are presented here to illustrate the impact environmental pollution poses on human life.

For example, in the eighteenth and nineteenth centuries, industrial workers used mercury nitrate, as part of the process of the animal furs, to produce felt for hats. Back then, the hazard of mercury nitrate was not well recognized, and the concept of occupational safety was nonexistent. The "hatters" thus developed a variety of physical and mental ailments, including speech difficulties, tremors, emotional problems, and hallucinations. The expression "mad as a hatter" was made as the problem became commonplace. Even earlier, there is evidence that Romans used lead acetate as a sweetener.

Cadmium was discovered by the German chemist Friedrich Stromeyer in 1817, and it changed the world of paintings forever, as this heavy metal produced vibrant yellow, orange, and red colors. Artists such as Paul Gauguin, Max Ernst, and Henri Matisse made extensive use of cadmium paints by taking advantage of the colors that cadmium made possible. It was also resistant to fading due to exposure to sunlight. However, artists have certainly inhaled the toxic cadmium powders and that have put them in health risks.

During the nineteenth century, another poisonous metal, arsenic, was used in wallpapers in bedrooms. Arsenic was known to create an array of bright colors, such as emerald green hue known as Scheele's Green. The use of such bedroom wallpapers resulted in illnesses and deaths, especially of young children.

Mycotoxin from toxigenic fungi has been a serious problem for humans since agriculture and storage of grains started approximately 10,000 years ago. Best documented mycotoxin cases are probably ergotism from Claviceps purpurea in rye and have been known over 2000 years. Ergotism caused the deaths of many thousands of people, mainly in Europe for the past 1000 years.

Beriberi is usually associated with vitamin B1 (thiamine) deficiency. However in Japan, a disease known since the seventeenth century as acute cardiac beriberi is caused by a mycotoxin from *Penicillium citreonigrum*. When the fungus infects rice grains, it makes them yellow. Since the sale of such moldy yellow rice was banned in 1910, the disease practically disappeared in Japan.

From these cases, we understand that industrial pollutions in a broad sense were part of the human history.

1.1.2 Modern Days

However, the scale of pollution is reaching a mass scale in every part of the world. Many mining activities caused environmental havoc all over the world, polluting the soils and waters with toxic heavy metals. Besides, mercury used to extract gold is a widespread source of pollution. In Bougainville, Papua New Guinea, a copper and gold mine called Panguna contaminated the surrounding environments and caused illnesses among the people who rely on water from the streams and crops grown on the local gardens. It eventually led to a civil war, resulting in the deaths of 10% of the population of Bougainville Province, where the mine was located. The pollution persisted long after the mine was permanently shut down. Besides, artisanal and small-scale gold mining make use of mercury and are being practiced long after the shutdown, This poses serious health and environmental problems.

Urbanization and industrialization complexed with advanced material production technology increased the consumption of materials and as a result waste discharge. To reduce the health risks stemming from these pollutions, first we must understand the nature of the polluting substances.

Environmental pollutants pose various hazards to health depending on the property of the material. The harmful effects include infant mortality, respiratory disorders, allergy, cancer, mental disorders, to mention a few. Long-term effects are more subtle to notice, but some pollutants increase the chance of certain cancers and other disorders.

1.1.3 Future Perspectives

With the development of new chemical and biological technologies, new environmental pollutants will arrive in our daily life. It is now easy to edit genes and genomes and even to create novel microorganisms and viruses. Such technology can be utilized to synthesize biological weapons. Inadvertent release of experimental materials of biological nature also constitutes risks and can be easily spread among the public.

Hazard of inhalation of nanomaterials is recently speculated. Animal studies demonstrated pulmonary effects such as inflammation, fibrosis, and carcinogenicity regarding some nanomaterials. Risk studies must be conducted to assess the potential danger, and if necessary, restriction on the use of nanomaterials must be considered.

In this chapter, we focus on most important environmental pollutants and explain their sources and mode of action, summarized from recent literatures. It is hoped that this chapter helps reduce the health risks by providing accurate accounts of pollutants.

1.2 Heavy Metals

By consuming various foods, we are inadvertently absorbing heavy metals into our body. The heavy metals react with our cell machinery in the manner that is harmful to the normal functions. Heavy metal poisoning occurs by natural causes. However, in modern days, it is usually associated with various forms of economic activities.

Heavy metals are present in soils and waters, as a result of industrial pollution, or mining activities. However, some soils and waters are naturally high in heavy metal content. Plants absorb heavy metals from roots and accumulate in their tissues. Subsequently, animals consume the plants and the metals are passed on to the meat that humans eat. Seas and freshwater are also contaminated with heavy metals. Fish accumulate heavy metals to dangerous levels for human consumption by the process called biological concentration. Therefore, careful monitoring of heavy metals in fish is required for dietary safety. The following is a list of heavy metals that are of concern to human health, with information on the mode of action and symptoms. In addition, we will describe a method to remediate heavy-metal polluted environment using live plants, termed phytoremediation, which is increasingly gaining popularity as it is both environmentally friendly and cost-effective.

1.2.1 Cadmium (Cd)

Cd is found in the earth's crust at about 0.1 part per million [1] and exists as impurity in zinc or lead deposits, thus is produced primarily as a by-product of zinc or lead smelting [2]. It is one of the most toxic elements that exist in nature. Cd is efficiently retained and accumulates in the human body throughout life [3]. The ionic radius of Cd is almost identical to that of Ca, which is an essential metal for many cellular functions. When the cell mistakes Cd for Ca, it blocks the signaling that Ca normally provides, causing toxic effects in our body.

Cd is primarily toxic to the kidney and it is the main site of accumulation [3]. Cd also causes bone demineralization [3]. Pollution by Cd is common in industrialized areas, and some foods, notably rice, can naturally accumulate Cd. Therefore, such long-term exposure to Cd can adversely affect the kidneys and bones of humans.

1.2.2 Arsenic (As)

Arsenic occurs in many minerals, usually in combination with sulfur and other metals. Arsenic is used mainly in alloying with lead. Lead components in car batteries are fortified by the presence of a very small amount of arsenic [4, 5]. The largest source of arsenic intake into human body is from food. Rice is known to accumulate As efficiently. Considering that rice is a staple food of most Asian populations, it is an important source of As exposure in the area. Long-term exposure to arsenic may alter the ability of cell functions. It could play a role in the development of diabetes, cancer, vascular disease, and lung disease.

Another source of As is water. The Environmental Protection Agency limits the amount of arsenic in U.S. public drinking water to 10 parts per billion (ppb). Water coming from wells may contain higher levels of arsenic, if the groundwater flows over arsenic-rich bedrock. In Bangladesh, As-contaminated well water caused mass sufferings of the people.

1.2.3 Chromium (Cr)

While trivalent chromium is a trace mineral that is essential to human nutrition, hexavalent chromium and its compounds are toxic when inhaled or ingested. Hexavalent chromium can be found in dyes and paints and in some products used in leather tanning. It is hemotoxic, genotoxic, and carcinogenic [6]. The signs and symptoms of chromium toxicity are fever, diarrhea, nausea, vomiting, renal failure, and severe gastrointestinal irritation or ulcers.

1.2.4 Nickel (Ni)

Nickel is a widely utilized metal as it is tolerant to corrosion. Cooking utensils, cellular phones, medical instruments, transportation, construction, power generation are only a few examples. As such, exposure to nickel occurs commonly in our daily activities.

The most common harmful effect of nickel is an allergic reaction to the skin of the sensitive people [7]. Nickel is also a potential immunomodulatory and immunotoxic agent in humans [8]. International Agency for Research on Cancer (IARC) classified nickel compounds except the nickel in the metallic form as carcinogenic to humans [9].

1.2.5 Mercury (Hg)

Mercury represents one of the four most hazardous heavy metals in the environment, along with cadmium, arsenic, and lead. In the1950s–1960s in Japan, mercury polluted seawater and fish, thereby causing massive poisoning of thousands of people in Kumamoto Prefecture (Minamata Disease). The symptoms of mercury poisoning may include muscle weakness, poor coordination, numbness in the hands and feet, skin rashes, anxiety, memory problems, trouble speaking, trouble hearing, or trouble seeing [10]. Mercuric compounds are more toxic than either the elemental form or the salts causing brain and liver damage. The most dangerous mercuric compound is dimethylmercury. A few microliters of dimethylmercury spilled on the skin can cause death [11, 12].

Currently, the use and sale of mercury is restricted by an international treaty, Minamata Convention on Mercury. However, illegal use still continues in small-scale gold mining operations (Fig. 1.1) to separate the gold from other materials. This process generates mercury vapor and contaminates atmosphere, soils, and water, and directly harm humans who handle the process.

Fig. 1.1 "Panning" the gold in defunct Panguna Mine, Bougainville, Papua New Guinea

1.2.6 Lead (Pb)

Lead is one of the four most poisonous heavy metals commonly found in the environment, other three being mercury, cadmium, and arsenic. The most sensitive human organ to lead is brain and the exposure to lead causes intellectual disability and behavioral problems [13]. Lead(II) acetate ($Pb(CH_3COO)_2$), also known as lead acetate, or Goulard's powder, is a white crystalline compound that tastes sweet. Therefore, it is especially hazardous to children who used to put lead acetate-containing paint in the mouth. Lead acetate is no longer used in the paint for this reason.

1.2.7 Aluminum (Al)

Aluminum is a common metal that is present in earth's crust. Nonetheless, aluminum has no known function in biology. Aluminum is not as toxic as other metals and not listed as a carcinogen by the United States Department of Health and Human Services.

The ingestion of aluminum was once suspected as a cause of Alzheimer's disease. However, no evidence has been found to prove the connection of aluminum with the disease [14].

Aluminum, however, can infrequently cause vitamin D-resistant osteomalacia, erythropoietin-resistant microcytic anemia, and central nervous system alterations. People with renal insufficiency are particularly at a risk [15].

1.2.8 Zinc (Zn)

Zinc is an essential trace element for humans, other animals, plants, and for micro-organisms. Zinc acts as a cofactor of over 300 enzymes and 1000 transcription factors. Although zinc is an essential element, excessive intake of zinc is toxic in humans by competing with copper and iron, causing the deficiency of these metals [16]. For example, a report has been published that elderly men taking 80 mg daily were hospitalized for urinary complications more often than those taking a placebo [17]. Some dietary supplements contain high amount of zinc. Therefore, dosage must be carefully monitored to avoid any adverse effects.

1.2.9 A Method of Decontamination of Polluted Soils: Phytoremediation

Soils and water polluted with heavy metals must be cleaned up before agriculture and other human activities are resumed. However, cleaning up contaminated soil is expensive. The estimated cost to remediate polluted sites in the EU alone is somewhere between €59 and €109 billion [18]. Besides, conventional methods of soil cleanup, such as replacing the contaminated soils with clean ones sourced from other locations, do not completely solve the problem. The removed soils must be stored elsewhere, which must be taken care of in the future. For this end, phytoremediation provides a cost-effective alternative, which is at the same time environmentally more friendly.

Phytoremediation is the technologies that use living plants to clean up soil, air, and water contaminated with hazardous materials, including heavy metals [19]. It is defined as "the use of green plants and the associated microorganisms along with proper soil amendments and agronomic techniques to either contain, remove, or render toxic environmental contaminants harmless" [20].

Phytoremediation utilizes a special group of plants termed hyperaccumulators. Hyperaccumulators are plants that are capable of growing in soil or water containing very high concentrations of metals, taking up these metals from their root system, and accumulating very high concentrations of metals in their tissues [21]. These plants have a selective advantage over ordinary (non-hyperaccumulators) plants on the soils with high metal content. Besides, it has been suggested that it is an effective strategy to avert herbivory or pathogen defenses by making the plants toxic to insects and animals feeding on the plants [22]. In hyperaccumulators, genes that are responsible for absorbing metals are upregulated. Expression of hyperaccumulation (HA) genes confers plants the ability to uptake and sequester metals such as As, Co, Fe, Cu, Cd, Pb, Hg, Se, Mn, Zn, Mo, and Ni, 100–1000 times the concentration found in non-hyperaccumulators [23, 24].

One good example of a hyperaccumulator is a fern *Pteris vittata* L., which can accumulate 27,000 mg of As per kg of tissue (fronds) [25]. In one experiment, 26% of As in the soil was removed after 20 weeks' plantation [26].

To remediate heavy-metal contaminated soils, first, appropriate hyperaccumulators are chosen based on the nature of the pollution. After the plants are grown and allowed to absorb heavy metals from the soils, the shoot systems are removed and processed elsewhere. In consideration of the safety of the wildlife, the hyperaccumulators should not be edible for local fauna. The advantage of phytoremediation is its cost-effectiveness and environmental friendliness. However, it is a long process and the problem of processing the shoot system containing high levels of heavy metals still exists.

1.3 Pesticides

Pesticide use is on the increase, as agriculture seeks to increase its outputs in response to the demand. Relatively safe pesticides have been developed in recent years. However, hazardous pesticides are still widely used. Organic farming is increasingly becoming popular, because it is a more sustainable way to produce agricultural products. Pesticides tend to kill the beneficial microorganisms in the soil and insects. As a result, continuous input of fertilizers and more pesticides is required, making the practice unsustainable. More targeted use of pesticides and the development of pesticides that are less harmful to environment will reduce the risk. Here, we list some of the most problematic pesticides and describe their effects on the health of human and other organisms.

1.3.1 Organochlorine

Organochlorine pesticides are synthetic pesticides used widely in the chemical industry and in agriculture. Organochlorines are known for their high toxicity, slow degradation, and bioaccumulation. Some examples of organochlorine are DDT, DDD, Dicofol, Eldrin, Dieldrin, Chlorobenzilate, Lindane, BHC, Methoxychloro Aldrin, Chlordane, Heptachlor, Endosulfan, Isodrin, Isobenzan, Toxaphene, and Chloro propylate.

Organochlorine toxicity is mainly due to stimulation of the central nervous system. Cyclodienes inhibit the calcium ion influx and Ca- and Mg-ATPase causing release of neurotransmitters [27]. Epidemiological studies suggest the etiological relationship between Parkinson's disease and organochlorine pollutants (for example, see [28]). These chemicals can be harmful to agricultural workers.

1.3.2 Organophosphates

Organophosphates are used in insecticides used in agriculture and chemical warfare (nerve agents) [29]. Some examples are Dimefox, Mipafox, Methyl Parathion, Ronnel, enitrothion, Bidrin, Phorate, Fenthion, coumaphos, Abate, Dichlorvos, Diptrex, Phosphamidon, Demetox, Oxydemeton-methyl, Malathion, Dimethoate, and Trichlorofan. These chemicals can be harmful to agricultural workers. Symptoms include increased saliva and tear production, diarrhea, vomiting, small pupils, sweating, muscle tremors, and confusion [30].

Organophosphates degrade readily on exposure to air and light. For this reason, they have been considered relatively safe to consumers [31]. However, on fruits and vegetables, the pesticides may remain undegraded. Some nerve agents, such as sarin and tabun, are organophosphates and extremely poisonous. Sarin has a notorious history of use as a chemical weapon. In 1995, a Japanese cult, Aum Shinrikyo used sarin in Tokyo subway, causing 12 deaths and 6200 injuries.

1.3.3 Carbamates

A carbamate is a chemical compound that is formally derived from carbamic acid (NH_2COOH). Carbamate insecticides target human melatonin receptors [32], along with inhibiting acetylcholinesterase [33].

1.3.4 Neonicotinoid

Recently neonicotinoids appear in the news as impacting the wildlife negatively. Neonicotinoid is a class of neuro-active insecticides chemically similar to nicotine [34]. The neonicotinoid family of pesticides includes acetamiprid, clothianidin, imidacloprid, nitenpyram, nithiazine, thiacloprid, and thiamethoxam. Imidacloprid is the most widely used insecticide in the world. Neonicotinoids, including some breakdown products, are particularly toxic to insects.

Neonicotinoids bind to nicotinic acetylcholine receptors (nAChRs) of a cell and initiate a response by that cell. In mammals, nicotinic acetylcholine receptors are found in the cells of the central nervous system and peripheral nervous systems, while in insects these are found only in the central nervous system. Nicotinic acetylcholine receptors are activated by the neurotransmitter acetylcholine. High levels overstimulate and block the receptors that results in paralysis and death.

Neonicotinoid was linked to honey-bee colony collapse disorder (CCD) and loss of birds as the direct result of the reduction in insect populations. It has been proposed that neonicotinoids reduce the survival ability of a bee colony during the winter. It is generally agreed that neonicotinoids have had a negative influence on bee populations. Accordingly, in 2018, the EU banned the three major neonicotinoids (clothianidin, imidacloprid, and thiamethoxam) for all outdoor uses.

Recently, the population decline of eels in Japan was linked to the use of neonicotinoid pesticide. The study, conducted in Lake Shinji in Shimane Prefecture, hinted at the reduction of feed for eels that were sensitive to neonicotinoid pesticide caused the decline of the eels Neonicotinoids disrupt aquatic food webs and decrease fishery yields [35].

1.3.5 Pyrethroids

A pyrethroid is an organic compound analogous to the pyrethrins, naturally occurring chemicals extracted from flowers of pyrethrums (*Chrysanthemum cinerariaefolium* and *C. coccineum*), which has a long history of use in China since Chou Dynasty. Pyrethroids are common household insecticides available commercially. Besides, they may have insect repellent properties and are generally considered harmless to humans [36]. The toxic effects of pyrethroids are mediated by preventing the closure of the voltage-gated sodium channels in the axonal membranes in the neuron cells.

Although being harmless to humans, pyrethroids are toxic to beneficial insects such as bees and dragonflies and other invertebrates, including those that constitute the base of aquatic and terrestrial food webs [37]. Notably, they are very toxic to fish and other aquatic organisms [38].

1.4 Persistent Organic Pollutants

Among chemical pollutants that adversely affect human and environmental well-being, a group of chemicals termed persistent organic pollutants, often abbreviated as POPs, are particularly problematic. For this reason, the United Nations Environment Programme Governing Council investigated POPs in 1995 and designated twelve POPs for the detrimental effects on human and the environmental health and banned the use of these compounds and required the member countries to take action to eliminate or reduce the release of POPs in the environment.

POPs are halogenated organic compounds. Because of this nature, they exhibit high lipid solubility. As a notable consequence, they bioaccumulate in fatty tissues. Halogenated compounds are also chemically very stable due to the nonreactivity of C–Cl bonds against hydrolysis and photolytic degradation, making them particularly problematic once released into the environment.

The 12 initial POPs ("dirty dozen") under the Stockholm Convention are classified into the following three categories:

Pesticides: aldrin, chlordane, DDT, dieldrin, endrin, heptachlor, hexachlorobenzene, mirex, toxaphene;

Industrial chemicals: hexachlorobenzene, polychlorinated biphenyls (PCBs); and

By-products: hexachlorobenzene; polychlorinated dibenzo-p-dioxins and polychlorinated dibenzofurans (PCDD/PCDF), and PCBs.

1.4.1 Aldrin

Aldrin is applied to soils to kill insect pests such as termites, grasshoppers, and corn rootworm. However, it can affect fish, birds, and humans as well. The lethal dose of aldrin for rats is estimated at 30–60 mg/kg. The toxicity for fish is much more potent with the lethal dose of 0.006–0.01 mg/kg [39]. In the soils, plant surface, and in the digestive tracts of insects, aldrin is oxidized to form more potent insecticidal epoxide called dieldrin.

1.4.2 Chlordane

Chlordane is a white and solid organochlorine compound used as a pesticide. In the USA, chlordane was used to control termites in homes until it was banned in 1988 [40]. Ten years prior, it had been banned for food crops such as corn and citrus and on the turf [39]. It is well documented that the chemical is persistent in the environment for an extended time. For example, in Japan, chlordane was detected 5 years after it was applied to homes to kill termites [41].

1.4.3 Dichlorodiphenyltrichloroethane (DDT)

Dichlorodiphenyltrichloroethane, or DDT, is a colorless and almost odorless organochlorine. DDT was first synthesized by the Austrian chemist Othmar Zeidler in 1874. However, the utility was not noticed until Paul Hermann Müller discovered that it can act as an insecticide in 1939. He was awarded a Nobel Prize in Physiology or Medicine in 1948 for this work. During World War II, the U.S. lost the supply of pyrethrum from Japan and started to use DDT to control mosquitos. It was also used to control malaria and typhus among civilians and troops. However, in 1962, Rachel Carson popularized the environmental damages by DDT in her book Silent Spring. Following this, the danger of DDT became known to the public and eventually it was banned in 1968. It is nonetheless a fact that DDT saved millions of lives especially in developing countries. In 2006, the World Health Organization endorsed the limited use of DDT for the control of malaria.

1.4.4 Dieldrin

Dieldrin was originally synthesized in 1948 by J. Hyman & Co, Denver. It is an insecticide closely related to aldrin, which reacts further to form dieldrin. it is an extremely persistent organic pollutant as it does not easily break down. Long-term exposure to dieldrin is harmful to a very wide range of animals including humans. For this reason, dieldrin is banned in most countries in the world.

Dieldrin is associated with an array of human diseases such as Parkinson's, breast cancer, and immune, reproductive, and nervous system damage. Besides, it is also an endocrine disruptor. It can also adversely affect testicular descent in the fetus if a pregnant woman is exposed to it [42].

1.4.5 Endrin

Endrin is an organochloride used as an insecticide, rodenticide, and piscicide, which takes a form of a colorless, odorless solid. Endrin was manufactured as an emulsifiable solution sold commercially as Endrex [43]. It is a persistent organic pollutant. In 2009, EPA released data indicating that the endrin in soil could last up to 14 years or more [44]. It is therefore banned in many countries.

Organochlorine pesticides such as endrin are resistant to degradation and are highly soluble in lipids [45]. This results in bioaccumulation in fats in fish tissues. A bioconcentration factor as high as 1335–10,000 has been reported [46].

Some symptoms of endrin poisoning in humans are headache, dizziness, nervousness, confusion, nausea, vomiting, and convulsions [47].

1.4.6 Heptachlor

Heptachlor is an organochlorine compound that was commonly used as an insecticide in the past. Heptachlor, along with other chlorinated insecticides, appears in Rachel Carson's Silent Spring published in 1962, as destructive to the environment. Heptachlor has a highly stable structure and persists in the environment for decades. In the USA, the Environmental Protection Agency limits the use to the control of fire ant control. The United States Environmental Protection Agency classifies heptachlor as a possible human carcinogen.

1.4.7 Hexachlorobenzene

Hexachlorobenzene is an organochloride used as a fungicide formerly used for seed treatment, especially on the bunt of wheat, which is caused by a fungus. Hexachlorobenzene is carcinogenic to animals and is also possibly a human carcinogen [48]. After it was introduced as a fungicide in 1945, it has been used widely to treat seeds. The use of hexachlorobenzene was banned in the USA in 1966.

1.4.8 Mirex

Mirex, a white crystalline odorless solid, is an organochloride that was used as an insecticide in the past, but now banned in many countries in the world. Mirex is an insecticide active in the insect's stomach. Therefore, it must be ingested by the organism to effect as a poison. Mirex was extensively used as an insecticide in Southeastern United States to control the imported fire ants *Solenopsis saevissima richteri* and *Solenopsis invicta*.

The 1995 report of the Agency for Toxic Substances and Disease Registry (ATSDR) states that Mirex caused fatty changes in the livers, hyperexcitability and convulsion, and inhibition of reproduction in animals. It is also a potent endocrine disruptor and interferes with estrogen-mediated functions such as ovulation, pregnancy, and endometrial growth [49].

1.4.9 Toxaphene

Toxaphene was a widely used pesticide mainly used for cotton in the Southern United States in 1960s and 1970s [50]. It is a mixture of more than 670 chemicals, produced by reacting chlorine gas with camphene [51]. Exposure to toxaphene stimulates the central nervous system and induces morphological changes in the thyroid, liver, and kidneys [52].

It is a persistent chemical that can remain in the environment such as in the soil for 1–14 years without degradation [53]. For this reason, it was banned in the USA in 1990 and by the 2001 Stockholm Convention on Persistent Organic Pollutants.

1.4.10 Polychlorinated Biphenyls (PCBs)

A polychlorinated biphenyl (PCB) is an organic chlorine compound which used to be widely used as dielectric and coolant fluids in electrical apparatus, carbonless copy paper, and in heat transfer fluids [54]. On the other hand, it has a potent biotoxicity and accumulates in fat tissues. It is carcinogenic and causes skin, organ, and hormonal disorders.

PCBs accumulate primarily in the hydrosphere, in the organic fraction of soil, and in living organisms, of which the hydrosphere is the primary reservoir. PDBs become heavier than water in the high pressure of the deep sea, they sink to the ocean trenches, where they accumulate in a concentrated form [55].

In 1968 in Japan, a mixture of dioxins and PCBs accidentally contaminated rice bran oil produced in Northern Kyushu. The contaminated cooking oil severely affected the health of over a thousand people. The disease is known as Kanemi Yusho Disease [56].

1.4.11 Polychlorinated Dibenzo-p-Dioxins (PCDDs) and Polychlorinated Dibenzofurans (PCDFs)

Polychlorinated dibenzo-p-dioxins (PCDDs) and polychlorinated dibenzofurans (PCDFs) are often called dioxins for short. According to a World Health

Organization (WHO) fact sheet, "the chemical name for dioxin is 2,3,7,8-tetrachlorodibenzo para dioxin (TCDD). The name 'dioxins' is often used for the family of structurally and chemically related polychlorinated dibenzo para dioxins (PCDDs) and polychlorinated dibenzofurans (PCDFs). Certain dioxin-like polychlorinated biphenyls (PCBs) with similar toxic properties are also included under the term 'dioxins.' Some 419 types of dioxin-related compounds have been identified but only about 30 of these are considered to have significant toxicity, with TCDD being the most toxic [57]."

Dioxins are very stable chemically and have the ability to be absorbed by fat tissue. Once it enters the body, dioxins stay there for many years. Their half-life is estimated to be 7–11 years [57].

Polychlorinated dibenzodioxins (PCDDs) are a group of polyhalogenated organic compounds recognized widely as environmental pollutants. PCDDs bioaccumulate in humans and wildlife due to their lipophilic properties. It may cause developmental disturbances and cancer. Dioxins are produced as by-products during the manufacturing of some organochlorides, in the incineration of materials that contain chlorine, such as polyvinyl chloride (PVC), and in the bleaching of paper with chlorine [58].

In Vietnam War (1961–1971), a herbicide and defoliant called Agent Orange was used in Vietnam by the United States Forces to destroy the plants that provided cover and food to opposition forces [59]. It is a mixture of equal parts of two herbicides, 2,4,5-T and 2,4-D. Besides it contained traces of dioxin (mainly TCDD, the most toxic of its type) [60] and these contaminants caused disastrous health problems for those who were involved. According to an estimate by the Red Cross, 3 million Vietnamese have been affected by dioxin and at least 150,000 children were born with serious birth defects.

Polychlorinated dibenzofurans (PCDFs) are a group of organic compounds with one or several of the hydrogens in the dibenzofuran replaced by chlorines. PCDFs are produced by incineration of chlorine-containing materials, such as PVC and PCBs at temperatures below 1200 °C [61]. PCDFs persist in the environment for an extended period of time and are possible human carcinogens. By consuming mainly animal products, humans are exposed to PCDFs. PCDFs were also detected in breastfed infants [62].

Besides the 12 initial POPs, the Stockholm Convention added the following 16 chemicals as new POSs [63]:

Alpha hexachlorocyclohexane,

Beta hexachlorocyclohexane,

Chlordecone,

Hexabromobiphenyl,

Hexabromocyclododecane,

Hexabromodiphenyl ether and heptabromodiphenyl ether (commercial octabromodiphenyl ether),

Hexachlorobutadiene,

Lindane,

Pentachlorobenzene,

Pentachlorophenol and its salts and esters,
Perfluorooctane sulfonic acid (PFOS), its salts and perfluorooctanesulfonyl fluoride (PFOSF),
Polychlorinated naphthalenes,
Technical endosulfan and its related isomers,
Tetrabromodiphenyl ether and pentabromodiphenyl ether (commercial pentabromodiphenyl ether),
Decabromodiphenyl ether (commercial mixture, c-DecaBDE), and
Short-chain chlorinated paraffins (SCCPs).

1.5 Antibiotics

An antibiotic is an antimicrobial substance targeted against bacteria used to treat bacterial infections. Since the discovery of the first antibiotic penicillin in 1928 by Alexander Fleming (1881–1955), numerous antibiotics have been in widespread use, which led to antibiotic resistance. Despite their specific and potent bactericidal effects, they have nonetheless side effects as well. Here we describe some more common antibiotics and their health effects on humans.

1.5.1 Clarithromycin

Clarithromycin, often sold as Biaxin, is an antibiotic used to treat bacterial diseases such a strep throat, pneumonia, skin infections, *H. pylori* infection, and Lyme disease, among others. The most common side effects of clarithromycin are gastrointestinal in nature, such as nausea, diarrhea, abdominal pain, and vomiting [64]. Clarithromycin also causes potential hazard to the fetus and therefore should be avoided during pregnancy.

1.5.2 Metronidazole

Metronidazole, often sold as Flagyl, Filmet, or Metro, is an antibiotic and antiprotozoal medicine. It is effective against pelvic inflammatory disease, endocarditis, and bacterial vaginosis. Its side effects include nausea, diarrhea, weight loss, abdominal pain, vomiting, headache, dizziness, and metallic taste in the mouth [65].

1.5.3 Ciprofloxacin

Ciprofloxacin is a broad-spectrum antibiotic and can be used to treat a number of bacterial infections including conditions such as bone and joint infections, intraabdominal infections, infectious diarrhea, respiratory tract infections, skin infections, typhoid fever, and urinary tract infections, among others [66]. Common side effects are nausea, vomiting, diarrhea, and rash. It is a widely prescribed antibiotic and therefore led to the development of resistance among the targeted bacteria.

1.5.4 Clindamycin

Clindamycin, often sold under trade names such as Cleocin, Clinacin, or Dalacin, is an antibiotic used for the treatment of a number of bacterial infections. It is effective for bone or joint infections, pelvic inflammatory disease, strep throat, pneumonia, middle ear infections, and endocarditis [67]. It is effective against Gram-positive bacteria, anaerobic bacteria, and mycoplasmas, but not effective against Gram-negative bacteria. Common side effects include diarrhea, pseudomembranous colitis, nausea, vomiting, abdominal pain, or cramps or rash.

1.5.5 Ampicillin

Ampicillin is an antibiotic in the beta-lactam group of antibiotics. It is similar to amoxicillin in terms of activity ampicillin [68]. Ampicillin is less toxic than most antibiotics. Very rare cases of side effects include angioedema and anaphylaxis.

Ampicillin has been contraindicated in those with a hypersensitivity to penicillins, due to the potential to cause anaphylactic reactions. Hypersensitivity reactions include skin rashes and hives, exfoliative dermatitis, erythema multiforme, and a temporary decrease in both red and white blood cells.

1.6 Biological Toxins

Plants, animals, and microbes produce some of the most sophisticated arsenal of poisons, to protect themselves from attacks by other biological entities. Popular belief is instilled in our society that natural foods are better than industrial foods. This concept is often in error, as in our agricultural history that dates back to thousands of years ago, we have been selecting varieties that are nonpoisonous and of high nutritional content. Therefore, without proper knowledge, these biological poisons can affect human health, especially because of the recent trend of valuing natural foods and beverages.

Here we describe some of the most important biological contaminants, commonly found in foods and the environment.

1.6.1 Mycotoxins

Mycotoxins are toxic secondary metabolites synthesized by fungi. They cause diseases in humans and other animals. There are numerous mycotoxins known to cause serious diseases. Some of the most prominent examples are ergotamine, aflatoxin, ochratoxin A, patulin, citrinin, fumonisins, zearalenone, and trichothecenes. One fungal species may produce more than one mycotoxins.

The tropical part of the world is especially high in the risk of swarm of mycotoxin contamination. For example, in the Philippines, seven mycotoxigenic *Aspergillus* species, four *Fusarium* species, and one *Penicillium* species have been isolated from various agricultural crop commodities. Five mycotoxin groups (aflatoxin, fumonisin, ochratoxin, nivalenol, and zearalenone) have been detected in both the raw form and the by-products of major crops grown in the Philippines. New information has been generated on mycotoxins and mycotoxigenic fungi since the first report of aflatoxin contamination in 1972, but very little is known about other mycotoxins. Despite the increased information accumulated on mycotoxigenic fungi and mycotoxins in the country, practices and measures that control both the fungi and the toxins are next to nonexistent [69]. The situation is practically the same in other tropical developing countries.

Of many different mycotoxins, aflatoxins pose a serious threat to humans. Aflatoxins are potent carcinogens and mutagens produced by molds (*Aspergillus flavus* and *Aspergillus parasiticus*). Aflatoxins are found in decaying foods such as cassava, peanuts, rice, sesame seeds, sunflower seeds, corn, nuts, wheat, among others. Children are most affected by aflatoxin exposure, resulting in stunted growth [70] and delayed development [71].

Outbreaks of aflatoxin contamination occur frequently. For example, 120 people died in Kenya in 2003 from acute aflatoxin poisoning [72]. In Nepal and Bangladesh, since 2014, unacceptable levels of aflatoxin have been detected in the bloodstream of pregnant women [73].

1.6.2 Ergotism

Ergotism is a disease caused by eating grain products, particularly rye, contaminated with the fungal plant pathogen *Claviceps purpurea* that produces toxic alkaloids. When the infected grains are milled, the ergot is reduced to a red powder, which may be missed in dark colored flour such as rye flour. Although not common in modern days, it still occurs in less developed countries. For example, in mid-2001, ergotism outbreak occurred in Ethiopia from contaminated barley.

There are two types of ergotism symptoms. Convulsive ergotism is characterized by nervous dysfunction. The victim twists and contorts their body in pain, with trembling and shaking. Gangrenous ergotism victim may lose parts of their extremities, such as toes and fingers.

To prevent ergotism, ergots must be removed by placing the yield in a brine solution; the ergots float in the brine, while the healthy grains sink [74].

1.6.3 Cyanotoxins

Cyanotoxins are toxins produced by cyanobacteria (formerly referred to as blue-green algae) that occur in lakes, ponds, rivers, and other surface waters. Cyanobacteria can cause harmful algal bloom (HAB) when the condition is conducive with abundant nutrients. If water supply is sourced from the bodies of water contaminated with cyanobacteria, the water treatment plant can remove the contaminants. However, during severe HAB events, it may face challenges.

Cyanotoxins can be divided into two main criteria: [1] on the basis of their mechanism of action on terrestrial vertebrates, especially mammals—e.g., hepatotoxins, neurotoxins, dermatotoxins, etc., and [2] according to their chemical structure—e.g., amino acid, cyclic peptides, alkaloids, or polyketides.

Some examples of cyanotoxins are guanitoxin, saxitoxins, *β*-*N*-methylamino-l-alanine (BMAA), nodularin, and aplysiatoxin.

Guanitoxin (an alkaloid), formerly Anatoxin-a(S), is a naturally occurring cyanotoxin commonly isolated from cyanobacteria of the genus *Anabaena*, and in a mouse study, it induced clinical signs of salivation, lacrimation, urinary incontinence, defecation, convulsion, fasciculation, and respiratory arrest [75].

Saxitoxin (STX, an alkaloid) is a neurotoxin and the best-known paralytic shellfish toxin (PST). Its symptom, paralysis, is caused by consuming shellfish contaminated with STX. Saxitoxin is a neurotoxin that blocks sodium channels [76]. It acts on the voltage-gated sodium channels of neurons, disturbing normal cellular function and resulting in paralysis.

***β*-Methylamino-*L*-alanine** (BMAA, an amino acid) is a non-proteinogenic amino acid synthesized by cyanobacteria in marine, freshwater, and terrestrial environments [77, 78]. BMAA is a neurotoxin that may be responsible for various neurodegenerative disorders including amyotrophic lateral sclerosis (ALS), Alzheimer's disease, Parkinson's disease, Huntington's disease [79–82].

Nodularins (cyclic peptides) are potent toxins produced by the cyanobacterium *Nodularia spumigena* [83]. Nodularins cause gastroenteritis, allergic irritation reactions, and liver diseases [84]. Among ten nodularin variants known to date, nodularin-R is well known as a potent hepatotoxin that may cause serious damage to the liver of animals including human. The WHO drinking water concentration limit for nodularins is 1.5 μg/L [85]. The full extent of the hazard of these toxins to humans has not been fully elucidated, predominantly due to the lack of exposure data [86].

Aplysiatoxin (a polyketide) is a cyanotoxin produced by cyanobacteria species in the genera *Lyngbya*, *Schizothrix*, or *Planktothrix*. It is used as a defensive secretion against fish. It acts as a potent irritant and carcinogen, by activating protein kinase C [87–90].

1.6.4 Cyanide

Cyanide prevents the cells from using oxygen. The symptoms of the exposure to small amount of cyanide include dizziness, headache, nausea and vomiting, rapid breathing, rapid heart rate, and restlessness. However, when exposed to a large amount of cyanide more serious symptoms are observed such as slow heart rate, convulsions, loss of consciousness, low blood pressure, lung injury, and respiratory failure leading to death.

Some plants accumulate cyanide in their tissues. This is thought to be a defense mechanism against insect and other pests. Important food crops that can accumulate cyanide include almonds, millet sprouts, lima beans, soybean, spinach, bamboo shoots, cassava, apple seeds, and peach and apricot pits.

Cassava, especially under water stress conditions, produces dangerous amounts of cyanide in the roots and causes numerous deaths in Africa. This happens often in Africa as it is the main cassava production area, even though the crop originated in South America. The matter is worsened as in the drought conditions, cassava can be the only food that can grow in the dry weather, and cyanide accumulation is heightened in such a condition. Cyanide poisoning of cassava can be prevented by processing the roots or flours.

1.6.5 Tetrodotoxin (TTX)

Tetrodotoxin (TTX) is a potent neurotoxin. The name tetrodotoxin derives from Tetraodontiformes, an order that includes pufferfish (fugu), in which the toxin is commonly found. The toxin is produced by infecting or symbiotic bacteria such as Pseudoalteromonas, Pseudomonas, and Vibrio [91].

Tetrodotoxin is extremely toxic with the lethal dose (LD50) for mice as 334 μg per kg [92]. In Japan fugu is an expensive delicacy and there are specialized fugu restaurants in major cities across the country. Poisoning from tetrodotoxin is reported every year when fugu is prepared by unlicensed individuals. To prevent this, it must be prepared and sold only in special restaurants where licensed fugu chefs carefully remove the tissues that contain toxins to eliminate the risk of poisoning [93].

1.7 Food Additives

1.7.1 Artificial Sweeteners

Artificial sweetener is a food additive that provides a sweet taste in place of sugar that contains less energy than natural sugars such as sucrose, glucose, or fructose. As their calorie value is either very low or zero, they can be used for dieting purposes. In North America, common artificial sweeteners are color-coded. Typical colors are blue for aspartame, pink for saccharin, yellow for sucralose, orange for monk fruit extract, and green for stevia [94].

As of 2017, sucralose was the most popular artificial sweetener for the production of foods and beverages, occupying 30% of the global market [95].

Most artificial sweeteners are considered safe. However, their long-term impact on human health is not well understood. Recently, a study in Canada did not show a consistent effect of artificial sweeteners on weight loss. It also showed a correlation between consumption of artificial sweeteners and higher risks of obesity, high blood pressure, diabetes, and heart disease [96]. More studies on long-term effect of artificial sweeteners on human health are thus warranted.

We describe some of the most commonly used artificial sweeteners below.

1.7.1.1 Saccharin

Saccharin is the first artificially synthesized sweetener developed in 1879 by Remsen and Fahlberg. It is 300–500 times as sweet as sucrose and used in toothpastes, diet foods, and diet beverages.

In 1960, a study showed that high levels of saccharin may cause bladder cancer in laboratory rats. In 1977, Canada banned saccharin based on the animal research. In the same year in the USA, the sales of saccharin required a warning label, and further study of saccharin safety was mandated.

Currently, the International Agency for Research on Cancer, part of the World Health Organization, states that "Saccharin and its salts were downgraded from Group 2B, possibly carcinogenic to humans, to Group 3, not classifiable as to carcinogenicity to humans." However, many countries still ban the use of saccharin as a sweetener.

1.7.1.2 Sucralose

Sucralose is the most commonly used artificial sweetener in the world that is about 600 times as sweet as sucrose.

The following regulatory agencies, among others, accept sucralose as a safe sugar alternative: the FDA, The Joint FAO/WHO Expert Committee Report on Food Additives, the European Union's Scientific Committee on Food, Health Protection

Branch of Health and Welfare Canada, and Food Standards Australia New Zealand. Canadian Diabetes Association reports that the amount of sucralose that can be consumed over a person's lifetime without any adverse effects is 900 mg per kg of body weight per day [97, 98].

There could be a possible link between sucralose and a reduction in beneficial gut microflora that might result in detrimental health effects, based on an animal study [99].

1.7.1.3 Aspartame

Aspartame is an artificial sweetener about 200 times sweeter than sucrose and is commonly used as a sugar substitute in various foods and beverages [100]. Its brand names include under the trade names Equal, NutraSweet, and Canderel.

Aspartame has been studied intensively since its discovery [101]. The United States Food and Drug Administration (FDA), UK Food Standards Agency, the European Food Safety Authority (EFSA), and Health Canada, among other agencies world over, consider aspartame to be safe for human consumption.

Aspartame is a methyl ester of the dipeptide of the natural amino acids *L*-aspartic acid and *L*-phenylalanine [102]. In 2009, the largest manufacturer of aspartame, Ajinomoto, re-branded it as AminoSweet to reflect its dual amino acid structure.

Aspartame is used in thousands of foods and beverages under the trade names such as Equal, NutraSweet, and Canderel. However, it is not suitable for baked foods as it breaks down when heated and loses its sweet taste [103].

1.7.2 Melamine

Melamine is a trimer of cyanamide, with a 1,3,5-triazine skeleton. Melamine can be used to manufacture melamine resins, which are durable thermosetting plastic used in various high pressure decorative laminates.

LD_{50} of melamine is 1–3 g/kg in rats studies and its acute toxicity is relatively low [104]. However, animal studies showed stone formation and bladder carcinogenicity at high doses of melamine [105–110].

In 2008, a scandal broke out in China that involved milk and infant formula along with other food materials that were adulterated with melamine. Melamine was used in the milk because the addition increased the nitrogen content of the milk, making it appear protein rich. In the scandal, out of approximately 300,000 victims in China [111, 112], six babies died from kidney stones and other kidney damages. Additionally, over 50,000 babies were hospitalized [113].

The 2008 event in China revealed that melamine by itself is able to cause toxicity when it is consumed at an excessive dosage [109]. It is likely because the melamine, when combined with endogenous urate, can produce calculi [114].

1.7.3 Azo Dyes

An array of azo dyes are added as food colors to make them attractive. It is also used to reinstate the natural colors lost during the production process [115]. However, many azo dyes are carcinogenic and mutagenic and can cause allergic reactions [116]. For this reason, azo dyes are banned in many countries and the use of most of the azo dyes in food is regulated.

1.7.4 Monosodium Glutamate (MSG)

Monosodium glutamate (MSG), also known as its brand name Ajinomoto, is the sodium salt of glutamic acid. It is one of the most common natural non-essential amino acids [117]. MSG was first prepared in 1908 by the Japanese chemist Kikunae Ikeda, who was interested to isolate the special taste of kombu, a kelp used for Japanese soups and other numerous traditional foods. MSG "balances, blends, and rounds the perception of other tastes" [118, 119].

Despite a popular belief, monosodium glutamate naturally occurs in tomatoes, grapes, cheese, mushrooms, and other foods [120]. MSG adds "umami" or pleasant taste to foods.

MSG is completely safe for humans to consume [121]. Anecdotal stories have been propagated that MSG can cause headaches and discomfort but controlled blinded tests have found no evidence to support this popular belief [122]. Therefore, international and national bodies overseeing food additives consider MSG safe for human consumption as a flavor enhancer [123].

1.7.5 Preservatives

A preservative is a chemical added to products such as food and beverages to slow degradation by microbes or by chemical changes in order to prevent spoilage. Some ancient methods of preservation use honey or salt as preservatives, which are completely safe for consumption.

However, modern preservatives include some toxic chemicals. For example, sodium nitrate is found in processed meats such as canned tuna and sausages. Studies demonstrated that nitrate and nitrite are precursors of N-nitroso compounds and induce tumors of the pancreas in animals [124].

Salts of benzoic acid are used as food preservatives in soy sauce and soft drinks to name a few. In the presence of ascorbic acid (vitamin C) and a transition-metal catalyst (iron and copper), benzene is produced from decarboxylation of benzoic acid [125]. Benzene is known to be carcinogenic.

Benzene in soft drinks has provoked public concerns in many occasions. For example, in 1990, a study reported the presence of benzene in bottles of Perrier sold in the USA, and the product was recalled [126]. In 2006, the Food Standards Agency of the UK reported that out of 150 beverages tested, 43 beverages contained benzene, four of which contained levels above 10 ppb, the World Health Organization drinking water standards [127]. These four were subsequently withdrawn from the market [128, 129].

1.8 Conclusion

We live in a world surrounded by more chemicals and substances that were unthinkable in the past. Every day, new chemicals are synthesized and introduced into the market. This requires the understanding of their properties, and potential risks to human health and environment must be assessed carefully. The risk of any chemical or material depends on the dose and also how it is used. Therefore, public awareness about most commonly utilized and occurring substances must be promoted to reduce the damaging effects for human and environmental well-being.

References

1. Wedepohl KH (1995) The composition of the continental crust. Geochim Cosmochim Acta 59 (7):1217–1232
2. Bernhoft RA (2013) Sci World J 2013:394652
3. Bernard A (2008) Indian J Med Res 128(4):557–564. Cadmium & its adverse effects on human health
4. The main use of arsenic is in alloying with lead. Lead components in car batteries are strengthened by the presence of a very small percentage of arsenic
5. Bagshaw NE (1995) Lead alloys: past, present and future. J Power Sources 53(1):25–30. Bibcode:1995JPS....53...25B. https://doi.org/10.1016/0378-7753(94)01973-Y
6. Barceloux DG, Barceloux D (1999) Chromium. Clin Toxicol 37(2):173–194. https://doi.org/10.1081/CLT-100102418. PMID 10382554
7. Das KK, Das SN, Dhundasi SA (2008) Nickel, its adverse health effects & oxidative stress. Indian J Med Res 128:412–425
8. Das KK, Buchner V (2007) Effect of nickel exposure on peripheral tissues: role of oxidative stress in toxicity and possible protection by ascorbic acid. Rev Environ Health 22:133–149
9. IARC (International Agency for Research on Cancer) (1990) IARC Monograph on the evaluation of carcinogenic risks to humans, vol 49. IARC, Lyans, pp 318–411
10. "Mercury" (2016) NIEHS. Archived from the original on 19 November. Retrieved 19 Nov 2016
11. The Karen Wetterhahn story Archived 2012-05-30 at the Wayback Machine—University of Bristol web page documenting her death. Retrieved 9 Dec 2006
12. OSHA update following Karen Wetterhahn's death Archived 2015-07-11 at the Wayback Machine
13. "Lead poisoning and health". WHO. September 2016. Archived from the original on 18 October 2016. Retrieved 14 Oct 2016

14. "Aluminum and dementia: Is there a link?". Alzheimer Society Canada. 24 August 2018
15. Dolara P (2014) Occurrence, exposure, effects, recommended intake and possible dietary use of selected trace compounds (aluminium, bismuth, cobalt, gold, lithium, nickel, silver). Int J Food Sci Nutr 65(8):911–924
16. Fosmire GJ (1990) Zinc toxicity. Am J Clin Nutr 51(2):225–227
17. Johnson AR, Munoz A, Gottlieb JL, Jarrard DF (2007) High dose zinc increases hospital admissions due to genitourinary complications. J Urol 177(2):639–643
18. Towards a Thematic Strategy for Soil protection. COM (2002)179 final. Brussels, Belgium: European Commission
19. Reichenauer TG, Germida JJ (2008) Phytoremediation of organic contaminants in soil and groundwater. ChemSusChem 1(8–9):708–717
20. Das PK (2018) Phytoremediation and Nanoremediation: emerging techniques for treatment of acid mine drainage water. Defence Life Sci J 3(2):190–196
21. Rascio N, Navari-Izzo F (2011) Heavy metal hyperaccumulating plants: how and why do they do it? And what makes them so interesting? Plant Sci 180(2):169–181
22. Poschenrieder C, Tolrá R, Barceló J (2006) Can metals defend plants against biotic stress? Trends Plant Sci 11:288–295
23. Pagliano C et al (2006) Evidence for PSII-donor-side damage and photoinhibition induced by cadmium treatment on rice (Oryza sativa L.). J Photochem Photobiol B Biol 84:70–78
24. Lange B, van der Ent A, Baker AJM, Echevarria G, Mahy G, Malaisse F, Meerts P, Pourret O, Verbruggen N (2017) Copper and cobalt accumulation in plants: a critical assessment of the current state of knowledge. New Phytol 213(2):537–551
25. Wang J, Zhao F-J, Meharg AA, Raab A, Feldmann J, McGrath SP (2002) Mechanisms of arsenic Hyperaccumulation in Pteris vittata. Uptake kinetics, interactions with phosphate, and arsenic speciation. Plant Physiol 130(3):1552–1561
26. Tu C, Ma LQ, Bondada B (2002) Arsenic accumulation in the Hyperaccumulator Chinese brake and its utilization potential for phytoremediation. J Environ Qual 31(5):1671
27. Mathew LL (2012) Organochloride Pesticide toxicity. Drugs, diseases and procedures. Medscape References
28. Chhillar N, Singh NK, Banerjee BD, Bala K, Mustafa M, Sharma D, Chhillar M (2013) Organochlorine pesticide levels and risk of Parkinson's disease in north Indian population. ISRN Neurol 2013:371034. https://doi.org/10.1155/2013/371034
29. Balali-Mood M, Shariat M (1998) Treatment of organophosphate poisoning. Experience of nerve agents and acute pesticide poisoning on the effects of oximes. J Physiol Paris 92 (5–6):375–378
30. Eddleston M, Buckley NA, Eyer P, Dawson AH (2008) Management of acute organophosphorus pesticide poisoning. Lancet 371(9612):597–607
31. Pesticide application and safety training for applicators of public health pesticides. Archived from the original on 2010-08-29. Retrieved 2013-03-25
32. Popovska-Gorevski M, Dubocovich ML, Rajnarayanan RV (2017) Carbamate insecticides target human melatonin receptors. Chem Res Toxicol 30:574–582
33. Colović MB, Krstić DZ, Lazarević-Pašti TD, Bondžić AM, Vasić VM (2013) Acetylcholinesterase inhibitors: pharmacology and toxicology. Curr Neuropharmacol 11:315–335
34. "Neonicotinoid pesticides & adverse health outcomes". ntp.niehs.nih.gov
35. Yamamuro M, Komuro T, Kamiya H, Kato T, Hasegawa H (2019) Yutaka Kameda Sci 366 (6465):620–623
36. Metcalf RL (2000) Ullmann's Encyclopedia of industrial chemistry. Wiley-VCH, Weinheim
37. Zaveri M (2010) Study links pesticides to river contamination. The Daily Californian
38. Thatheyus AJ, Gnana S, Deborah A (2013) Synthetic Pyrethroids: toxicity and biodegradation. Appl Ecol Environ Sci 1(3):33–36
39. Robert L (2002) Metcalf "insect control" in Ullmann's Encyclopedia of industrial chemistry. Wiley-VCH, Weinheim

40. Toxicological profile for chlordane, U.S. department of health and human services, agency for toxic substances and disease registry
41. Uemura S, Kawamura H, Tsuji M, Tomita S, Maeda S (2002) In Encyclopedia on the toxicity of agricultural pesticide. Revised. Sanseido Press
42. Andersen R, Helle, Vinggaard AM, Rasmussen H, Thomas, Gjermandsen IM, Cecilie Bonefeld-Jørgensen E (2002) Effects of currently used pesticides in assays for estrogenicity, androgenicity, and aromatase activity in vitro. Toxicol Appl Pharmacol 179(1):1–12
43. van Esch GT, van Heemstra-Lequin EAH (1992) Environmental health criteria 130: endrin. International programme on chemical safety. World Health Organization
44. "Technical Factsheet on: Endrin" (PDF). www.epa.gov. United States Environmental Protection Agency
45. Zitko V (2003) Persistent organic pollutants (PDF). Springer-Verlag, Berlin, Heidelberg, pp 47–90
46. "Technical Factsheet on: Endrin". www.epa.gov. United States Environmental Protection Agency
47. "Toxicological Profile for Endrin" (1996) Agency for toxic substances and disease registry. United States Department of Health and Human Services. August
48. "Report on Carcinogens, Eleventh Edition"
49. Faroon O, Kueberuwa S, Smith L, DeRosa C (1995) ATSDR evaluation of health effects of chemicals. II. Mirex and chlordecone: health effects, toxicokinetics, human exposure, and environmental fate. Toxicol Ind Health 11(6):1–203
50. "Toxaphene" (2014) Report on carcinogens. National Toxicology Program, Department of Health and Human Services. 13. October 2
51. Saleh MA (1983) Capillary gas chromatography-electron impact chemical ionization mass spectrometry of toxaphene. J Agric Food Chem 31(4):748–751
52. "Toxaphene". Technology transfer network—air toxics web site. United States Environmental Protection Agency
53. "Technical Factsheet on: TOXAPHENE". National primary drinking water regulations. United States Environmental Protection Agency
54. Rossberg M, Lendle W, Pfleiderer G, Tögel A, Dreher E-L, Langer E, Rassaerts H, Kleinschmidt P, Strack (2006) Chlorinated hydrocarbons. In: Ullmann's encyclopedia of industrial chemistry. Wiley-VCH, Weinheim
55. "Nasty chemicals abound in what was thought an untouched environment". Economist. 2017-02-18
56. Aoki Y (2001) Polychlorinated biphenyls, Polychloronated Dibenzo-p-dioxins, and polychlorinated Dibenzofurans as endocrine disrupters—what we have learned from Yusho disease. Environ Res 86(1):2–11
57. World Health Organization fact sheet: dioxins and their effects on human health 4 October 2016
58. Beychok MR (1987) A data base for dioxin and furan emissions from refuse incinerators. Atmos Environ 21(1):29–36
59. Buckingham WH Jr (1982). Operation ranch hand: the air force and herbicides in southeast Asia 1961–1971. Office of Air Force History, United States Air Force
60. "Agent orange and cancer". American Cancer Society. February 11, 2019
61. "Proceedings of the subregional awareness raising workshop on persistent organic pollutants (POPs), Bangkok, Thailand". United Nations Environment Programme. November 25–28, 1997
62. WHO web site. http://chm.pops.int/TheConvention/ThePOPs/The12InitialPOPs/tabid/296/Default.aspx
63. Stockholm convention on persistent organic pollutants (POPs) June 2017. The 16 New POPs: an introduction to the chemicals added to the Stockholm convention as persistent organic pollutants by the conference of the parties)
64. Clarithromycin side effects in detail—Drugs.com

65. Rossi S (ed) (2013) Australian medicines handbook. The Australian Medicines Handbook Unit Trust, Adelaide
66. "Ciprofloxacin hydrochloride". The American society of health-system pharmacists
67. "Clindamycin hydrochloride". The American society of health-system pharmacists
68. The American society of health-system pharmacists
69. Balendres et al (2019)
70. Khlangwiset P, Shephard GS, Wu F (2011) Aflatoxins and growth impairment: a review. Crit Rev Toxicol 41(9):740–755
71. Abbas HK (2005) Aflatoxin and food safety. CRC Press. isbn:978-0-8247-2303-3
72. "Eastern and Southern Africa 2011 Highlights" (PDF). ICRISAT. 2012
73. "Aflatoxin threat in Nepal, Bangladesh. SciDev.Net South Asia. 2014-12-17
74. Wegulo SN, Carlson MP (2011) Ergot of small grain cereals and grasses and its health effects on humans and livestock. University of Nebraska–Lincoln Extension
75. Mahmood NA, Carmichael WW, Pfahler D (1988) Anticholinesterase poisonings in dogs from a cyanobacterial (blue-green algae) bloom dominated by Anabaena flos-aquae. Am J Vet Res 49(4):500–503
76. Huot RI, Armstrong DL, Chanh TC (1989) Protection against nerve toxicity by monoclonal antibodies to the sodium channel blocker tetrodotoxin. J Clin Investig 83(6):1821–1826
77. Cox PA, Banack SA, Murch SJ, Rasmussen U, Tien G, Bidigare RR, Metcalf JS, Morrison LF, Codd GA, Bergman B (2005) Diverse taxa of cyanobacteria produce b-N-methylamino-L-alanine, a neurotoxic amino acid. PNAS 102(14):5074–5078
78. Esterhuizen M, Downing TG (2008) β-N-methylamino-L-alanine (BMAA) in novel south African cyanobacterial isolates. Ecotoxicol Environ Saf 71(2):309–313
79. Murch SJ, Cox PA, Banack SA (2004) A mechanism for slow release of biomagnified cyanobacterial neurotoxins and neurodegenerative disease in Guam. PNAS 101 (33):12228–12231
80. Murch SJ, Cox PA, Banack SA, Steele JC, Sacks OW (2004) Occurrence of b-methylamino-L-alanine (BMAA) in ALS/PDC patients from Guam. Acta Neurol Scand 110(4):267–269
81. Pablo J, Banack SA, Cox PA, Johnson TE, Papapetropoulos S, Bradley WG, Buck A, Mash DC (2009) Cyanobacterial neurotoxin BMAA in ALS and Alzheimer's disease. Acta Neurol Scand 120(4):215–225
82. Bradley WG, Mash DC (2009) Beyond Guam: the cyanobacterial/BMAA hypothesis of the cause of ALS and other neurodegenerative diseases. ALS 10:7–20
83. Sivonen K, Kononen K, Carmichael WW, Dahlem AM, Rinehart KL, Kiviranta J, Niemela SI (1989) Occurrence of the hepatotoxic cyanobacterium Nodularia spumigena in the Baltic Sea and structure of the toxin. Appl Environ Microbiol 55(8):1990–1995
84. Dawson RM (1998) The toxicology of microcystins. Toxicon 36(7):953–962
85. "Nodularin". Substances of Biological Interest, Bacterial Toxin, Natural Toxin. SelfDecode
86. Chen Y, Shen D, Fang D (2013) Nodularins in poisoning. Clin Chim Acta 425:18–29
87. Kato Y, Scheuer PJ (1974) Aplysiatoxin and debromoaplysiatoxin, constituents of the marine mollusk Stylocheilus longicauda (Quoy and Gaimard, 1824). J Am Chem Soc 96 (7):2245–2246
88. Weinstein IB, Arcoleo J, Backer J, Jeffrey A, Hsiao WL, Gattoni-Celli S, Kirschmeier P, Okin E (1983) Molecular mechanisms of tumor promotion and multistage carcinogenesis. Princess Takamatsu Symp 14:59–74
89. Arcoleo JP, Weinstein IB (1985) Activation of protein kinase C by tumor promoting phorbol esters, teleocidin and aplysiatoxin in the absence of added calcium. Carcinogenesis 6 (2):213–217
90. Nagai H, Yasumoto T, Hokama Y (1996) Aplysiatoxin and debromoaplysiatoxin as the causative agents of a red alga Gracilaria coronopifolia poisoning in Hawaii. Toxicon 34 (7):753–761

91. Lago J, Rodríguez LP, Blanco L, Vieites JM, Cabado AG (2015) Tetrodotoxin, an extremely potent marine neurotoxin: distribution, toxicity, origin and Therapeutical uses. Mar Drugs 13 (10):6384–6406
92. "Material Safety Data Sheet Tetrodotoxin ACC# 01139". Acros Organics N.V
93. Warin RH, Steventon GB, Mitchell SC (2007) Molecules of death. Imperial College Press, London, p 390
94. "Artificial sweeteners. What's the difference?". Tribunedigital-chicagotribune
95. "Sweetener Market Projected to Be Worth USD 2.84 Billion by 2021: Technavio". Yahoo Finance. Archived from the original on 25 April 2017. Retrieved 10 Jan 2018
96. Azad MB, Abou-Setta AM, Chauhan BF, Rabbani R, Lys J, Copstein L, Mann A, Jeyaraman MM, Reid AE, Fiander M, MacKay DS, McGavock J, Wicklow B, Zarychanski R (2017) Nonnutritive sweeteners and cardiometabolic health: a systematic review and meta-analysis of randomized controlled trials and prospective cohort studies. Can Med Assoc J 189(28):E929. https://doi.org/10.1503/cmaj.161390
97. Canadian diabetes association 2008 clinical practice guidelines for the prevention and management of diabetes in Canada (2008) Can J Diabetes 32(Supplement 1):S41
98. Goldsmith LA (2000) Acute and subchronic toxicity of sucralose. Food Chem Toxicol 38 (Suppl 2):S53–S69
99. Schiffman SS, Rother KI (2013) Sucralose, a synthetic organochlorine sweetener: overview of biological issues. J Toxicol Environ Health B 16(7):399–451
100. "Aspartame". PubChem, National Library of Medicine, US National Institutes of Health. 17 August 2019. Retrieved 24 August 2019
101. EFSA National Experts (2010) Report of the meetings on aspartame with national experts. EFSA
102. "Aspartame". PubChem, National Library of Medicine, US National Institutes of Health. 17 August 2019
103. Struck S, Jaros D, Brennan CS, Rohm H (2014) Sugar replacement in sweetened bakery goods. Int J Food Sci Technol 49(9):1963–1976
104. MSDS Melamine
105. Mast RW, Jeffcoat AR, Sadler BM, Kraska RC, Friedman MA (1983) Metabolism, disposition and excretion of [14C]melamine in male Fischer 344 rats. Food Chem Toxicol 21(6):807–810
106. Melnick RL, Boorman GA, Haseman JK, Montali RJ, Huff J (1984) Urolithiasis and bladder carcinogenicity of melamine in rodents. Toxicol Appl Pharmacol 72(2):292–303
107. IUCLID dataset substance ID: 108-78-1. Melamine. In. 18 Feb ed: European Commission, European Chemicals Bureau; 2000
108. National Toxicology Program (1983) NTP carcinogenesis bioassay of melamine (CAS no. 108-78-1) in F344/N rats and B6C3F1 mice(feed study). Natl Toxicol Program Tech Rep Ser 245:1–171
109. Ogasawara H, Imaida K, Ishiwata H, Toyoda K, Kawanishi T, Uneyama C et al (1995) Urinary bladder carcinogenesis induced by melamine in F344 male rats: correlation between carcinogenicity and urolith formation. Carcinogenesis 16(11):2773–2777
110. Okumura M, Hasegawa R, Shirai T, Ito M, Yamada S, Fukushima S (1992) Relationship between calculus formation and carcinogenesis in the urinary bladder of rats administered the non-genotoxic agents thymine or melamine. Carcinogenesis 13(6):1043–1045
111. Branigan T (2008) Chinese figures show fivefold rise in babies sick from contaminated milk. The Guardian, London. Archived from the original on 5 December 2008
112. J Med Toxicol (2010) 6:50–55
113. Scott McDonald (2008) Nearly 53,000 Chinese children sick from milk. Associated Press. Archived from the original on 10 February 2014
114. Workshop on Melamine, 2008 November 18–19. Beijing, People's Republic of China
115. Food Chem (2016) 192:813–824

116. Gičević A, Hindija L, Karačić A (2020) Toxicity of Azo dyes in pharmaceutical industry. In: Badnjevic A, Škrbić R, Gurbeta Pokvić L (eds) CMBEBIH 2019. CMBEBIH 2019. IFMBE proceedings, vol 73. Springer, Cham
117. Ninomiya K, Technical Committee, Umami Manufacturers Association of Japan (1998) Natural occurrence. Food Rev Intl 14(2 & 3):177–211
118. Loliger J (2000) Function and importance of glutamate for Savory foods. J Nutr 130 (4s Suppl):915s–920s
119. Yamaguchi S (1991) Basic properties of umami and effects on humans. Physiol Behav 49 (5):833–841
120. "Questions and Answers on Monosodium glutamate (MSG)". www.fda.gov. U.S. Food and Drug Administration. 19 November 2012
121. Barry-Jester AM (2016) How MSG Got A Bad Rap: Flawed Science And Xenophobia
122. Obayashi Y, Nagamura Y (2016) Does monosodium glutamate really cause headache? : A systematic review of human studies. J Headache Pain 17:54
123. Walker R, Lupien JR, School of Biological Sciences, University of Surrey, UK, and Food and Nutrition Division, FAO of the United Nations, Italy (2000) The safety evaluation of monosodium glutamate. J Nutr 130(4S Suppl):1049S–1052S
124. Am J Epidemiol (2011) 174(3):305–315. https://doi.org/10.1093/aje/kwr092. Epub 2011 Jun 17
125. Aschebrook-Kilfoy B, Cross AJ, Stolzenberg-Solomon RZ, Schatzkin A, Hollenbeck AR, Sinha R, Ward MH. Pancreatic cancer and exposure to dietary nitrate and nitrite in the NIH-AARP Diet and Health Study
126. Lalita K, GardnerGlen D (1993) Lawrence, benzene production from decarboxylation of benzoic acid in the presence of ascorbic acid and a transition-metal catalyst. J Agric Food Chem 41(5):693–695
127. Food and Drug Administration (1990) U.S. public health service, Department of Health and Human Services FDA Enforcement Report February 28
128. Food Standards Agency (2006) Survey of benzene in soft drinks. March
129. Elliott V (2006) Soft drinks pulled from shelves over cancer fear. The Times

Chapter 2
Gut Microbiota and Health

Chunjiang Zhang, Amanpreet Kaur Virk, Israr Khan, and Haoran Qin

2.1 Gut Microbiota

2.1.1 Introduction

Microbiota is a sophisticated community of microorganisms comprising of bacteria, viruses, protozoa, and fungi, dwelling in various zones of human body, for example, mouth, respiratory framework, skin, gastroenteric tube, and vagina [1]. More than 70% of microbiota resides within the gastrointestinal (GI) tract in a mutually beneficial association with its host, spreading continuously from gastric lumen to colon/rectum, where it arrives at its most severe concentration.

The human gastrointestinal tract (GIT) constitutes the largest interfaces (250–400 m^2) among the host, ecological elements, and antigens within the human body. Approximately, 60 tons of food runs through the human GIT in an average lifespan, along with an abundance of environmental microorganisms that pose a major threat to the integrity of gut [2]. Assortment of bacteria, eukarya, and archaea occupying the GIT is named as "gut microbiota" and has co-developed with the host to establish a complex, and mutually beneficial connection [3, 4]. The mammalian GIT has higher and varied amount of microbes, best-known as intestinal microbiota. Archaea, bacteria, protozoa, fungi, and viruses live together and associate with the host, especially immune and epithelial cells [5]. The quantity of microorganisms

Chunjiang Zhang and Amanpreet Kaur Virk contributed equally to this work

C. Zhang (✉) · I. Khan · H. Qin
School of Life Science, Lanzhou University, Lanzhou, Gansu, PR China
e-mail: chjzh@lzu.edu.cn

A. K. Virk
Faculty of Applied Sciences and Biotechnology, Shoolini University of Biotechnology and Management Sciences, Solan, India

X. Li, P. Liu (eds.), *Gut Remediation of Environmental Pollutants*,
https://doi.org/10.1007/978-981-15-4759-1_2

living in GIT has been evaluated to surpass 10^{14} that include approximately 10 times more bacterial cells than the quantity of human cells and more than 100 times the quantity of genomic material (microbiome) as the human genome [3, 6]. Nevertheless, an amended estimate has recommended that the proportion of bacterial: human cells is probably close to 1:1 [1]. Because of the immense quantity of bacterial cells in the body, the host and microorganisms occupying it are often mentioned as a "superorganism" [6, 7].

Microbiota provides numerous advantages for the host, by means of physiological roles, for example, reinforcing the integrity of gut or forming the epithelium of intestine [8], extracting energy [9], guarding from pathogens [10], and controlling immunity of host [11]. Because of a modified microbial composition, known as dysbiosis, there is possibility for the disruption of above-mentioned mechanisms. With the development of progressively advanced methods to characterize sophisticated biological systems, a function of the microbiota in an enormous number of intestinal and extra-intestinal diseases has become consistently evident [12, 13]. This chapter summarizes our present comprehension of the human GI microbiota composition and development, and its effect on host health and gut integrity.

2.1.2 Structure and Composition of the Human GI Microbiota

An adult gut microbiota contains 10 to 100 trillion microbes, which is 10 times the quantity of total somatic and germ cells of humans [14]. Gut microbiome contain 100- to 150-times more genes than human genome [15]. The gut microbiota has co-developed with humans and has demonstrated significant consequences for different host reactions. The modified composition of gut microbiota has been connected to metabolic diseases, like obesity, diabetes, or non-alcoholic fatty liver diseases. Such studies have shown the significance of gut in modulating metabolic disorders and host metabolism.

Intestinal microbiota comprises autochthonous individuals occupying the gut mucosa, as well as transitory microbiota that is component of the food consumed. Gut microbiota has been assessed to include more than 100 distinct species in every organism. Around 1500 unique species were described as component of the human gut microbiota. Intestinal microbiota is established by a total of 10^{13}–10^{14} microbial cells and is generally expected to represent ten times more cells than eukaryotic cells of humans. Large intestine is the site of the body with highest abundance of microbes, with 10^{11}–10^{12} cells/g of intestinal matter [16]. Bacteria rule the gut microbiota, which is mainly portrayed by Firmicutes and Bacteroidetes, Actinobacteria, Fusobacteria, Proteobacteria, Synergistetes, and Verrucomicrobia [17]. Fungi and archaea account for up to 1% of the human gut microbiota species [18]. Among the major typical genera of the above-mentioned phyla, *Bacteroides* sp., *Prevotella* sp., *Blautia* sp., *Clostridium* sp., *Ruminococcus* sp., *Faecalibacterium* sp., and *Bifidobacterium* sp., (in breast-fed infants) are important because of their high abundance [15, 19].

It has been recently suggested that all of the inter-individual variation of intestinal microbiota could be categorized into enterotypes, characterized as a system of co-abundant microbial communities controlled by the salient existence of one of these three genera: *Bacteroides, Ruminococcus*, and *Prevotella* [20]. Some authors found enterotypes to be a very simplified theory, thus, decreasing the complexity of intestinal microbiota into three groups [21]. For example, only two of these enterotypes [22] have been identified by some authors, as two perpetual clusters of microbiota configurations isolated by a gradient of bacterial species with varied abundances [23]. Classifying the intestinal microbiota into enterotypes or other classes, having strong connections with dietary patterns, could be very useful in customizing the cure of diseases continuing with microbial dysbiosis [24]. This will necessitate the advancement of mathematical models capable of consolidate the entire complexity, and subsequently more experimental data will be required [25].

Evolution of next-generation DNA sequencing technologies over the last 10 years has permitted a profound comprehension of microbial composition of species living in the gut, upper airways of the respiratory tract, vagina, skin, or mouth. Research was conducted to study about the improvement of diversity of gut microbiota because of the advent of culture-independent methodologies, for example, low-cost and high-throughput sequencing strategies. Focusing on 16S ribosomal RNA (rRNA) gene of bacteria is a well-known methodology [26, 27] as this gene occurs in all archaea and bacteria and comprises nine highly variable (V1–V9) regions, thus permitting the easy recognition of species. Previous strategies focused on sequencing the whole 16S rRNA gene. By utilizing this strategy, the strong insensitivity and bias of culturing techniques were featured in an early investigation, as 76% of the sequences of rRNA acquired from an adult male fecal sample belonged to new and uncharacterized species [28]. Lately, the focal point of 16S rRNA sequencing has moved towards more prominent depth investigation of shorter subregions of gene [27]; even so, the usage of shorter read lengths will lead to errors [26]. More accurate estimation of microbiota composition and diversity might be given by entire genome shotgun metagenomics because of the sensitivity, and high resolution of these methods [26]. The most detailed perspective of human-related microbial selection to date has been provided by combined knowledge from the human microbiome project and MetaHit [29, 30]. Accumulated information from these investigations grouped 2172 species, isolated from humans, into twelve separate phyla, out of which 93.5% species belonged to Proteobacteria, Firmicutes, Actinobacteria, and Bacteroidetes. Three of the twelve distinguished phyla enclosed just a single species isolated from humans, along with an intestinal species, *Akkermansia muciniphila*, the sole recognized representative of Verrucomicrobia phyla. 386 species known in humans are anaerobic and are located mostly in mucosal habitats, for example, GIT and oral cavity [29].

2.1.3 Metabolic Roles of Microbiota

Mammals have a restricted inherent ability to process polysaccharides, but they can assimilate simple sugars in the small intestine. The primary substratum for the growth and maintenance of intestinal flora is all the indigestible elements, which represents the main energy source in colon [31, 32]. Since the microbiota's genetic and species diversity gives various host-related enzymatic, metabolic, and biochemical pathways, the outcome is energy extraction, digestible substrates for the host, and an energy and nutrients supply for the expansion of particular inhabitant species of bacteria [33]. Thus, the microbiota is known as an important metabolic organ [34].

Intestinal bacteria, primarily Firmicutes, Bacteroidetes, and Actinobacteria, obtain energy from the transformation and fermentation of indigestible food substrates, especially from carbohydrate fermentation. Indigestible polysaccharides break down into monosaccharides, and later into bacterial fermentation products, particularly gases (CO_2 and H_2) and short-chain fatty acids (SCFAs) [35, 36]. For adults, the average supply of substrates is around 5–20 g of carbohydrates and 20–60 g of proteins. The fermentation process achieves high levels with an abundant generation of SCFAs in the ascending colon and cecum, where the pH is relatively acidic (in the range of 5 and 6) and the growth of bacteria is rapid. The supply of substrates reduces in the distal colon (having neutral pH), where the activity of bacterial community reduces dramatically and putrefactive procedures become quantitatively more crucial. Therefore, the generation of SCFAs (butyrate, propionate, acetate in the proportion 15:25:60) portrayed metabolic endpoint, which employ a strong trophic and energetic activity in the intestinal lumen [37]. Bacteroidetes generates acetate and propionate by degrading the undigested polysaccharides, and Firmicutes creates butyrate [38]. Acetate is ingested and afterward transferred to the peripheral level, and there it serves as a substratum for cholesterol synthesis, while propionate takes an active part in gluconeogenesis. Butyrate, as a primary energy source for colonocytes, enhances the sensitivity to insulin in mice and has a potential anti-obesogenic activity and also an anti-inflammatory effect [39]. Butyrate and different SCFAs have a major role in controlling intestinal cell proliferation and growth of obesity [40]. Butyrate encourages the constancy of cellular heritage, preferring the transformation of cells from neoplastic to non-neoplastic phenotype. Production of SCFAs is also induced by the anaerobic metabolism of protein substrates and/or peptides that may produce harmful components such as ammonia, thiols, amines, indoles, and phenols. SCFAs are responsible for performing various biological activities, such as modulation of glycemia [41], action on glucose homeostasis [42], inhibitory control of excessive production of cholesterol [36], regulation of satiety through peptides [43], increasing intake of energy without increasing the peptide YY or glucagon-like peptide 1 concentration in humans and rodents [44, 45], management of bowel kinetic activity, transport of fluid, muco-protective action [46], anti-carcinogenic action [47], and anti-inflammatory action [48] (Fig. 2.1).

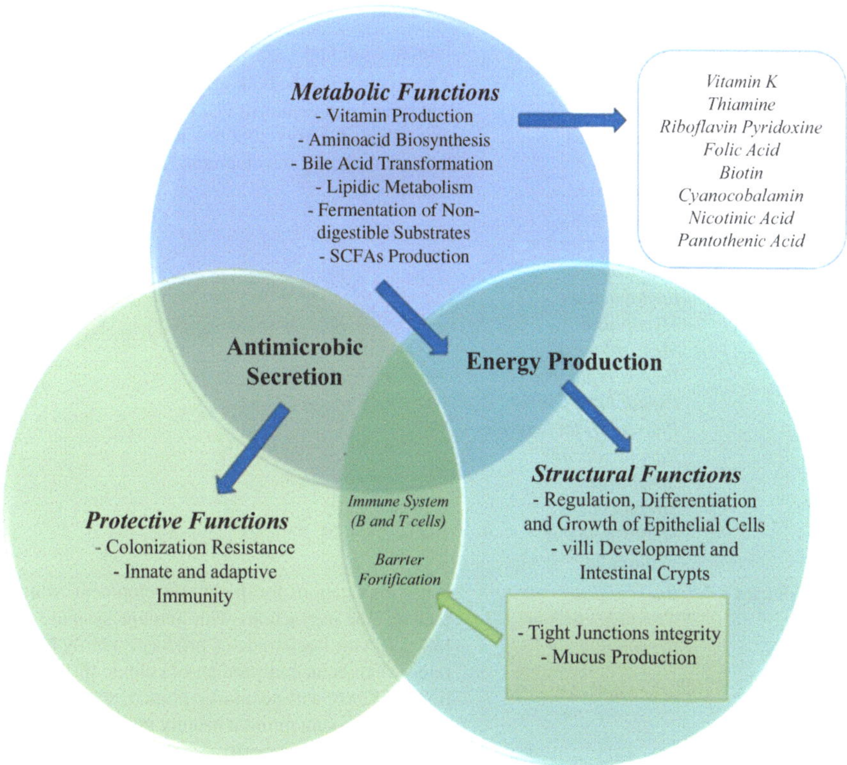

Fig 2.1 Functions of intestinal microbiota [49]

Microbiota can influence its own composition as well. The production of SCFAs differs depending on the fermentable carbohydrates existing in the bowel lumen [50, 51] that can alter the microbiota composition itself. Furthermore, starch resistant to digestion has been reported to directly enhance levels of butyrate in humans [52], and arabinoxylan, formed by the prebiotic arabinoxylan oligosaccharides, enhances levels of propionate in transversal colon [53]. Also, the microbiota conducts another significant metabolic functions, for instance, at the intestinal level it is necessary for synthesis of certain enzymatic co-factors and vitamins (folic acid, pantothenic acid, vitamin B1, B2, B6, B12, PP, H, K) and for the assimilation of iron, calcium, and magnesium [38]. It is additionally accountable for bile acids deconjugation in the liver catalyzed by an enzyme bile salt hydrolase that exists in numerous species of bacteria. Hydrolysis hinders the reuptake of these molecules by enterocytes while promoting their elimination and blocking their enterohepatic recirculation [54]. The interference of intestinal bacteria in hepatic transformation of cholesterol into bile acids, with significant implications in fat assimilation, has therefore been proven (Table 2.1).

Table 2.1 Phyla and their genera in gut: putative relationships with metabolic and gut functions

Phyla	Genera	Functions in Gut
Firmicutes	*Anaerostipes* *Bacillus* *Coprococcus* *Clostridium* *Eubacterium* *Enterococcus* *Faecalibacterium* *Lactococcus* *Lactobacillus* *Mycoplasma* *Megasphaera* *Peptostreptococcus* *Pseudobutyrivibrio* *Phascolarctobacterium* *Ruminococcus* *Roseburia* *Streptococcus* *Staphylococcus* *Veillonella*	These constitute the bulk of human gut microbiome and have been demonstrated to be associated with extraction of energy, and possibly related to diabetes and obesity development [16, 55–57].
Bacteroidetes	*Bacteroides* *Corynebacterium* *Prevotella*	It has implications for the gut development, which includes the interactions with immune system [58, 59]. Gut Bacteroidetes produces mainly butyrate, which is an end-product of colonic fermentation, and have anti-neoplastic properties and play a function in maintaining a healthy gut [60], with implications in the obesity development [57].
Actinobacteria	*Eggerthella* [61] *Olsenella* [62]	They are present in the human colon and feces, and are responsible for causing liver and anal abscesses, ulcerative colitis, and systemic bacteremia [61, 63].
Cyanobacteria	*Spirulina*	Spirulina (*Arthrospira platensis*) has hypolipidemic, hypoglycemic, and anti-hypertensive properties [64].
Proteobacteria	*Citrobacter* *Klebsiella* *Escherichia* *Shigella* *Helicobacter* *Salmonella* *Sutterella*	Proteobacteria is most unstable in host life among the four major represented gut microbiota phyla and its irregularity is proposed as a possible diagnostic reference for gut-associated ailments [65].
Spirochaetes	*Brachyspira*	The most popular species is swine dysentery, *Brachyspira hyodysenteriae*, which induces broad and extreme mucohemorrhagic colitis in rising pigs [66].
Verrucomicrobia	*Akkermansia*	*A. muciniphila* is a common inhabitant of human intestinal tract, containing up to 1% of total intestinal bacteria. It develops ideally at 37 °C and is able of fermenting glucose, N-acetyl galactosamine, and N-acetyl glucosamine [67].
Fusobacteria	*Fusobacterium* (Five species in GIT)	Fusobacteria have an impact on CRC development by interaction with innate immune system or host factors [68].

2.1.4 Development of the Human GI Microbiota

The human GIT begins from the mouth, spreading through the anatomical regions—the esophagus, stomach, small intestine, colon, rectum, and terminating at the anus [69]. The structural and functional growth of GIT is a pivotal component of human growth, since the gut must harbor the heterogeneity of dietary inputs and external antigens which are incorporated along with food into human body across various phases of life [70]. Human GIT maturation begins in utero and proceeds after birth with certain roles, for example, epithelial barrier systems, intestinal immune system, and accessory structures [70]. The primitive gut is formed about 22 days after conception from the dorsal portion of yolk sac, directing towards the emergence of foregut, midgut, and hindgut, around 25 days after conception [71]. The midgut increases quickly in length so far that it cannot fit within the developing abdominal cavity and herniates into the vitelline sac before experiencing complex turns and coming back to the abdominal cavity after gestation period of around 10 to 12 weeks [71].

It is assumed that the production of microbiota starts from birth, despite the fact that this dogma is confronted by a confined various investigations in which microorganisms have been found in womb tissues, such as placenta [72, 73]. GIT is quickly colonized after birth, with life events, for example, sickness, changes in diet, and antibiotic treatment causing disordered microbiota shifts [73, 74]. Mode of delivery seems to affect the microbiota composition, with microbiota of infants delivered vaginally possessing higher number of *Lactobacilli* during the initial days, as a result of elevated *Lactobacilli* load in the flora of vagina [75, 76]. The microbiota of infants born by C-section is insufficient and deferred in the colonization of *Bacteroides* genus, but are colonized by facultative anaerobes like *Clostridium* species [77–79]. The microbiota is commonly low in diversity in the initial stages of development, and is governed by two fundamental phyla, Actinobacteria and Proteobacteria [73, 80]. Microbial abundance increases during the first year of development, and the composition of microbiota changes to adult-like microbial profile with time-related patterns specific to each newborn child [81]. At around 2.5 years old, the newborn child microbiota's composition, diversity, and functional capabilities are close to those of adult microbiota [73, 74]. Despite the fact that the composition of gut microbiota is generally steady in adulthood, it remains exposed to perturbation by life events [82]. The microbial community shifts in people aged over 65 years, with an elevated prevalence of Bacteroidetes and *Clostridium* cluster IV, in comparison to young individuals with more prevalent cluster XIVa [83]. Another report discovered the similarity of microbiota of young generation and an elderly population (70 years), and a significant decline of microbiota diversity from a cohort of centenarians [84]. A notable relationship among diversity and living arrangements has been identified in the older population, like group dwelling or long-term residential care [85]. Microbiota's ability to perform metabolic processes, such as SCFA synthesis, and amylolysis, is typically decreased in elders, while there is an increase in proteolytic activity [86]. With increasing evidence of the role of

SCFAs as metabolic and immune mediators, the decline in SCFAs was believed to support the inflammation-ageing process in aged people's intestine [87].

Advances in metagenomic technologies have revealed the composition of human gut microbiota from early infancy [81] to old age [88]. The human intestine after birth is quickly occupied by a variety of factors and microbes considered to impact colonization which involves gestational age, delivery mode, sanitation, diet, and antibiotic treatment [89, 90]. Facultative anaerobes are the first colonizers, which builds a new environment promoting the colonization of anaerobes such as *Bacteroides*, *Bifidobacterium*, and *Clostridium* sp. Low diversity and relative abundance of Proteobacteria and Actinobacteria define the intestinal microbiota of neonates, which becomes more complicated with the growth and abundance of Firmicutes and Bacteroidetes as time period after birth increases [91–93]. At the end of first year of development, infants have an individually defined microbial profile, converging towards the distinctive microbiota of an adult, so that by the age of 25, the microbiota completely matches the composition and diversity of an adult [74, 81, 94]. The initial three years of life serves as the most important phase for dietary interventions to promote child growth and development. At this time, the intestinal microbiota, a crucial tool for health and neuro-development [95] is developed and its modification during this phase can significantly influence health and development of host. Development of gut microbiota is influenced by various factors such as delivery mode, genetics, diet, health status, gestational age, etc. (Fig. 2.2).

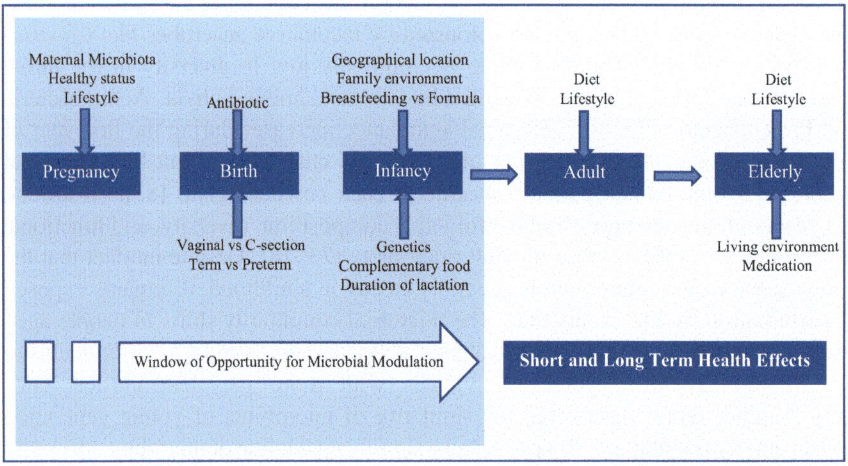

Fig 2.2 Factors that affect the development of infant, adult, and elderly gut microbiota [96]

2.1.5 Biogeography of the Human Microbiota in GIT

The microbiota composition in GIT represents the physiological properties of a particular part and is formed on both a longitudinal and transverse axis [97]. Chemical, metabolic, and immunological gradients along the intestine affect the microbiota density and composition. There are usually elevated concentrations of acids, oxygen and anti-microbials in the small intestine, and a limited transition time [98]. These characteristics restrict the development of bacteria to such an extent that only quickly growing, facultative anaerobes having the capacity to bind to mucus/ epithelia are thought to be enduring [98]. *Lactobacillaceae* dominates the microbial community of small intestine of mice [99]. Colonic environment supports a dense and abundant bacterial community, predominantly anaerobes having the capacity to use complex carbohydrates that are indigestible in the small intestine. The colon has been reported to be dominated by *Lachnospiraceae*, *Prevotellaceae*, and *Rikenellaceae* [98, 99]. Contrary to the different composition of microbiota within different GI organs, the microbiota of various colorectal mucosal areas in the same organism is conserved structurally in terms of diversity and composition [100, 101]. This property is evident even at the time of localized inflammation [101]. However, fecal/luminal and mucosal composition is significantly different [100, 101]. For instance, Bacteroidetes concentration is reported to be high in fecal/ luminal samples than in the mucosal [19, 100]. Conversely, Firmicutes, primarily *Clostridium* cluster XIVa, are augmented in the mucus layer relative to the lumen [19]. Many experiments in mice colonized with pathogen-free microbiota demonstrated a distinct microbial niche formed by the large intestine's outer mucus, and the bacterial species existing in the mucus exhibit differential proliferation and resource utilization relative to the same species in intestinal lumen [102].

Inter-individual differences in the arrangement of species and subspecies are suggested to overcome the variations in the organization of community in an individual [100, 103, 104]. The concept of a core microbiota has been projected, suggesting to be a group of the similar abundant species found in all individuals. In the set of microbial genes present between organisms, however, greater comparability can be seen than the taxonomic profile, indicating that the "core microbiota" might be best characterized at a functional rather than organismal level [103]. Individual microbiota arrangements have been recently classified into "community types" that are related with background and can be predictive of one another [105]. Multi-dimensional study of thirty-three samples from various nationalities uncovered the existence of three enterotypes recognizable by differences in the level of one of three genera: *Bacteroides* (enterotype 1), *Prevotella* (enterotype 2), and *Ruminococcus* (enterotype 3) [106]. Nevertheless, there is conflicting data encompassing the presence and development of these enterotypes [21].

2.1.6 Factors Influencing the GI Microbiota

The microbial community's complexity and richness progress via a number of stages of development spanning from neonatal phase before the apparent stabilization after weaning. In combination with individuality, there are essential inter-linked factors that assume a significant part in forming the microbial composition of human GI. Those factors involve age [107, 108], diet [109, 110], genetics of host [109–111], infections, antibiotic usage [108–110], physiology of colonization site [69], birth mode [109, 110, 112], feeding type [109, 112], and the birth environment of infants [112].

Technical variation also influences the form of developing microbial composition. For instance, culture-dependent microbe identification procedures are subject to biases that emerge from: (1) sensitivity to oxygen; (2) intractability of some species of bacteria to culturing media; and, (3) competitiveness among fast-growing and slow-growing bacteria. It restricts the existing culture-dependent techniques to be effective for the isolation of only 70% of intestinal microbes in a sample relative to culture-independent methodologies [113].

2.1.6.1 Age

The infant's microbiota is seeded during childbirth and is at first undifferentiated over the different body habitats. The predominance of aerobic bacteria at time of birth is changed during perinatal and postnatal development. During initial weeks of life, the microbiota diversifies to form a diverse microbial population dominated by anaerobes. This early stage of colonization corresponds with the stimulation of hypothalamic pituitary adrenal (HPA) axis that affects the enteric nervous system thus innervating the GIT 123]. Enteroendocrine cells of gut release a number of metabolically linked peptides, all of which are associated with food consumption, lipid accumulation, energy equilibrium and may be regulated by microbial metabolites, for example, SCFAs. Some investigations have shown that young people have a greater concentration of *Bifidobacteria* and *Clostridia* than adults; however, the gut microbiota is more stable during adult life. During old age, a final set of age-related changes in gut microbiota's composition and function occurs [114]. Aging is related with modified physiological functions, involving function of immune system, which influence the makeup of the gut microbiota. Age-related differences detailed in composition of gut microbiota include rise in the total amount of facultative anaerobes, changes in the proportion of Bacteroidetes to Firmicutes, and a pronounced reduction of *Bifidobacteria* in humans > 60 years old, during which the immune system begins to weaken. Metabolic shifts that correlate with the development and maturation of gut microbiota can be seen in the excretion profiles of bacterial products of amino acid metabolism and in energy-linked metabolites [115].

2.1.6.2 Diet

Current research indicates that diet affects the gut microbiota enormously [98]. Meta-transcriptomic research has shown the ideal microbiota to be driven by the ability of microbial individuals to metabolize simple sugars, indicating microbiota's adjustment to the abundance of nutrients in the small intestine [116]. Formation of colonic microbiota depends upon the accessibility of microbiota-accessible carbohydrates (MACs) present in dietary fiber. "Animal-based" or "plant-based" diets result in widespread modifications of human gut microbiota [117]. A crossover study showed the impact of fiber, indicating that otherwise balanced diets high in resistant starch or in non-starch polysaccharide fiber (wheat bran) lead to a powerful and reproducible augmentation of various species of bacteria in the human gut [118].

The role of food-consumed bacteria in gut microbiome had previously been underestimated, potentially as a result of methodological restrictions [119]. Various investigations have indicated that high-calorie diet brings obesity and type-2 diabetes (T2D) both in humans and mice [120–124]. Many evidences propose that the connection among diet and obesity is related to gut microbiota [125–131]. Changes in diet bring significant and rapid changes in gut microbiome composition, as indicated by various interventional studies [22, 132]. High-fat diet (60% fat) reduces the quantity of bacterial species in the gut microbiome of mice, and the composition of gut microbiome between mice on a high-fat (unpurified) diet and on a regular unpurified diet is totally different. Another study in obese mice having T2D revealed that the abundance of *A. muciniphila* was reduced and prebiotic feeding of *A. muciniphila* normalized its abundance, improved metabolic profiles, decreased fat mass, inflammation, and insulin resistance elicited by a high-fat diet [133]. It has been demonstrated that a fiber-rich diet is favorable to health, as it balances the gut microbiome [134]. Studies of 16S rRNA sequencing in humans have categorized the gut microbiota of humans into various enterotypes recognized by the kinds of bacteria present [106]. Enterotypes have been connected with long-term diets, especially those with protein and animal fat. Wu et al. [22] indicated that *Bacteroides* were related with protein and animal fat, while *Prevotella* was related to carbohydrates. The authors also examined controlled feeding in ten subjects and discovered that microbiome composition altered within 24 h of starting a low-fat and high-fiber diet or high-fat and low-fiber, and remained stable throughout the 10-d study [22]. The outcomes suggested the strong connection of diet with partitioning of enterotypes. In another study, a plant-based diet rich in legumes, grains, fruits, and vegetables, or an animal-based diet consisting of eggs, meat, and cheese was consumed ad libitum by six male and four female volunteers (aged 21 to 33 years with BMI (in kg/m^2) ranging from 19 to 32) for five consecutive days. The subject's fecal samples were cultured or directly analyzed by 16S rRNA gene sequencing [132]. It was indicated that microbiota changes in the high-fat animal-based diet, and was hypothetically connected to modified fecal bile acid profiles and microorganisms development able of activating inflammatory bowel disease (IBD) [132]. The outcomes demonstrated that a high-fat diet can change the bacteria in the gut and contribute to dysbiosis and eventually disease.

2.1.6.3 Host Genetics

The quantity of different bacteria present in the gut microbiota is affected by the host's genetic constitution in manners that influence host metabolism and can eventually affect health [135]. It has been found that family members have more comparable microbiota communities than unrelated individuals, and the gut microbiota is more comparable in mono-zygotic than in di-zygotic twins [135]. At present, there are no genome-wide investigations characterizing the specific genes and pathways to determine the gut microbiome composition [136], although some genes of the immune system are related with IBD [137, 138].

The microbiota can also be formed by the immune system of host. This impact is generally constrained to compartmentalization of bacteria in order to prevent opportunistic colonization of host tissue, while species-specific impacts are less likely because of the high levels of functional redundancy in the microbiota [16, 139–142]. Both anti-microbials collected from the host and administered have a central role in forming the gut microbiota. Paneth cells in GIT produce anti-microbials, for example, angiogenin 4, α-defensins, cathelicidins, collectins, histatins, lipopolysaccharide (LPS)-binding protein, lysozymes, secretory phospholipase A2, and lectins [143]. Such proteins are confined in the mucus layer and are almost absent from the lumen, most likely because of poor mucus dispersion or luminal degradation [144, 145]. Attenuated expression of mucosal α-defensin was observed in ileal Crohn's disease (CD) patients, featuring the significance of these proteins [146, 147]. Secretory IgA (SIgA), another part of the immune system, co-localizes with gut bacteria in the outer mucus layer and helps with constraining the exposure of epithelial cell surface to bacteria [143, 148]. SIgA is suggested to intercede the shaping of bacterial biofilm by means of binding to SIgA receptors on bacteria [149]. In IgA-deficient individuals, the expression of SIgA receptors by bacteria is reduced [150]. Microbiotic dysbiosis, specifically an over-representation of segmented filamentous bacteria (SFB), arises in mice with IgA deficiency, an impact that might be especially harmful to the host because of the capacity of SFB to firmly bind the epithelium and trigger the immune system [151].

2.1.6.4 Infections

Even though the gut microbiota influences bacterial and viral infections, the opposite is likewise obvious [152–157]. One research explored the impact of an enteropathogenic infection caused by *Citrobacter rodentium* on mice microbiota and discovered that some gut bacterial groups are altered because of *C. rodentium* infection, including a decrease in the relative abundance of *Lactobacillus* [158]. A human investigation of *Clostridium difficile* patients and asymptomatic carriers with the utilization of 16S rRNA gene pyrosequencing revealed that both had decreased microbial richness and diversity relative to healthy individuals [159]. *C. difficile* infection is characteristic of severe gut microbiota dysbiosis [160, 161]. Transplantation of gut microbiome from healthy donors to infected patients have increased the microbial richness and diversity, and it is, at present, applied clinically [162–165]. By utilizing a mouse model of hepatitis B virus infection, Chou et al. [152] demonstrated that the clearance of hepatitis B virus infection demands the formation of gut microbiota. It is apparent that the change in gut microbiota of host influences both pathogenesis and clearance of bacterial and viral infections.

2.1.6.5 Antibiotic Usage

Increasing evidence proposes that numerous non-antibiotic drugs including the medications used to treat T2D affect the gut microbiota [166–169]. The gut microbiota also influences drug efficacy [170, 171]. Antibiotics are ordinarily endorsed drugs that profoundly affect the normal microbiota of gut and their impact is fast, and relentless at times. Broad-spectrum antibiotics decrease the diversity of bacteria while increasing the concentration of certain bacteria that can be utilized by pathogens and reducing the number of beneficial bacteria [172]. The utilization of wide range antibiotics in infants and young children, for example, clindamycin, has been revealed to have the longest-enduring consequences on gut microbiota composition [173–175]. Early exposure to antibiotic in neonates can prompt microbial dysbiosis, which might be a predisposing factor for IBD [176]. There is also an association between diet and antibiotic administration. Research in mice and humans has discovered that the utilization of antibiotics early in life can promote obesity later in life, mediated by the modification of gut microbiota [177–179]. However, those studies do have limitations. Most of the mice studies on obesity are instigated by a high-fat diet with or without antibiotic treatment utilized by only male mice since they gain more weight than female mice, although no obvious sex bias is observed in human obesity. One study demonstrated that antibiotics modified the gut microbiota of host without altering the host metabolism [180, 181]. Many studies showed that antibiotics lower body weight and improve sensitivity to insulin [182, 183]. Berberine, the primary component of a Chinese herbal extract used for the treatment of bacterial diarrhea, has an anti-diabetic impact by balancing the gut microbiota and reducing glucose and insulin resistance [184, 185].

2.1.6.6 Physical and Biochemical Barriers

Intestinal mucus provides the gut microbiota a source of carbohydrates [186, 187]. The layers of intestinal mucus are made-up around the large, highly glycosylated gel-forming mucin MUC2 (Muc2 in mice), which is secreted by goblet cells [188]. The glycan structures in mucins are different and dependent on four core mucin-type O-glycans including N-acetyl galactosamine, N-acetyl glucosamine, and galactose. O-glycans represent up to 80% of the total molecular mass of Muc2/MUC2 [189]. Mucus is present throughout GIT and is thickest in the colon where it is important to mediate the relationship between host and microbiota [190]. Normalization of layers of host's intestinal mucus needs long-term microbial colonization [191]. Colonic mucus is separated into two layers comprising of a dense and impermeable internal layer and a loose external coating that is penetrable by bacteria [190]. While the internal layer is almost sterile, the mucin proteins in the external layer, embellished with a rich and diverse collection of O-glycans, provide an energy source and preferential binding sites for commensal bacteria [189, 192, 193]. The type of mucin O-glycosylation depends on the expressed glycosyl transferases and their location in the Golgi apparatus [187], modifications of which influence the composition of microbiota. For example, the presence or absence of H and ABO antigens in GI mucosa, as dictated by the genotype FUT2 (a gene that expresses an α1,2-fucosyl transferase), influences the abundance of numerous bacterial species [194]. Mucus and mucin glycosylation are consequently a key in defining the microbiota and for allowing the selection of most ideal microbial species to mediate host health [195–197]. A loss of MACs from mice diet can lead to narrow mucus in the distal colon, increased expression of the inflammatory marker, REGIIIβ, and increased microbe proximity to epithelium [198]. Colonic mucus barrier erosion under dietary fiber deficiency is related with shifting of gut microbiota towards the usage of secreted mucins as a nutrient source [199]. In contrast, administration of *A. muciniphila* (a mucin degrader) to mice avoids the development of high-fat diet-induced obesity and strengthens metabolic endotoxemia-induced inflammation by restoring the gut barrier [133, 200]. The protective function of *A. muciniphila* could be recapitulated by utilizing its purified membrane protein or the pasteurized bacterium [201]. It has been recently shown that supplementation of *A. muciniphila* reduces fat mass and alleviates body weight gain in chow diet-fed mice by mitigating metabolic inflammation [202]. The capability of *A. muciniphila* was therefore suggested as an alternative therapy to target human obesity and related disorders.

The ability of gut bacteria to use dietary or mucin glycans is directed by the collection of polysaccharide lyases (PLs) and glycoside hydrolases (GHs) encoded by their genomes [187]. Many species serve as generalists capable of degrading many polysaccharides, while others are specialists in targeting specific glycans [203]. Bacteroidetes encode a lot more glycan-cleaving enzymes than members of Firmicutes [204]. The genome of *Bacteroides thetaiotaomicron* contains 260 GHs, relative to 97 hydrolases encoded by humans [205]. The most represented family in the gut microbiota is GH13 family, which includes enzymes associated with the starch breakdown [204]. The biochemical and structural characterization of extensive degrading assembly of prominent gut species like *B. thetaiotaomicron* or *Bacteroides ovatus* uncovered that the identification and breakdown of complex carbohydrates by the human gut microbiota is considerably more complex than previously recommended [206–211]. Firmicutes members also show some unique and complex highlights, such as the recent discovery of amylosomes in the resistant starch using *Ruminococcus bromii* bacterium [212].

Mutations and lateral gene transfer can lead to diversification of microbial population [213, 214]. New bacterial functions encourage niche variation, making it a positive feedback loop where more diversification can occur [215, 216]. Additionally, interaction between gut microbes permits colonization by a diverse set of microorganisms, shaping the gut microbiota community. One mechanism proposed to intervene this impact is microbial cross-feeding. Several products of carbohydrate fermentation, including succinate, lactate, and 1,2-propanediol, do not generally aggregate to higher levels in the healthy adult human's colon, because they can act as substrates for other bacteria, including propionate and butyrate producers [217]. For instance, acetate produced by *R. bromii* (fermentation of resistant starch) [218] or lactate produced by lactic acid bacteria (*Lactobacilli* and *Bifidobacteria*) provides substrate for other microbiota members such as *Eubacterium hallii* and *Anaerostipes caccae* which convert it into butyrate [219, 220]. *B. ovatus* has recently been shown to conduct extra-cellular insulin digestion at its own expense, but to the benefit of other species that provide reciprocal advantages [221]. Such association is especially obvious in the outer mucus layer where mucin-degrading bacteria give mono- or oligo-saccharides to bacteria lacking specialized mucolytic ability [102]. For instance, the limit of cleaving sialic acid off mucins is confined to bacterial groups encoding GH33 sialidases. Numerous bacteria, including pathogens, for example, *Salmonella typhimurium* or *C. difficile*, lack a sialidase but harbor a "nan cluster" dedicated to the metabolism of sialic acid, and hence depend on other members of gut microbiota to supply them with this carbon source [222]. Intramolecular trans-sialidase, new class of sialidases is recently recognized in strains of *Ruminococcus gnavus* that can help the gut commensal bacteria to adapt to the niche of mucosa [186, 223, 224]. This action may give such bacteria a competitive nutritional advantage over other species in the gut mucosal environment, particularly in IBD which are rich in short, sialylated mucin glycans [186, 225]. Accessibility of sulfated compounds in the colon, either organic (host mucins and dietary amino acids) or inorganic (sulfites and sulfates), may impact specific bacterial groups like sulfate-reducing bacteria, which are gut microbiota occupants involved in the etiology of intestinal disorders, for example, IBS, IBD, or colorectal cancer [226].

As extensively reviewed, the bile acids distribution in small and large intestine can influence the dynamics of bacterial community within the gut [227, 228]. Essential bile acids, like taurocholate, can give homing signals to gut bacteria and encourage spore germination, as well as alleviate microbiota recovery after antibiotics or toxin-induced dysbiosis [113]. In addition, decreased concentration of bile acid in gut can play a significant part in permitting pro-inflammatory microbial taxa to expand [229].

2.1.6.7 Mode of Birth

Birth mode determines the microbial population to which babies are exposed at time of birth. For example, vaginal birth exposes infants to the microbes that are presently colonizing the birth canal of mother. Infants born via vaginal delivery have a comparative microbiota to that of their own mother as compared to other mothers [77, 230]. On the other hand, no substantial difference has been found between the microbiota of mothers and children delivered by C-section [77, 230]. Environmental factors (air, delivery and surgical equipment, other infants and health care workers) seem to affect the infant's microbiome delivered by C-section [69, 231]. Recent results for C-section-delivered infants showed that a time of labor before surgery was related to infants with a microbiota that looked like that of vaginally delivered infants, while infants born without any duration of labor had a microbiota that resembled that of the skin of mother [232]. C-section is recommended to be a reason for microbial disruption at early stages of life and this disturbance in microbial colonization influences host-microbial interaction that can prompt long-term metabolic results in the host [233–235]. Furthermore, C-section infants have higher chances of developing atopic diseases in the initial two years after birth, when compared to vaginally born infants based on data collected from 2500 full-term healthy newborns in LISA-Study [236].

The birth mode effect on acquiring *Lactobacillus* in infant's GIT is a good example of birth mode impact on the gut microbiota. In the maternal vagina, *Lactobacillus* is exceptionally common with an IndVal index of 0.922 [232]. Infants delivered through the mother's birth canal contain *Lactobacillus* as part of their microbiome profile, but those delivered by C-section do not [234]. One more study detected less *Lactobacillus* genus in the infant's microbiome profile delivered by C-section ($n = 17$, detection rate = 6%) versus vaginal ($n = 134$, detection rate = 37%) [237]. This variation in *Lactobacilli* detection rates, however, disappeared by the age of three [237].

The level of bacteria within an individual's microbiota in the genera of *Bacteroides* and *Clostridium* (*Bacteroides fragilis* and *Clostridium difficile*) is also connected with birth mode [77, 230, 231, 238–241]. In the Netherlands study of KOALA Birth Cohort ($n = 1032$), diverse bacterial species from stool samples obtained at one month of age were identified by real-time quantitative PCR assays [238]. Infants delivered by unassisted vaginal mode ($n = 826$) had reduced quantity of *C. difficile* and relatively high quantity of *B. fragilis* in comparison to C-section infants [238]. On the other hand, the inverse relationship was indicated by stool samples of infants delivered by C-section ($n = 108$) [238]. Identification of *C. difficile* on the hands and in the stools from healthy hospital personnel could be connected to ecological factors rather than with the mother [238, 242]. *C. difficile* was regarded a microorganism that only exists in hospitals [243] and was absent in women's vaginal swabs before delivery [244, 245]. This could clarify the *C. difficile* levels in the infants born in hospital and by C-section [238]. A study of 24 infants has further indicated the low abundance of Bacteroidetes ($p = 0.002$) in C-section-delivered infants ($n = 9$) in comparison to vaginally delivered infants [77]. Remarkably, this decrease in Bacteroidetes abundance continued for the first two years following birth [77]. The above studies are consistent with earlier studies that illustrate deferred formation of *Bacteroides* in first six months [231] and one year of life [246] of C-section infants.

Not all investigations have discovered a relationship between birth mode, the development and inheritance of GI microbiota. For instance, an investigation of 21 infants discovered that birth mode did not influence population of microbes in premature babies during the initial three months after birth [247, 248]. Studies have shown that infants delivered via C-section appear to have: less quantity of anaerobes; less diverse microbiota [77, 231, 249]; slower colonization of microbial population [239]; and, they develop atopic diseases [249] and metabolic disorders [235] more often than infants delivered by unassisted vaginal mode.

2.1.6.8 Type of Feeding

Methods of feeding may also influence the concentration of certain bacterial groups in infant's gut microbiota. The primary food, added into GIT postpartum is milk and its composition is known to have a direct influence on the development of early GI microbiota [250, 251]. This effect can occur by providing: fundamental nutrients for proliferation of bacteria [250]; immuno-modulatory molecules [252]; and, microbes able to colonizing the infant [253]. The form of feeding contributes towards the early post-natal growth of GI flora which is confirmed by a reported closeness between microbial composition in colostrum and the meconium of infants that were breast-fed from the first hour after birth [254]. Shared bacterial DNA has been found in human breast milk and infant's fecal samples [255]. This association is increasingly articulated between infants, their mother's milk, and areolar skin as compared to a random mother ($p < 0.001$) [256]. Such outcomes, together, are associated with the vertical movement of microbial species to the infant's gut, mediated by breast milk [256].

Methodologies focused on culture have detected more assorted microbiomes in formula-fed infants as opposed to breast-fed infants [246]. This finding has been confirmed by culture-independent studies [257, 258]. For instance, Lee et al. [257] described the impact of feeding type on the microbiota of 20 vaginally born Korean infants. Fecal samples from 10 predominantly breast-fed and 10 formula-fed babies were collected at age of four weeks. Relatively limited quantities of formula supplementation (once every 24 h in the first week after birth) to breast-fed infants changed the microbial profile to motif close to that found for formula-fed infants exclusively [259]. Some formula-fed infants were fed a diet containing 70 to 100% of formula milk, and they were also exposed to breast milk [257]. In this analysis, five bacterial species were found to be present in the fecal samples of all infants (both formula- and breast-fed groups contained *Bifidobacterium longum*, *Streptococcus lactarius*, *Streptococcus salivarius*, *Lactobacillus gasseri*, and *Streptococcus pseudopneumoniae*). Lee et al. [257] argued that the existence of these bacterial species in these babies' intestines must be independent of the feeding type, and therefore these species represent specific commensal bacteria found in 4 week-old Korean infants. The higher abundance of *B. longum*, *L. gasseri*, and *S. pseudopneumoniae*, and lesser abundance of *S. lactarius*, and *S. salivarius* were observed in breast-fed babies as compared to formula-fed babies. These outcomes are consistent with the predictions that disclosure to varied feeding types, breast or formula milk changes the relative abundance of certain commensal bacteria.

On the whole, formula-fed infants have more stable and diverse GI microbial populations with high levels of facultative and strict anaerobes as compared to breast-fed infants [257, 260–262]. Fecal samples of breast-fed infants are less complex, have high quantity of aerobic bacteria, and have shown more changes in the microbial composition in the first year following birth [257, 261, 262]. Studies recommend that once the introduction of solid foods into the diet begins, the distinctions in microbial population among breast and formula-fed infants are lost and microbial communities migrate towards an intricate adult microbiome [69, 250].

2.1.6.9 Birth Environment of the Infants

Disclosure of multiple extra-uterine disorders at time of early development of gut adds to the colonization and development of infant's GI microbiota. It is known that infants delivered by C-section are more vulnerable to ecological factors [263, 264]. It is especially valid for premature infants having high possibility of developing a flora that reflects NICU (Neonatal Intensive Care Unit), owing to the immatureness of their GIs and extended vulnerability to the environment [251].

The path of microbial transmission from surroundings to neonates is difficult to confirm yet investigations have demonstrated that microbes from the surroundings can be separated from fecal samples of neonates [265, 266]. However, cross-transference among patients and spread of a multi-drug resistant (MDR) strain, *Acinetobacter baumannii* additionally prompted an outburst in a Tunis NICU. 31 infants (26–41 weeks gestational age) got pneumonia induced by MDR *A. baumannii* and 10 deaths occurred because of infection after the transfer of MDR *A. baumannii* from an infant to another hospital's epidemic-associated surgical ward [266]. Such outcomes are agreeable with reviews that infants belonging to different geographical regions/hospitals harbor diverse microbial communities [261, 265]. Despite the fact, the PiPS experiment, a double-blind randomized placebo-controlled trial of probiotic treatment with *Bifidobacterium breve* was conducted to prevent sepsis and necrotizing enterocolitis in 1310 premature babies (born in the range of 23–30 weeks period of gestation) from 24 hospitals. The probiotic strain of *B. breve* was reported to be recognized in the feces of 37% of infants in the placebo arm, in comparison to 85% of the intervention arm, showing that ecological-associated parameters lead to cross-colonization of *B. breve* in infants [267]. Interestingly, this PiPS trial indicated no distinction in the microbial diversity of babies microbiome in two arms of the study [268].

The environment of hospital, handling, feeding, and treatment mechanisms can improve microbial transference to infants [265]. Nonetheless, information of transmission mechanisms, dominating microbial communities in the environment of hospitals and the strains of bacteria with high probability of effectively colonizing the infant's GI remain subtle and are worth investigating in further studies.

2.1.6.10 Other Factors

Various ecological parameters have been involved in forming the microbiota that involves surgery, geographic location, depression, smoking, and living arrangements (rural/urban) [73, 269–271].

2.2 Gut Microbiota Balance and Health

Bacteria are colonized in the human GIT from the time of human birth. The species and quantity of the flora are dynamically changing with conditions such as life, diet, and environment until a stable adult microbiota is established. Total number of bacteria in the intestinal tract of normal people is as many as 10^{14} [272]. This bacterial community is mainly composed of obligate anaerobic bacteria, aerobic bacteria, and facultative anaerobic bacteria. Among them, anaerobic bacteria are more prevalent than aerobic bacteria, and 60% of anaerobic bacteria are thick-walled bacteria, more than 20% *Bacteroides* [273]. The intestinal flora of healthy people can be roughly divided into three categories: (1) Intestinal dominant bacteria, mainly obligate anaerobic bacteria that are symbiotic with the host, including *Bifidobacterium, Bacteroides, Lactobacillus, Clostridium* genus, with nutrition, immune regulation and metabolism; (2) pathogenic bacteria coexisting with the host, mainly facultative anaerobic bacteria, when the intestinal flora is disordered, can cause disease; (3) Pathogens, such as *Proteus* and *Pseudomonas*, due to the small number of bacteria and long-term colonization opportunities, once the body's immunity is low, the number is beyond the normal range, causing disease. Due to the bactericidal action of gastric acid and intestinal peristalsis, the number of bacteria in the stomach is very small, the small intestine acts as a transition zone, the jejunum is dominated by a small amount of aerobic bacteria, and the number of ileal bacteria is large, mainly gram-negative anaerobic bacteria in the colon. The number and type are obviously increased, the concentration can reach 10^{12} cfu/mL, mainly composed of anaerobic bacteria such as *Bifidobacterium*, *Lactobacillus*, *Bacteroides*, and *Clostridia* [274]. The terminal colon is very different and is regulated by pathophysiological conditions. In healthy individuals, the host maintains a steady state symbiotic relationship with the microbe, the host provides a nutritious and stable environment, and the microbes participate in the protective barrier of the intestinal mucosa. The gut microbiota provides a broad, anaerobic or hypoxic, constant temperature environment, which helps the host to improve the decomposition efficiency of nutrients, increase the absorption of beneficial substances, synthesize nutrients and essential vitamins needed by the body, and maintain the nervous system. Stability promote the immune system. In an unbalanced state, dysbacteriosis affects host growth, development, health and disease, and can also affect drug treatment [275].

2.2.1 Gut Microbiota and Gut Barrier

A layer of polarized columnar epithelial cells and epithelial area, including lamina propria, enteric nervous system, connective tissue, and muscular layer, are present in intestinal mucosa. There are four types of intestinal barriers: mechanical barriers, immune barriers, chemical barriers, and biological barriers. First is the mechanical barrier which is tightly connected. The intestinal mucosa is not only an anatomical physical structure, but more importantly it is an intestinal barrier. The energy through the intestinal epithelial cells is primarily through an extra-cellular pathway, with specific membrane channels and pumps, as well as a para-cellular pathway that is regulated by tight junctions. Under the microscope, they look like discrete contacts of a series of adjacent cells. Eventually a complex tight junction is formed that maintains the normal structure of the intestinal mucosal cells. Next to the immune barrier, this barrier helps the intestinal cells to secrete IgA normally. The third barrier is a chemical barrier in which microorganisms and antigens in the gut are degraded in a non-specific manner through the gastric acid environment, pancreatic fluid, and biliary secretions. Digestive enzymes are mainly proteases, lipases, amylases, and nucleases that kill microorganisms by destroying the cell walls of bacteria [276]. A large amount of digestive juice produced by the intestine can adulterate the toxin and clean out the intestinal lumen, making it hard for potential pathogenic bacteria to bind to the intestinal epithelium, thereby shortening the presence of potentially toxic or pathogenic substances in the intestinal lumen. It can stimulate the secretion of gastric acid protease. Finally, the biological barrier, the intestinal flora is located in the outermost layer of the mucus, is an important part of the metabolism, proliferation, and maintenance of the intestinal barrier of the epithelial barrier [4]. However, the interaction between microorganisms and intestinal epithelial cells is twofold. Some are considered pathogens, while others are considered symbiotic. The symbiotic flora limits the colonization of pathogens by competing for nutrients and niches, changing pH, releasing antibacterial substances that allow exchanges between species, and optimizing the number of beneficial microorganisms. Of course, the gut flora also provides other important functions for the host. The results indicate that the native bacteria can regulate gene expression involved in a variety of crucial intestinal functions that includes absorption of nutrients, mucosal barrier enhancement, angiogenesis, xenogeneic metabolism, and postnatal intestinal maturation [277]. The intestinal barrier plays a significant role in maintenance of human health. The destruction of intestinal barrier can cause dysfunction of the body and lead to a variety of disorders.

2.2.2 Gut Microbiota in Metabolism

The mixed oxygen in the food is consumed by aerobic and facultative bacteria in the upper part of the intestine, and the closure of the intestinal wall makes the large intestine meet the anaerobic environment required by the obligate or facultative anaerobic bacteria fermentation. In the large intestine, crude fibers and non-starch polysaccharides (NSP), which cannot be decomposed and used by the host, become the raw materials for its fermentation and eventually produce volatile fatty acids, thus providing energy for the host. At the same time, volatile fatty acids can also promote the growth of intestinal epithelial cells, accelerate the repair of intestinal damaged mucosa, and even regulate the gene expression of epithelial cells, inhibit the occurrence of enteritis and colon cancer, thereby promoting the health of the host. In addition to producing beneficial substances, intestinal microbial fermentation in the body also produces metabolites that inhibit host growth. Intestinal microorganisms degrade tyrosine and tryptophan into highly toxic phenol and aromatic compounds in the intestinal tract and expel them from the urine, but these phenol compounds are not found in the urine of sterile mice. Ammonia is another toxic waste produced by microbial urease fermentation of amino acids in the intestinal tract. However, urea hydrolysis in sterile animals cannot take place, and the concentration of ammonia in the colon of normal animals is several times the concentration required for cell damage, which inhibits the growth of the host. Therefore, the main mechanism of using antibiotics to promote growth may be to reduce the inhibiting effect of toxic and harmful substances produced by intestinal microbial fermentation on the growth of animals [278].

2.2.2.1 Lipid Metabolism

Fiaf is an endocrine signal expressed in intestinal epithelium, liver, and adipose tissue that activates the Tie2 receptor and initiates intracellular signal transduction to inhibit lipoprotein lipase (LPL) activity and reduce triglyceride deposition in adipose cells. Backhed et al. found that the total body fat content, weight of epididymal fat pad, and LPL activity of aseptic fed Fiaf+/+ mice and conventionally fed Fiaf−/− and Fiaf+/+ mice were higher than those of aseptic fed Fiaf+/+ mice. The inhibition of intestinal microorganisms on the expression of Fiaf and the deletion of the mutation of Fiaf gene will lead to the decrease of the expression of Fiaf in intestinal epithelial cells, weakening the inhibition of Fiaf on IPL activity and promoting the storage of triglycerides in fat cells. The fat precipitation effect caused by Fiaf gene deletion is consistent with the effect of microbial inhibition of Fiaf. Srebp-1 and ChREBP are transcription factors mediating the lipid response of liver cells to insulin and glucose. Acetyl CoA carboxylase (Acc) and fatty acid synthase (Fas) genes are the target sequences of srebp-1 and ChREBP, which can promote the synthesis and storage of fat. Studies have proved that ChREBP mRNA in liver of conventionally fed mice was significantly increased ($p < 0.01$), and srebp-1 mRNA was also significantly increased ($p < 0.05$).

2.2.2.2 Protein Metabolism

The proteins ingested by the host are mainly broken down into amino acids that can be absorbed and utilized by protease and peptidase. Studies have shown that although only a few bacteria contain protease, almost all bacteria have peptidase. As a result, intestinal microbes are able to independently break down the proteins taken by the host to meet their own needs. The proteins degraded and utilized by intestinal microorganisms cannot be utilized by the host. Amino acids that are broken down by gut microbes but not used can be used by the host to help digest proteins. Intestinal microbes can not only break down proteins but also use ammonia in the intestine to synthesize bacterial proteins. Microorganisms in the rumen of cattle are able to synthesize bacterial proteins from ammonia and provide proteins to the host. In the case of protein deficiency, ammonia formed by the degradation of amino acids by intestinal microorganisms can enter the host and recycle to synthesize amino acids, which makes up for the deficiency of protein and is beneficial to the growth of the host.

2.2.2.3 SCFAs Production

At least four different pathways allow the SCFAs to signal to the host. First, SCFAs, particularly butyrate, serve as an energy substrate for colonocytes [67, 68], and in retaliation to decreased availability of energy, germ-free mice slow down the transportation through small intestine to permit more time for nutrient absorption [69]. Second, propionate act as a substrate for gluconeogenesis and can stimulate intestinal gluconeogenesis, by signaling through the central nervous system (CNS) to defend the host from diet-induced obesity and glucose intolerance [64]. Third, acetate and butyrate, can act as inhibitors of histone deacetylase [70, 71]. Fourth, SCFAs signal through G-protein-coupled receptors like GPR41 and GPR43, and thus affecting various crucial processes including inflammation [72] and enteroendocrine regulation [73]. SCFAs generation is, however, just one feature of microbial metabolism in the gut [279].

2.2.2.4 Bile Acid Conversion

Bile acids are generated in the liver, stored in the gall bladder, and secreted into the duodenum after consumption. Bile acids have long been known as single emulsifiers for absorption of lipids, and have also been found to be effective signaling molecules regulating other metabolic pathways. Intestinal flora is a significant controller of bile acid metabolism. Intestinal flora can not only regulate the synthesis of bile acid but also promote it to produce secondary metabolites. Therefore, the diversity of bile acids in germ-free mice is much less than that in colonized mice [280].

Bile acids can bind to cell receptors like farnesoid X receptors (NR1H4) and G-protein-coupled receptor (GPCRs, TRG5), and are involved in regulating lipid metabolism and maintaining homeostasis of the body's internal environment. Activation of FXR has a crucial role in modulating bile acid equilibrium in the body. Studies have shown that the activation of the ileum FXR receptor can promote the increase of the expression level of the growth factor (FGF)19 gene in fibroblasts and the homologous FGF15 gene in mice, thereby inhibiting the synthesis of bile acid. In addition, the activation of FXR receptor also promotes the expression of small heterodimer (SHP) genes, the transcription level of ileum bile acid-binding protein (IBABP) gene, and the expression level of organic solute transporter-ost beta gene, thereby regulating the absorption and transport of bile acid in the terminal ileum. Activation of TRG5 receptor induces glp-1 secretion by intestinal L cells, which improves liver and pancreas role and enhances glucose tolerance in mice suffering from obesity. Studies have shown that TGR5, which activates brown fat tissue and muscle, enhances expenditure of energy and prevents diet-induced obesity. The intestinal flora, therefore, can be used to regulate the metabolism of bile acid pool of FXR and TGR5 receptors to adjust and control signal, and regulate the body fat metabolism and sugar metabolism, and finally play a decisive role for diabetes and obesity. In addition, the study of Baghdasaryan et al. on the mouse model of bile duct sclerosis showed that inhibiting the absorption of intestinal bile acid can effectively improve the cholestatic liver and bile duct injury in mice. Molecular concatenates (anti-apoptotic protein Bcl2, long non-coding rna-hi9, and nuclear receptor Shp) can maintain normal liver function by regulating the balance of bile acids in the body. Therefore, maintaining bile acid homeostasis is an important prerequisite for improving body health [281].

2.2.3 Gut Microbiota and Host Immunity

Firmicutes and *Bacteroides* are the most important intestinal bacteria in animals. Firmicutes are mainly gram-positive bacteria, such as *Clostridium*, *Streptococcus*, and *Lactobacillus*. Bacteroidetes are mainly gram-negative bacteria, including *Bacteroidetes multiformis* and ovalis. An important role of intestinal bacteria is to improve the host's digestion and utilization efficiency of nutrients. However, in the process of co-evolution with the host, animal intestinal microbes have evolved more functions. For example, intestinal microbes can regulate intestinal development, angiogenesis, and lymphocyte development as signal molecules. In addition, intestinal bacteria also has an extremely crucial role in protecting the host from pathogens. By competing with bacterial pathogens for dietary nutrients, intestinal bacteria limit the rapid colonization of pathogens in the intestinal tract. Gut microbes can also stimulate host immune responses. However, the association between microbes in the intestinal tract and the host is not always mutually beneficial. For example, *Enterococcus faecalis*, one of the most important flora in the human intestinal tract, can invade mucosal tissues and increase the incidence of bacteremia and infectious endocarditis in humans.

Intestinal microorganisms are rebooting and regulating factors of host innate immunity and adaptive immunity. When the body is exposed to pathogenic factors, the body will activate related receptors, for example, Toll-like receptors (TLRs) and nod-1ike receptors (NLRs), to activate inflammatory response and kill pathogenic factors. Therefore, it is important to comprehend the association of gut microbes and immune system [282].

2.2.4 Gut Microbiota and Innate

Intestinal microbes are known as "superorganisms" that encode genes for breaking down dietary fiber, amino acids, and drugs. Intestinal microbes can promote the formation of immune function and influence the composition of T-cell subsets. Wu et al. established a sterile chicken model, indicating that intestinal microorganisms can promote the development of spleen and improve immunity. Gut microbes can adjust the immune function of the immune system, for example, *Bifidobacterium* stimulating immune cells to secrete IL-6, IL-1, that promote differentiation of mature B-lymphocytes and T-lymphocyte proliferation, enhance the killing ability of NK cells. In addition to this, some strains of *Bifidobacterium* having anti-inflammatory activity increase the secretion of intestinal IgA, and induction of mature dendritic cells [283].

Modulation of immune system is not only affected by microbial flora, but also the reaction in the microbial flora of immune system played a key function in shaping gut microbes group. SIg-A the secretion of intestinal lamina propria of gram-negative bacteria have special affinity, can pack by bacteria, inhibit bacteria and intestinal epithelial cells, specific binding to prevent bacteria in intestinal epithelial cell adhesion, shifting to avoid bacteria through intestinal epithelium [284].

2.2.5 Diet-Mediated Production of Beneficial or Detrimental Metabolites by the Gut Microbiome

Microbial metabolites are produced by microorganism–microorganism and host–microorganism interactions.

2.2.5.1 Polyamines

Putrescine, spermine, and spermidine are polycationic molecules present in all living cells and are essential to many biological functions that includes gene transcription and translation, growth of cell, and death. The intestinal tract comprises a large amount of polyamines, derived from diet and *de novo* by host and microbial cells. Polyamines are accountable for increasing the integrity of intestinal epithelial cells (IECs) barrier [285]. Polyamines, as demonstrated by the in vitro studies, can promote the generation of inter-cellular junction proteins, which are essential for controlling para-cellular permeability and reinforcing epithelial barrier function.

2.2.5.2 SCFAs

Bacterial fermentation in the colon produces SCFAs (acetic acid, butyric acid, and propionic acid) as their main metabolic end products by using undigested complex carbohydrates as substrates. SCFA concentrations in the gut [31] are dependent upon microbiota composition, intestinal transit time, microbiota-host metabolic flux of SCFAs, and fiber content of host diet [286]. These microbiota-generated metabolites are crucial sources of energy for gut microbiota and IECs. Apart from acting as substrates for energy production, SCFAs have various regulatory functions, and their impact on physiology and immunity of host is still apparent.

2.2.5.3 Formyl Peptides

Formyl peptide receptors (FPRs) can recognize conserved N-formyl peptide motifs that are present in bacteria, and their closely associated motifs present in mitochondria. Non-formylated endogenous ligands are also detected by FPRs, which includes serum amyloid A, protein annexin, cathelicidin anti-microbial peptide. Instigation of FPRs results in enlisting the leukocytes and generation of pro-inflammatory cytokines, super oxides, and enzymes to fight infections. FPRs are stated by innate immune cells, endothelial cells, epithelial cells, neural cells, and muscle cells, and many studies suggested the instigation of FPRs on non-phagocytic cells to be necessary to achieve tissue homeostasis after infection or injury [287].

2.3 Gut Microbiota Dysbiosis and Disease

Stability of the intestinal micro-ecology is an indispensable part of human health. The imbalance of intestinal micro-ecology may induce a series of diseases, such as T2D, autoimmune diseases, senile dementia, obesity, IBD, depression, IBS, Alzheimer's disease, cancer, etc. According to "China's adult diabetes prevalence and control status," the prevalence of diabetes in adults aged 18 and over in China has reached 11.6%. Diabetes has become one of the most important and difficult public health problems in China.

From the birth of the baby, the bacteria settle into the intestines. Under the influence of dietary intake and environmental conditions, the ratio of various intestinal microbes tends to be stable. Therefore, each individual's gut microbiota is unique in the genus and species level, but has a strong universality at the door level, such as *Bacteroides* and thick-walled bacteria. The microbiota colonizes for a long time and forms a gut micro-ecology with its living environment. These intestinal flora participate in the regulation of human health through various ways such as absorption of energy, alteration of intestinal permeability, production of SCFAs, choline metabolism, bile acid metabolism, and brain–gut axis. Therefore, the intestinal flora is closely related to the metabolism and immunity of the human body. In addition, the normal intestinal flora prevents the invasion of foreign pathogenic microorganisms by establishing mechanical, biological, and immune barriers, and maintains the stability and micro-ecological equilibrium of intestinal environment. Probiotics colonize the intestinal mucosa to create a biological barrier, reducing the infection and colonization of pathogenic microorganisms. Certain probiotics produce anti-bacterial substances that suppress the growth and reproduction of noxious bacteria [288].

When the internal or external environment causes imbalance of intestinal micro-ecology, it will lead to disease. In Gordon's study, the intestinal flora of obese mice was transplanted into sterile mice, which showed a significant increase in body weight [289]. Taiwanese scholars have found that WEGL can alleviate metabolic disorders caused by intestinal flora imbalance and obesity [290]. AIEC bacteria in the gut of CD patients can adhere to and invade IECs. AIEC releases macrophages and releases IFN-γ and TNF-α, which enhances its own value and aggravates inflammation [291]. A study by the Tokyo University of Science and the University of Tokyo pointed out that laminarin in seaweed can prevent the occurrence of IBD by increasing the number of *Lactobacilli* in the intestine [292].

Investigations have shown that gut microbiota diversity is the key to gut health. Some treatments can reduce the diversity of intestinal microbes, so the patient relapses after stopping the drug. Microbiota may also promote the resistance of pathogenic species to drugs, or lead to the expansion of disease-causing populations and enhance virulence [293]. Research on the gut microbiota has become the key to treating these diseases.

2.3.1 Gut Microbiota and Metabolic Disorders

The human's gut microbiome as a part of the digestive system, can participate in the body's digestion of nutrients, and can affect the body's own metabolic activities [294]. Among them, *Bacteroides* bacteria can degrade a large group of plant polysaccharides (such as cellulose, hemicellulose, pectin, resistant starch, etc.) that cannot be digested in the human body, thus providing additional energy to the host. For the extra energy provided by bacteria (mainly in the form of carbohydrates), the body combines it into fat storage in adipose tissue, making the effect *Bacteroides* on the body's sugar metabolism a major cause of obesity

[295]. Similarly, Phylum Firmicutes bacteria that degrade non-degradable polysaccharides in the body's digestive tract are also likely to be a major contributor to obesity.

In 2004, a study by Backhed et al. [296] found that gut microbes may affect the body's energy storage, suggesting that obesity may be associated with it. Studies have shown that gut microbes use the body's undigested polysaccharide metabolism to produce small molecular compounds that can be used by the body to increase their energy, and in mouse models, gut microbes can increase the host's metabolic rate, increase its ineffective circulation, and store excess energy in fat form. Intestinal microorganisms increase the density of capillaries under the intestinal microflu, which contributes to the absorption of nutrients; the intestinal microbe inhibits the expression of the intestinal epithelial to Fiaf and may promote the synthesis of fat in the liver.

Imbalances in the gut microbiome can lead to metabolic disorders, such as insulin resistance due to steady state imbalances [297], which cause abnormalities in the sugar metabolism of the TMA/FMO3/TMAO pathway regulation. The use of sugar-reducing lipid-adjusting side intervention after 3 months can significantly reduce blood sugar lipid levels in patients with combined hyperlipidemia in obese T2D, improve insulin resistance, and be equal to metformin, while regulating the patient's intestinal flora, increasing the beneficial bacteria represented by *Blautia* and *Faecalibacterium*. Changes in the structure of the flora were significantly related to an improvement in blood sugar lipid levels [298].

A new study has seen [299] a change in the composition of the fecal microbiome in postmenopausal obese women with low-calorie diet interventions, preserving the core microbiome and changing the structure of some functional microbiomes. At the same time, the concentration of fecal bile acid decreased significantly, which was related to the metabolic pathways of amino acids, radon, and lipids in plasma. Intestinal flora can also produce SCFAs by fermenting soluble dietary fiber [300, 301], and SCFAs can reduce serum triglyceride and cholesterol levels by inhibiting the activity of liver lip-creation enzymes, promoting the production of cholesterol oxidase that accelerates the degradation of cholesterol, improves liver utilization, and increases bile acid synthesis [29], lower serum cholesterol. Intestinal flora regulates fat cytokines, component binding proteins, and other genes and enzymes to regulate blood lipids [30–32]. There have been a large number of experiments and clinical studies which showed that the disorder of intestinal flora structure is related to metabolic syndrome.

2.3.2 Gut Microbiota and Hepatic Disorders (e.g. NAFLD and ALD)

Recent reports have indicated that gut microbiota is closely associated with alcoholic liver disease (ALD) and non-alcoholic fatty liver disease (NAFLD). ALD is a series of liver lesions due to long-term heavy drinking. According to pathological features, it is divided into mild alcoholic liver disease, alcoholic hepatitis, alcoholic fatty liver, alcoholic liver fibrosis, and alcoholic cirrhosis. One of its pathogenesis is the damage of the intestinal barrier. The damage of intestinal barrier results in intestinal micro-ecological disorders, enhanced permeability of intestinal mucosa, displacement of a large number of bacteria and endotoxin (LPS) in the intestinal tract, and excessive production of inflammatory factors, thereby accelerating the occurrence and development of the disease [302]. Inokuchi et al. found that alcohol favors the development of gram-negative bacteria such as Proteobacteria in intestine, thereby reducing the number of anaerobic bacteria such as *Bifidobacteria*. Since the Proteobacteria are considered to be important bacteria that initiate the innate immune system, an increase in the number of Proteobacteria can result in activation of immune system, which will promote the development of chronic inflammation of the liver [303]. Bull-Otterson et al. found that alcohol intake can cause damage to the local immune defense system of the GI tract, promote the growth of intestinal bacterial overgrowth (SIBO), and significantly reduce the number of thick-walled bacteria and *Bacteroides* in the intestine. Gram-positive (Actinomycetes) and gram-negative (Proteobacteria and Prevotella) increased in number, and LPS in the intestine was released in large quantities, causing liver damage [303]. NAFLD has become a reason of chronic liver disease (CLD), and its occurrence is the result of a combination of genetics, environment, and lifestyle. A growing number of reports have indicated that the imbalance of intestinal microecology is involved in the evolution and progression of NAFLD, mainly through the function of enteric axis, and elevated levels of bacterial lipopolysaccharide (LPS) in the systemic or portal or circulation in various CLDs [303]. The study found that there was a rise in the amount of SIBO and inflammatory factor, tumor necrosis factor alpha in NASH patients [303]. In summary, the relationship between microbial populations and NAFLD and ALD can be represented by the following figure:

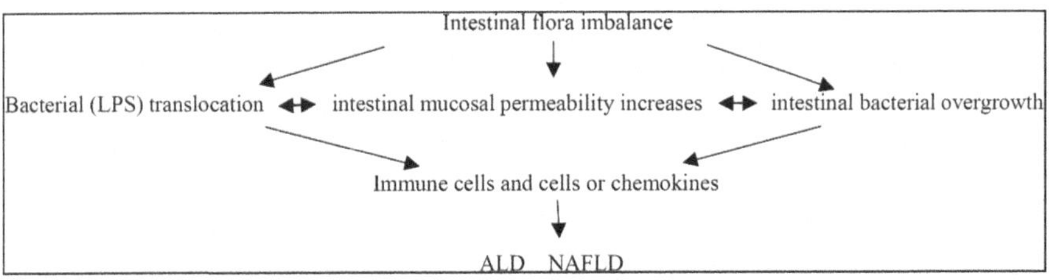

Regulating the intestinal flora becomes a new direction for the treatment of ALD and NAFLD. Use of probiotics and prebiotics can regulate the intestinal flora to prevent or treat NAFLD. Kirpich et al. found that ALD patients were supplemented with *Bifidobacterium* and germ lactic acid bacteria to maintain the integrity of the intestinal barrier, rebuild the balance of intestinal microbes, and prevent intestinal microbial translocation and harmful inflammatory reactions [304].

2.3.3 Gut Microbiota and Autoimmune Diseases: Inflammatory Bowel Diseases

The dysregulation of gut flora may lead to a variety of autoimmune diseases, including IBD. Autoimmune refers to the phenomenon that the body's immune system produces antibodies and sensitized lymphocytes against its own tissue components, causing an immune response. When autoimmunity causes dysfunction of its own tissues and organs and clinical symptoms appear, it is called autoimmune disease (AID). At present, there are more than 30 kinds of autoimmune diseases, most of which are primary and a few are secondary. The cause of primary autoimmune disease is unknown, closely related to genetic factors, and is divided into organ-specific and non-organ-specific. Target antigens and lesions of organ-specific AIDs are often restricted to a specific organ. Target antigens and lesions of non-organ-specific AIDs are often systemic or systemic, and secondary refers to other diseases or treatments. The dysregulation of intestinal microecology may lead to a variety of autoimmune-related diseases. Intestinal microorganisms can directly affect the body's innate immune system through TLRs and other related immune receptors, and have a significant function in the pathogenesis of a variety of autoimmune and inflammatory diseases [305]. Recent studies have shown that a variety of auto-immune diseases, for example IBD, metabolic syndrome, multiple sclerosis, rheumatoid arthritis, etc., are associated with abnormal changes in intestinal microecology [305]. Many studies have shown that small molecules secreted by intestinal bacteria can enter the cell through transporters or endocytosis on the surface of intestinal mucosal cells, and activate a series of signal pathways related to cell survival. It was found that patients with IBD have different degrees of intestinal microbial abnormalities, the most common is the reduction of thick-walled bacteria and the increase of Proteobacteria. Some people have suggested through clinical analysis that patients with active IBD have lower abundance of *Clostridium sphaeroides*, *Clostridium sp.*, *Bifidobacteria*, in the active period and remission period of ulcerative colitis. The abundance of *E. coli* and *Lactobacilli* did not differ between the active phase of IBD and the remission period [305]. By altering the population or community of microorganisms, reshaping the structure and function of intestinal microbes, and then regulating immunity, it is expected to provide new possibilities for the treatment of autoimmune diseases.

2.3.4 Gut Microbiota and Cardiovascular Disease

Community structure modifications in the gut microbiota are closely associated with cardiovascular disease (CVD). CVD is considered to be one of the major causes of death in contemporary human diseases, with the most common diseases including hypertension, coronary atherosclerosis, and heart failure. Trimethylamine N-oxide (TMAO), a metabolic derivative formed by the intestinal flora, can increase atherosclerosis and promote the risk of cardiovascular diseases such as chronic heart failure [306]. Yang et al. found that the abundance of intestinal flora in the hypertension group decreased significantly from both clinical observation and animal experiments. The main reason was the decrease in the number of probiotics such as *Bifidobacteria*. Some scholars believe that the intestinal flora metabolites may regulate blood pressure through the buffer system of SCFAs receptor-olfactory receptor 78 and G-protein coupled receptor orphan [306]. In recent years, gene sequencing has found that the intestinal flora of patients with coronary heart disease is disordered, and the content of *E. coli*, *Helicobacter pylori*, and *Streptococcus* is increased, and *Bifidobacteria*, *Lactobacillus* content is reduced [306]. The metabolite TMAO of the intestinal flora is also associated with atherosclerosis. Experiments have shown that plasma levels of TMAO are positively correlated to mouse atherosclerotic plaque load [306]. Patients with heart failure are often accompanied by gastrointestinal congestion, prone to loss of appetite, abdominal distension and other symptoms, decreased gastrointestinal motility leads to accumulation of gastrointestinal contents, a large number of bacteria can easily destroy intestinal homeostasis, causing dysbacteriosis. Further research found that the pathogenic bacteria such as *Salmonella* and *Shigella* in the intestinal flora of the patients increased significantly [306]. A study has shown that the severity of heart failure is also related to TMAO. Therefore, changing the intestinal ecology through probiotics will be a new entry point for the prevention and cure of cardiovascular diseases.

2.3.5 Intestinal Microflora and Nervous System Diseases

2.3.5.1 Microbiota–Gut–Brain Axis

At present, many mental diseases (autism, Parkinson's disease, and Alzheimer's disease) are highly related to intestinal flora, and our joys and sorrows may also be regulated by flora. Many of our desires and preferences may also be affected by intestinal flora, including appetite food preferences, and even sexual orientation. These connections involve an important chain of relationships: the bacteria–intestine–brain axis. Although our brain and intestines are located in two separate parts of our body, there is a very strong relationship between them. In fact, there may be three channels in the bacteria–intestine–brain axis. One is the nerve channel, the second is the blood channel, and the third is the immune channel. Some intestinal substances may pass through the intestinal barrier, through the blood, pass over the brain–blood barrier, thus affecting the brain. Some of the cells involved in the intestinal immune response may repeat the same immune response in the brain.

2.3.5.2 Intestinal Microbiology Group Is Closely Related to Neurological Diseases

One study found that many patients with Parkinson's disease suffer from severe constipation for a long time before they are diagnosed. Bacteria in the human gut decompose undigested proteins into toxic substances such as ammonia, mercaptan, indole, hydrogen sulfide, and histamine. These toxic substances can be excreted from the body through the stool. However, the intestinal function of the elderly is declining, especially in elderly patients with constipation. It is very difficult for elderly patients to rule out these toxic substances. Over time, toxic substances will accumulate in large quantities. When toxic substances accumulate to a certain extent, they will slowly enter the brain with blood circulation. Damage to the CNS can lead to Alzheimer's disease. For Parkinson's disease, higher the enterobacteriaceae in the intestinal tract of patients, more serious the symptoms will often be, and the pathogenic protein in the brain, α-synaptic nucleoprotein, is also closely related to the pathological changes of the enteric nervous system. [307].

The researchers first bred two groups of mice that produced too much α-synaptic nucleoprotein, which is thought to be one of the "culprits" of Parkinson's disease. The only difference between the two groups was that one group had a complete intestinal microflora and the other group was sterile. The results showed that aseptic mice not only did not show the symptoms of Parkinson's disease but also performed much better in running, pole climbing, and other motor performance tests. The researchers then fed some aseptic mice with SCFAs formed by the decomposition of food fiber by intestinal flora and transplants intestinal flora obtained from the feces of patients with Parkinson's disease. As a result, all of the mice developed symptoms of Parkinson's disease and it is concluded that intestinal microbiome is an important promoter of this disease. Changes in the composition of intestinal flora or intestinal bacteria themselves may contribute to or even lead to deterioration of motor function, which is the main symptom of Parkinson's disease.

In this framework, antibiotics, probiotics, diet, fecal bacteria transplants, and meditation, which may regulate flora, may be ideal tools and the best way to treat neurological or mental illness.

2.3.6 Intestinal Microflora and Cancer

2.3.6.1 Importance of Microorganisms in Human Cancer

Cancer is the number one killer of human health, but the complex relationship between the mechanism of cancer and environmental microorganisms has been difficult to prove. Since the partial success of William Coley's attempt to treat sarcomas with local injection of bacteria (Coley's toxin) in the late nineteenth century, the relationship between cancer and pathogens such as bacteria, viruses, and fungi has attracted worldwide attention [308]. Especially after the first discovery of microbial membrane on the surface of cancer cell mucosa by Christine et al., the study of the interaction between human microorganisms, especially intestinal microorganisms and cancer has become a hot topic.

2.3.6.2 Progress of Intestinal Microbiome in Cancer Research

Intestinal microflora is not only related to the formation of the immune system, but also to the interaction between the immune system. Under normal homeostasis conditions, intestinal symbiotic bacteria are recognized by TLRs and has a crucial role in maintaining the homeostasis of intestinal epithelial cells. In the experiment of chemical induction of intestinal epithelial cell injury in mice, Rakoff-Nahoum found that mice lacked a key connector molecule in the microbial ligand or linker protein pathway produced by pathogenic microorganisms and intestinal symbiotic bacteria which will aggravate the damage to the cells [309]. It can be seen that the health of the body and disease state is the outcome of interaction between pathogenic bacteria and intestinal flora. Upadhyay et al. demonstrated that the intestinal microbe group interacts with the immune response and forms the related lipid metabolism by affecting obesity. Russell et al. found that if *Candida albicans* mutates in intestinal flora, the specific chemicals produced will affect the immune response and make the immune system oversensitive and produce allergic diseases.

For example, related studies have shown that *Clostridium nucleatum* is a common bacteria living in human large intestine, and it is also considered to be a key leader in colon cancer. In addition, intestinal *Clostridium* and *Bacteroides* are also one of the pathogenic bacteria of colon cancer. The researchers have found that a group of probiotic bacteria in the intestinal tract can stimulate intestinal cells to activate the Nrf2 signaling pathway, which has a protective effect on small intestinal cells [310]. This finding is of great significance for the use of bacteria to treat intestinal diseases and to reduce the intestinal damage caused by cancer radiotherapy.

2.3.6.3 Achievements of Intestinal Microbiome in Cancer Prevention and Control

French scientist Sophie Viaud used a cyclophosphamide anticancer drug to change the composition of the intestinal microbial population, driving gram-positive bacteria into the secondary lymphatic system, triggering a special helper T cell attack on the tumor. In order to achieve the therapeutic effect of killing tumor, Chen et al. found that the intestinal microflora of individuals is dominated by bacteria that use different fibers, such as Plumeria and *Bacteroides* to ferment the fiber in food into SCFAs. Butyric acid, as the preferred energy source of colon cells, can promote intestinal barrier function and reduce inflammation. Therefore, feeding fiber can optimize the structure and function of intestinal flora, which is very important for the early prevention and control of the disease.

Researchers at Xin Zhou University and Tokyo Pharmaceutical University in Japan have used transgenic technology to develop a *Bifidobacterium* whose life activities can cut off the nutritional supply of cancer cells, thereby inhibiting the growth of tumor tissue, a technology that can be used to treat cancer. *Bifidobacterium* is a common bacteria in human intestinal tract, which is easy to survive in anoxic environment, and the interior of breast and chest cancer tissue belongs to anoxic state.

2.3.6.4 Research Prospect of Intestinal Microbiome

Intestinal flora plays a significant part in regulating anxiety, emotional disorders and other neurological diseases, and chronic diseases such as IBD, type I diabetes, obesity, cardiovascular disease, and cancer [311]. It is worth noting that, intestinal microbiome can maintain homeostasis in the human body, and may also produce potential carcinogenic toxins and metabolites through bacteria to have a negative impact on cancer prevention. Therefore, in the future, anti-tumor therapy can be carried out through the combination of intestinal microbiome and its metabolites with immunotherapy, or it can also be combined with the traditional method of directly targeting malignant cells for anti-tumor therapy. Based on the immune response induced by intestinal microorganism group and the mechanism of cancer induction, high efficient anticancer strains were screened to develop new and efficient anticancer agents.

2.3.7 *Renal Diseases*

Although intestinal flora lives in the gut, its role is not limited to the digestive system. The effect of intestinal flora on human body is systemic through its influence on human metabolism and immune function. The kidney is the main organ of excretion of metabolites in the body and also the important site of deposition of immune complex. Therefore, intestinal flora has a crucial role in the development and treatment of renal diseases. For example, Vaziri et al. found that the quantity of Firmicutes and Bacteroidetes in the intestinal tract of chronic renal failure rats was lower, especially that of *Lactobacillus* and *Prevotellaceae*. Wong et al. found that in patients having end-stage renal dirty disease, the abundance of bacteria producing ammonia, indole, cresol, and other harmful metabolites increased, while the abundance of bacteria producing SCFAs (including *Lactobacillus* and *Prevotellaceae*) decreased. IS, PCS, and PAG can be detected in the early stages of renal dysfunction. Meanwhile, kidney stone disease is closely related to changes in intestinal flora. The main pathological change of kidney stone disease is crystal formation in the kidney, and its incidence rate is increasing day by day. Stern et al. used 16sRNA test to find that intestinal *Bacteroides* in patients with kidney stones had a higher abundance, while *Prevotella* was lower. *Eubacterium* and *E. coli* were negatively correlated with urinary oxalic acid and citric acid content at 24 h, respectively. Calcium oxalate stone is a common type of KSD. Gnanandarajah et al. suggested that the lack of bacterial colonization in the intestine was a risk factor for calcium oxalate urolithiasis. Sadaf et al. found that oxalate *Bacillus* and *Lactobacillus* prevent stone deposition and formation in the kidney by producing enzymes conducive to oxalate degradation. Xiaoying et al. found that *Enterobacteriaceae* was significantly elevated in kidney stone disease. Recently, it was found that the fecal microbial diversity of patients with recurrent idiopathic calcium calculi was low, and the expression of oxalate degradation related bacteria gene was significantly reduced, which was negatively correlated with oxalate excretion. At the same time, it is also believed that kidney stone disease is not caused by the lack of oxalate formate bacteria or one kind of bacteria, but is related to the extensive changes of intestinal flora. IgAN is deposited in the glomeruli by a polyimmune complex containing IgA, causing kidney damage. DeAngelis et al. discovered that the composition of intestinal flora in IgAN patients changed, mainly manifested by the increase of *Streptococcus*, *Enterobacter* and the decrease of *Bifidobacteria* [312].

Ley et al. sequenced 16S ribosomal RNA genes in fat and lean mice and found that the number of *Bacteroides* in fat mice was relatively high. For obese and non-obese people, human trials also showed the same changes in bacteria as animal studies. T2D patients are also often associated with differences compared with the normal population. Larsen et al. compared the degree of abnormality in the types and quantities of intestinal flora of T2D group and non-T2D group, and it was found that *E. coli*, *Salmonella*, and *Vibrio cholerae* belong to proteobacteria are present in the intestines of T2D patients, and the proportion of bacterial flora change related to blood glucose concentration. Qin et al. found that T2D patients were accompanied by moderate-intensity bowel. The proportion of trace bacteria was unbalance, which was reflected by the benefit of producing butyric acid of Hoffmann-La Roche Inc. A large number of bacteria were lost, while the number of harmful bacteria such as *Clostridium* was increased. The diabetic patients were supplemented with probiotics, prebiotics, and other microecological preparations to make intestinal flora; after being regulated and reaching steady state, its blood glucose level will also improve. Intestinal flora structural changes (e.g. reduction of Bacteroidetes/Firmicutes ratio, butyric acid production, salt bacteria, etc.) is closely related to T2D and may pass through those involved in SCFAs, LPS, fence-induced fat factors and bile acids in vivo synthesis, induces the body to produce a variety of mechanisms (such as chronic inflammatory response, generation Endotoxemia, etc.), which then leads to the destruction of islet beta cells [313]. T2D reduces the body's sensitivity to insulin, and ultimate leads to death. Therefore, intestinal flora and T2D were actively studied to make full use of intestinal flora for better control of T2D patients' blood sugar.

References

1. Sender R, Fuchs S, Milo R (2016) Revised estimates for the number of human and bacteria cells in the body. PLoS Biol 14:e1002533. https://doi.org/10.1371/journal.pbio.1002533
2. Bengmark S (1998) Ecological control of the gastrointestinal tract. The role of probiotic flora. Gut 42:2–7. https://doi.org/10.1136/gut.42.1.2
3. Backhed F (2005) Host-bacterial mutualism in the human intestine. Science 307:1915–1920. https://doi.org/10.1126/science.1104816
4. Neish AS (2009) Microbes in gastrointestinal health and disease. Gastroenterology 136:65–80. https://doi.org/10.1053/j.gastro.2008.10.080
5. Hooper LV, Gordon JI (2001) Commensal host-bacterial relationships in the gut. Science 292:1115–1118
6. Gill SR, Pop M, DeBoy RT, Eckburg PB, Turnbaugh PJ, Samuel BS et al (2006) Metagenomic analysis of the human distal gut microbiome. Science 312:1355–1359. https://doi.org/10.1126/science.1124234
7. Luckey TD (1972) Introduction to intestinal microecology. Am J Clin Nutr 25:1292–1294
8. Natividad JMM, Verdu EF (2013) Modulation of intestinal barrier by intestinal microbiota: pathological and therapeutic implications. Pharmacol Res 69:42–51. https://doi.org/10.1016/j.phrs.2012.10.007
9. den Besten G, van Eunen K, Groen AK, Venema K, Reijngoud DJ, Bakker BM (2013) The role of short-chain fatty acids in the interplay between diet, gut microbiota, and host energy metabolism. J Lipid Res 54:2325–2340. https://doi.org/10.1194/jlr.R036012
10. Bäumler AJ, Sperandio V (2016) Interactions between the microbiota and pathogenic bacteria in the gut. Nature 535:85–93. https://doi.org/10.1038/nature18849
11. Gensollen T, Iyer SS, Kasper DL, Blumberg RS (2016) How colonization by microbiota in early life shapes the immune system. Science 352:539–544. https://doi.org/10.1126/science.aad9378
12. Chang C, Lin H (2016) Dysbiosis in gastrointestinal disorders. Best Pract Res Clin Gastroenterol 30:3–15. https://doi.org/10.1016/j.bpg.2016.02.001
13. Schroeder BO, Bäckhed F (2016) Signals from the gut microbiota to distant organs in physiology and disease. Nat Med 22:1079–1089. https://doi.org/10.1038/nm.4185
14. Backhed F, Ley RE, Sonnenburg JL, Peterson DA, Gordon JI (2005) Host-bacterial mutualism in the human intestine. Science 307:1915–1920
15. Qin J, Li R, Raes J, Arumugam M et al (2010) A human gut microbial gene catalogue established by metagenomic sequencing. Nature 464:59–65
16. Ley RE, Peterson DA, Gordon JI (2006) Ecological and evolutionary forces shaping microbial diversity in the human intestine. Cell 124:837–848
17. Eckburg PB, Bik EM, Bernstein CN, Purdom E et al (2005) Diversity of the human intestinal microbial flora. Science 308:1635–1638
18. Tannock GW (2007) What immunologists should know about bacterial communities of the human bowel. Semin Immunol 19:94–105
19. Van den Abbeele P, Belzer C, Goossens M, Kleerebezem M et al (2013) Butyrate-producing Clostridium cluster XIVa species specifically colonize mucins in an in vitro gut model. ISME J 7:949–961
20. Arumugam M, Raes J, Pelletier E, Le Paslier D et al (2011) Enterotypes of the human gut microbiome. Nature 12:174–180
21. Jeffery IB, Claesson MJ, O'Toole PW, Shanahan F (2012) Categorization of the gut microbiota: enterotypes or gradients? Nat Rev Microbiol 10:591–592
22. Wu GD, Chen J, Hoffmann C, Bittinger K et al (2011) Linking long-term dietary patterns with gut microbial enterotypes. Science 334:105–108
23. Huse SM, Ye Y, Zhou Y, Fodor AA (2012) A core human microbiome as viewed through 16S rRNA sequence clusters. PLoS One 7:e34242

24. Wu GD, Chen J, Hoffmann C, Bittinger K et al (2012) NIH public access. Science 334:105–108
25. Knights D, Ward TL, McKinlay CE, Miller H et al (2014) Rethinking enterotypes. Cell Host Microbe 16:433–437
26. Poretsky R, Rodriguez-R LM, Luo C, Tsementzi D et al (2014) Strengths and limitations of 16S rRNA gene amplicon sequencing in revealing temporal microbial community dynamics. PLoS One 9:e93827. https://doi.org/10.1371/journal.pone.0093827
27. Mizrahi-Man O, Davenport ER, Gilad Y, White BA (2013) Taxonomic classification of bacterial 16S rRNA genes using short sequencing reads: evaluation of effective study designs. PLoS One 8:e53608. https://doi.org/10.1371/journal.pone.0053608
28. Suau A et al (1999) Direct analysis of genes encoding 16S rRNA from complex communities reveals many novel molecular species within the human gut. Appl Environ Microbiol 65:4799–4807
29. Hugon P, Dufour JC, Colson P, Fournier PE et al (2015) A comprehensive repertoire of prokaryotic species identified in human beings. Lancet Infect Dis 15:1211–1219. https://doi.org/10.1016/S1473-3099(15)00293-5
30. Li J, Jia H, Cai X, Zhong H et al (2014) An integrated catalog of reference genes in the human gut microbiome. Nat Biotechnol 32:834–841. https://doi.org/10.1038/nbt.2942
31. Donohoe DR, Garge N, Zhang X, Sun W et al (2011) The microbiome and butyrate regulate energy metabolism and autophagy in the mammalian colon. Cell Metab 13:517–526. https://doi.org/10.1016/j.cmet.2011.02.018
32. Kobyliak N, Virchenko O, Falalyeyeva T (2016) Pathophysiological role of host microbiota in the development of obesity. Nutr J 15:1–12. https://doi.org/10.1186/s12937-016-0166-9
33. Guarner F, Malagelada JR (2003) Gut flora in health and disease. Lancet 361:512–519. https://doi.org/10.1016/S01406736(03)12489-0
34. Borre YE, O'Keeffe GW, Clarke G, Stanton C et al (2014) Microbiota and neurodevelopmental windows: implications for brain disorders. Trends Mol Med 20:509–518. https://doi.org/10.1016/j.molmed.2014.05.002
35. Morrison DJ, Preston T (2016) Formation of short chain fatty acids by the gut microbiota and their impact on human metabolism. Gut Microbes 7:189–200. https://doi.org/10.1080/19490976.2015.1134082
36. den Besten G, van Eunen K, Groen AK, Venema K et al (2013) The role of short-chain fatty acids in the interplay between diet, gut microbiota, and host energy metabolism. J Lipid Res 54:2325–2340. https://doi.org/10.1194/jlr.R036012
37. Macfarlane GT, Gibson GR, Cummings JH (1992) Comparison of fermentation reactions in different regions of the human colon. J Appl Bacteriol 72:57–64. https://doi.org/10.1111/j.1365-2672.1992.tb04882.x
38. LeBlanc JG, Chain F, Martín R, Bermúdez-Humarán LG et al (2017) Beneficial effects on host energy metabolism of short-chain fatty acids and vitamins produced by commensal and probiotic bacteria. Microb Cell Factories 16:1–10. https://doi.org/10.1186/s12934-017-0691-z
39. Chakraborti CK (2015) New-found link between microbiota and obesity. World J Gastrointest Pathophysiol 6:110–119. https://doi.org/10.4291/wjgp.v6.i4.110
40. Li X, Shimizu Y, Kimura I (2017) Gut microbial metabolite short chain fatty acids and obesity. Biosci Micro, Food Heal 36:135–140. https://doi.org/10.12938/bmfh.17-010
41. Pingitore A, Chambers ES, Hill T, Maldonado IR et al (2017) The diet-derived short chain fatty acid propionate improves beta-cell function in humans and stimulates insulin secretion from human islets in vitro. Diabetes Obes Metab 19:257–265. https://doi.org/10.1111/dom.12811
42. Schönfeld P, Wojtczak L (2016) Short- and medium-chain fatty acids in energy metabolism: the cellular perspective. J Lipid Res 57:943–954. https://doi.org/10.1194/jlr.R067629
43. Cooper DN, Martin RJ, Keim NL (2015) Does whole grain consumption alter gut microbiota and satiety? Healthc (Basel, Switz) 3:364–392. https://doi.org/10.3390/healthcare3020364
44. Everard A, Cani PD (2014) Gut microbiota and GLP-1. Rev Endocr Metab Disord 15:189–196

45. Frost G, Sleeth ML, Sahuri-Arisoylu M, Lizarbe B et al (2014) The short-chain fatty acid acetate reduces appetite via a central homeostatic mechanism. Nat Commun 5:3611. https://doi.org/10.1038/ncomms4611
46. Willemsen LEM, Koetsier MA, van Deventer SJH, van Tol EAF (2003) Short chain fatty acids stimulate epithelial mucin 2 expression through differential effects on prostaglandin E(1) and E(2) production by intestinal myofibroblasts. Gut 52:1442–1447
47. Cousin FJ, Jouan-Lanhouet S, Theret N, Brenner C et al (2016) The probiotic Propionibacterium freudenreichii as a new adjuvant for TRAIL-based therapy in colorectal cancer. Oncotarget 7:7161–7178. https://doi.org/10.18632/oncotarget.6881
48. Ni J, Wu GD, Albenberg L, Tomov VT (2017) Gut microbiota and IBD: Causation or correlation? Nat Rev Gastroenterol Hepatol 14:573–584
49. Pascale A, Marchesi N, Marelli C, Coppola A et al (2018) Microbiota and metabolic diseases. Endocrine 61:357–371. https://doi.org/10.1007/s12020-018-1605-5
50. Flint HJ, Bayer EA (2008) Plant cell wall breakdown by anaerobic microorganisms from the mammalian digestive tract. Ann New Y Acad Sci 1125:280–288
51. van den Abbeele P, Gérard P, Rabot S, Bruneau A et al (2011) Arabinoxylans and inulin differentially modulate the mucosal and luminal gut microbiota and mucin-degradation in humanized rats. Environ Microbiol 13:2667–2680. https://doi.org/10.1111/j.1462-2920.2011.02533.x
52. McOrist AL, Miller RB, Bird AR, Keogh JB et al (2011) Fecal butyrate levels vary widely among individuals but are usually increased by a diet high in resistant starch. J Nutr 141:883–889. https://doi.org/10.3945/jn.110.128504
53. Grootaert C, Van Den Abbeele P, Marzorati M, Broekaert WF et al (2009) Comparison of prebiotic effects of arabinoxylan oligosaccharides and inulin in a simulator of the human intestinal microbial ecosystem. FEMS Microbiol Ecol 69:231–242. https://doi.org/10.1111/j.1574-6941.2009.00712.x
54. Begley M, Hill C, Gahan CGM (2006) Bile salt hydrolase activity in probiotics. Appl Environ Microbiol 72:1729–1738. https://doi.org/10.1128/AEM.72.3.1729-1738.2006
55. Ley R, Turnbaugh P, Klein S, Gordon J (2006) Microbial ecology: human gut microbes associated with obesity. Nature 444:1022–1023. https://doi.org/10.1038/nature4441021a
56. Ley RE, Bäckhed F, Turnbaugh P, Lozupone CA et al (2005) Obesity alters gut microbial ecology. Proc Natl Acad Sci U S A 102:11070–11075. https://doi.org/10.1073/pnas.0504978102
57. Komaroff AL (2017) The microbiome and risk for obesity and diabetes. JAMA 317:355. https://doi.org/10.1001/jama.2016.20099
58. Mazmanian SK, Round JL, Kasper DL (2008) A microbial symbiosis factor prevents intestinal inflammatory disease. Nature 453:620–625. https://doi.org/10.1038/nature07008
59. Wen L, Ley RE, Volchkov PY, Stranges PB et al (2008) Innate immunity and intestinal microbiota in the development of Type 1 diabetes. Nature 455:1109–1113. https://doi.org/10.1038/nature07336
60. Kim YS, Milner JA (2007) Dietary modulation of colon cancer risk. J Nutr 137:2576S–2579S
61. Lau SKP, Woo PCY, Woo GKS, Fung AMY et al (2004) Eggerthella hongkongensis sp. nov. and Eggerthella sinensis sp. nov., two novel Eggerthella species, account for half of the cases of Eggerthella bacteremia. Diagn Microbiol Infect Dis 49:255–263. https://doi.org/10.1016/j.diagmicrobio.2004.04.012
62. Kraatz M, Wallace RJ, Svensson L (2011) Olsenella umbonata sp. nov., a microaerotolerant anaerobic lactic acid bacterium from the sheep rumen and pig jejunum, and emended descriptions of Olsenella, Olsenella uli and Olsenella profusa. Int J Syst Evol Microbiol 61:795–803. https://doi.org/10.1099/ijs.0.022954-0
63. Lau SK, Woo PC, Fung AM, Chan K et al (2004) Anaerobic, non-sporulating, gram-positive bacilli bacteraemia characterized by 16S rRNA gene sequencing. J Med Microbiol 53:1247–1253. https://doi.org/10.1099/jmm.0.45803-0

64. Finamore A, Palmery M, Bensehaila S, Peluso I (2017) Antioxidant, immunomodulating, and microbial-modulating activities of the sustainable and ecofriendly Spirulina. Oxidative Med Cell Longev 2017:3247528. https://doi.org/10.1155/2017/3247528
65. Shin NR, Lee JC, Lee HY, Kim MS et al (2014) An increase in the Akkermansia spp. Population induced by metformin treatment improves glucose homeostasis in diet-induced obese mice. Gut 63:727–735. https://doi.org/10.1136/gutjnl-2012-303839
66. Hampson DJ, La T, Phillips ND (2015) Emergence of Brachyspira species and strains: reinforcing the need for surveillance. Porc Heal Manag 1:8. https://doi.org/10.1186/s40813-0150002-1
67. Galperin MY (2008) New feel for new phyla. Environ Microbiol 10:1927–1933. https://doi.org/10.1111/j.1462-2920.2008.01699.x
68. Yamauchi M, Lochhead P, Morikawa T, Huttenhower C et al (2012) Colorectal cancer: a tale of two sides or a continuum? Gut 61:794–797. https://doi.org/10.1136/gutjnl-2012-302014
69. Mackie RI, Sghir A, Gaskins HR (1999) Developmental microbial ecology of the neonatal gastrointestinal tract. Am J Clin Nutr 69:1035S–1045S
70. Trahair J (2001) Digestive system. In: Harding R, Bocking AD (eds) . Cambridge University Press, Cambridge, pp 137–153. ISBN 0521645433
71. Trahair JF, Harding R (1994) Development of the gastrointestinal tract. In textbook of fetal physiology; Thorburn, G.D., Harding, R., Eds. Oxford University Press, New York, NY.. ISBN 0198577486
72. Aagaard K, Ma J, Antony KM, Ganu R et al (2014) The placenta harbors a unique microbiome. Sci Transl Med 6:237ra65. https://doi.org/10.1126/scitranslmed.3008599
73. Rodriguez JM et al (2015) The composition of the gut microbiota throughout life, with an emphasis on early life. Microb Ecol Health Dis 26:26050
74. Koenig JE, Spor A, Scalfone N, Fricker AD et al (2011) Succession of microbial consortia in the developing infant gut microbiome. Proc Natl Acad Sci U S A 108:4578–4585. https://doi.org/10.1073/pnas.1000081107
75. Avershina E, Storrø O, Øien T, Johnsen R et al (2014) Major faecal microbiota shifts in composition and diversity with age in a geographically restricted cohort of mothers and their children. FEMS Microbiol Ecol 87:280–290. https://doi.org/10.1111/1574-6941.12223
76. Aagaard K, Riehle K, Ma J, Segata N et al (2012) A metagenomic approach to characterization of the vaginal microbiome signature in pregnancy. PLoS One 7:e36466. https://doi.org/10.1371/journal.pone.0036466
77. Jakobsson HE, Abrahamsson TR, Jenmalm MC, Harris K et al (2014) Decreased gut microbiota diversity, delayed Bacteroidetes colonisation and reduced Th1 responses in infants delivered by caesarean section. Gut 63:559–566. https://doi.org/10.1136/gutjnl-2012-303249
78. Salminen S (2004) Influence of mode of delivery on gut microbiota composition in seven year old children. Gut 53:1388–1389. https://doi.org/10.1136/gut.2004.041640
79. Backhed F, Roswall J, Peng Y, Feng Q et al (2015) Dynamics and stabilization of the human gut microbiome during the first year of life. Cell Host Microbe 17:852. https://doi.org/10.1016/j.chom.2015.05.012
80. Bäckhed F (2011) Programming of host metabolism by the gut microbiota. Ann Nutr Metab 58:44–52. https://doi.org/10.1159/000328042
81. Palmer C, Bik EM, DiGiulio DB, Relman DA et al (2007) Development of the human infant intestinal microbiota. PLoS Biol 5:e177. https://doi.org/10.1371/journal.pbio.0050177
82. Dethlefsen L, Relman DA (2011) Incomplete recovery and individualized responses of the human distal gut microbiota to repeated antibiotic perturbation. Proc Natl Acad Sci U S A 108:4554–4561. https://doi.org/10.1073/pnas.1000087107
83. Claesson MJ, Cusack S, O'Sullivan O, Greene-Diniz R et al (2011) Composition, variability, and temporal stability of the intestinal microbiota of the elderly. Proc Natl Acad Sci U S A 108:4586–4591. https://doi.org/10.1073/pnas.1000097107

84. Biagi E, Nylund L, Candela M, Ostan R et al (2010) Through ageing, and beyond: gut microbiota and inflammatory status in seniors and centenarians. PLoS One 5:e10667. https://doi.org/10.1371/journal.pone.0010667
85. Claesson MJ et al (2012) Gut microbiota composition correlates with diet and health in the elderly. Nature 488:178
86. Woodmansey EJ, McMurdo MET, Macfarlane GT, Macfarlane S (2004) Comparison of compositions and metabolic activities of fecal microbiotas in young adults and in antibiotic-treated and non-antibiotic-treated elderly subjects. Appl Environ Microbiol 70:6113–6122. https://doi.org/10.1128/AEM.70.10.6113-6122.2004
87. Biagi E, Candela M, Turroni S, Garagnani P et al (2013) Ageing and gut microbes: perspectives for health maintenance and longevity. Pharmacol Res 69:11–20. https://doi.org/10.1016/j.phrs.2012.10.005
88. Claesson MJ, Cusack S, O'Sullivan O, Greene-Diniz R et al (2011) Composition, variability, and temporal stability of the intestinal microbiota of the elderly. PNAS 108:458691
89. Adlerberth I, Wold AE (2009) Establishment of the gut microbiota in western infants. Acta Paediatr 98:22938
90. Marques TM, Wall R, Ross RP, Fitzgerald G et al (2010) Programming infant gut microbiota: influence of dietary and environmental factors. Curr Opin Biotechnol 21:14956.12
91. Eckburg PB, Bik EM, Bernstein CN, Purdom E et al (2005) Diversity of the human intestinal microbial flora. Science 308:16358
92. Qin J, Li R, Raes J, Arumugam M et al (2010) A human gut microbial gene catalogue established by metagenomic sequencing. Nature 464:5965
93. Backhed F (2011) Programming of host metabolism by the gut microbiota. Ann Nutr Metab 58:4452
94. Yatsunenko T, Rey FE, Manary MJ, Trehan I et al (2012) Human gut microbiome viewed across age and geography. Nature 486:2227
95. Borre YE, Moloney RD, Clarke G, Dinan TG et al (2014) The impact of microbiota on brain and behavior: mechanisms & therapeutic potential. Adv Exp Med Biol 817:373403
96. Murphy JMRK, Stanton C, Ross RP, Kober OI et al (2015) The composition of the gut microbiota throughout life, with an emphasis on early life. Microb Ecol Health Dis 26:26050. https://doi.org/10.3402/mehd.v26.26050
97. Macpherson AJ, McCoy KD (2013) Stratification and compartmentalisation of immunoglobulin responses to commensal intestinal microbes. Semin Immunol 25:358–363. https://doi.org/10.1016/j.smim.2013.09.004
98. Donaldson GP, Lee SM, Mazmanian SK (2015) Gut biogeography of the bacterial microbiota. Nat Rev Microbiol 14:20–32. https://doi.org/10.1038/nrmicro3552
99. Gu S, Chen D, Zhang JN, Lv X et al (2013) Bacterial community mapping of the mouse gastrointestinal tract. PLoS One 8:e74957. https://doi.org/10.1371/journal.pone.0074957
100. Eckburg PB (2005) Diversity of the human intestinal microbial flora. Science 308:1635–1638. https://doi.org/10.1126/science.1110591
101. Lavelle A, Lennon G, O'Sullivan O et al (2015) Spatial variation of the colonic microbiota in patients with ulcerative colitis and control volunteers. Gut 64(10):1553–1561
102. Li H, Limenitakis JP, Fuhrer T, Geuking MB et al (2015) The outer mucus layer hosts a distinct intestinal microbial niche. Nat Commun 6:8292. https://doi.org/10.1038/ncomms9292
103. Turnbaugh PJ, Hamady M, Yatsunenko T, Cantarel BL et al (2009) A core gut microbiome in obese and lean twins. Nature 457:480–484. https://doi.org/10.1038/nature07540
104. Jakobsson HE, Jernberg C, Andersson AF, Sjölund-Karlsson M et al (2010) Short-term antibiotic treatment has differing long-term impacts on the human throat and gut microbiome. PLoS One 5. https://doi.org/10.1371/journal.pone.0009836
105. Ding T, Schloss PD (2014) Dynamics and associations of microbial community types across the human body. Nature 509:357–360. https://doi.org/10.1038/nature13178
106. Arumugam M, Raes J, Pelletier E, Le Paslier D et al (2011) Enterotypes of the human gut microbiome. Nature 473:174–180. https://doi.org/10.1038/nature09944

107. Odamaki T, Kato K, Sugahara H, Hashikura N et al (2016) Age-related changes in gut microbiota composition from newborn to centenarian: a cross-sectional study. BMC Microbiol 16:90
108. Jandhyala SM, Talukdar R, Subramanyam C, Vuyyuru H et al (2015) Role of the normal gut microbiota. World J Gastroenterol 21:8787–8803
109. Wen LL, Duffy A (2017) Factors influencing the gut microbiota, inflammation, and type 2 diabetes. J Nutr 147:1468S–1475S
110. Voreades N, Kozil A, Weir TL (2014) Diet and the development of the human intestinal microbiome. Front Microbiol 5:494
111. Spor A, Koren O, Ley R (2011) Unravelling the effects of the environment and host genotype on the gut microbiome. Nat Rev Microbiol 9:279–290
112. Nagpal R, Tsuji H, Takahashi T, Nomoto K et al (2017) Ontogenesis of the gut microbiota composition in healthy, full-term, vaginally born and breast-fed infants over the first 3 years of life: a quantitative bird's-eye view. Front Microbiol 8:1388
113. Browne HP, Forster SC, Anonye BO, Kumar N et al (2016) Culturing of 'unculturable' human microbiota reveals novel taxa and extensive sporulation. Nature 533:543–546
114. Mitsou EK, Kirtzalidou E, Oikonomou I, Liosis G et al (2008) Fecal microflora of Greek healthy neonates. Anaerobe 14(2):94–101
115. Samuel BS, Shaito A, Motoike T, Rey FE et al (2008) Effects of the gut microbiota on host adiposity are modulated by the short-chain fatty-acid binding G protein-coupled receptor, Gpr41. Proc Natl Acad Sci 105(43):16767–16772
116. Zoetendal EG, Raes J, van den Bogert B, Arumugam M et al (2012) The human small intestinal microbiota is driven by rapid uptake and conversion of simple carbohydrates. ISME J 6:1415–1426. https://doi.org/10.1038/ismej.2011.212
117. David LA, Maurice CF, Carmody RN, Gootenberg DB et al (2013) Diet rapidly and reproducibly alters the human gut microbiome. Nature 505:559–563. https://doi.org/10.1038/nature12820
118. Walker AW, Ince J, Duncan SH, Webster LM et al (2011) Dominant and diet-responsive groups of bacteria within the human colonic microbiota. ISME J 5:220–230. https://doi.org/10.1038/ismej.2010.118
119. Veiga P, Pons N, Agrawal A, Oozeer R et al (2014) Changes of the human gut microbiome induced by a fermented milk product. Sci Rep 4:6328
120. Field AE, Willett WC, Lissner L, Colditz GA (2007) Dietary fat and weight gain among women in the Nurses Health Study. Obesity (Silver Spring) 15(967–76)
121. Shai I, Schwarzfuchs D, Henkin Y, Shahar DR et al (2008) Weight loss with a low carbohydrate, Mediterranean, or low-fat diet. N Engl J Med 359:29–41
122. Sacks FM, Bray GA, Carey VJ, Smith SR et al (2009) Comparison of weight-loss diets with different compositions of fat, protein, and carbohydrates. N Engl J Med 360:859–873
123. Mozaffarian D, Hao T, Rimm EB, Willett WC et al (2011) Changes in diet and lifestyle and long-term weight gain in women and men. N Engl J Med 364:2392–2404
124. Winzell MS, Ahren B (2004) The high-fat diet-fed mouse: a model for studying mechanisms and treatment of impaired glucose tolerance and type 2 diabetes. Diabetes 53:S215–S219
125. Musso G, Gambino R, Cassader M (2010) Obesity, diabetes, and gut microbiota: the hygiene hypothesis expanded? Diabetes Care 33:2277–2284
126. DiBaise KK, Frank DN, Mathur R (2012) Impact of the gut microbiota on the development of obesity: current concepts. Am J Gastroenterol Suppl 1:22–27
127. Cani PD, Bibiloni R, Knauf C, Waget A et al (2008) Changes in gut microbiota control metabolic endotoxemia induced inflammation in high-fat diet-induced obesity and diabetes in mice. Diabetes 57:1470–1481
128. De Filippo C, Cavalieri D, Di Paola M, Ramazzotti M et al (2010) Impact of diet in shaping gut microbiota revealed by a comparative study in children from Europe and rural Africa. Proc Natl Acad Sci U S A 107:14691–14696

129. Scott KP, Gratz SW, Sheridan PO, Flint HJ et al (2013) The influence of diet on the gut microbiota. Pharmacol Res 69:52–60
130. Graf D, Di Cagno R, Fak F, Flint HJ et al (2015) Contribution of diet to the composition of the human gut microbiota. Microb Ecol Health Dis 26:26164
131. Sonnenburg ED, Smits SA, Tikhonov M, Higginbottom SK et al (2016) Diet-induced extinctions in the gut microbiota compound over generations. Nature 529:212–215
132. David LA, Maurice CF, Carmody RN, Gootenberg DB et al (2014) Diet rapidly and reproducibly alters the human gut microbiome. Nature 505:559–563
133. Everard A, Belzer C, Geurts L, Ouwerkerk JP et al (2013) Cross-talk between Akkermansia muciniphila and intestinal epithelium controls diet-induced obesity. Proc Natl Acad Sci U S A 110:9066–9071
134. Koh A, De Vadder F, Kovatcheva-Datchary P, Backhed F (2016) From dietary fiber to host physiology: short-chain fatty acids as key bacterial metabolites. Cell 165:1332–1345
135. Goodrich JK, Waters JL, Poole AC, Sutter JL et al (2014) Human genetics shape the gut microbiome. Cell 159:789–799
136. Blekhman R, Goodrich JK, Huang K, Sun Q et al (2015) Host genetic variation impacts microbiome composition across human body sites. Genome Biol 16:191
137. Thompson-Chagoyàn OC, Maldonado J, Gil A (2005) Aetiology of inflammatory bowel disease (IBD): role of intestinal microbiota and gut-associated lymphoid tissue immune response. Clin Nutr 24:339–352
138. Rehman A, Sina C, Gavrilova O, Hasler R et al (2011) Nod2 is essential for temporal development of intestinal microbial communities. Gut 60:1354–1362
139. Hooper LV, Littman DR, Macpherson AJ (2012) Interactions between the microbiota and the immune system. Science 336:1268–1273. https://doi.org/10.1126/science.1223490
140. Macpherson AJ (2000) A primitive T cell-independent mechanism of intestinal mucosal IgA responses to commensal bacteria. Science 288:2222–2226. https://doi.org/10.1126/science.288.5474.2222
141. Macpherson AJ, Uhr T (2004) Induction of protective IgA by intestinal dendritic cells carrying commensal bacteria. Science 303:1662–1665. https://doi.org/10.1126/science.1091334
142. Cash HL (2006) Symbiotic bacteria direct expression of an intestinal bactericidal lectin. Science 313:1126–1130. https://doi.org/10.1126/science.1127119
143. McGuckin MA, Lindén SK, Sutton P, Florin TH (2011) Mucin dynamics and enteric pathogens. Nat Rev Microbiol 9:265–278. https://doi.org/10.1038/nrmicro2538
144. Hooper LV, Macpherson AJ (2010) Immune adaptations that maintain homeostasis with the intestinal microbiota. Nat Rev Immunol 10:159–169. https://doi.org/10.1038/nri2710
145. Meyer-Hoffert U, Hornef MW, Henriques-Normark B, Axelsson LG et al (2008) Secreted enteric antimicrobial activity localises to the mucus surface layer. Gut 57:764–771. https://doi.org/10.1136/gut.2007.141481
146. Wehkamp J (2004) NOD2 (CARD15) mutations in Crohn's disease are associated with diminished mucosal alpha-defensin expression. Gut 53:1658–1664. https://doi.org/10.1136/gut.2003.032805
147. Wehkamp J, Salzman NH, Porter E, Nuding S et al (2005) Reduced Paneth cell alpha-defensins in ileal Crohn's disease. Proc Natl Acad Sci U S A 102:18129–18134. https://doi.org/10.1073/pnas.0505256102
148. Rogier EW, Frantz A, Bruno M, Kaetzel C (2014) Secretory IgA is concentrated in the outer layer of colonic mucus along with gut bacteria. Pathogens 3:390–403. https://doi.org/10.3390/pathogens3020390
149. Bollinger RR, Everett ML, Palestrant D, Love SD et al (2003) Human secretory immunoglobulin A may contribute to biofilm formation in the gut. Immunology 109:580–587. https://doi.org/10.1046/j.1365-2567.2003.01700.x
150. Friman V et al (1996) Decreased expression of mannose-specific adhesins by Escherichia coli in the colonic microflora of immunoglobulin A-deficient individuals. Infect Immun 64:2794–2798

151. Suzuki K, Meek B, Doi Y, Muramatsu M et al (2004) Aberrant expansion of segmented filamentous bacteria in IgA-deficient gut. Proc Natl Acad Sci U S A 101:1981–1986. https://doi.org/10.1073/pnas.0307317101
152. Chou HH, Chien WH, Wu LL, Cheng CH et al (2015) Age-related immune clearance of hepatitis B virus infection requires the establishment of gut microbiota. Proc Natl Acad Sci U S A 112:2175–2180
153. Singh P, Teal TK, Marsh TL, Tiedje JM et al (2015) Intestinal microbial communities associated with acute enteric infections and disease recovery. Microbiome 3:45
154. Qin N, Zheng B, Yao J, Guo L et al (2015) Influence of H7N9 virus infection and associated treatment on human gut microbiota. Sci Rep 5:14771
155. Zaiss MM, Rapin A, Lebon L, Dubey LK et al (2015) The intestinal microbiota contributes to the ability of helminths to modulate allergic inflammation. Immunity 43:998–1010
156. Yang L, Poles MA, Fisch GS, Ma Y et al (2016) HIV induced immunosuppression is associated with colonization of the proximal gut by environmental bacteria. AIDS 30:19–29
157. Zilberman-Schapira G, Zmora N, Itav S, Bashiardes S et al (2016) The gut microbiome in human immunodeficiency virus infection. BMC Med 14:83
158. Hoffmann C, Hill DA, Minkah N, Kirn T et al (2009) Community-wide response of the gut microbiota to enteropathogenic Citrobacter rodentium infection revealed by deep sequencing. Infect Immun 77:4668–4678
159. Zhang L, Dong D, Jiang C, Li Z et al (2015) Insight into alteration of gut microbiota in Clostridium difficile infection and asymptomatic C. difficile colonization. Anaerobe 34:1–7
160. Seekatz AM, Young VB (2014) Clostridium difficile and the microbiota. J Clin Invest 124:4182–4189
161. Blanchi J, Goret J, Megraud F (2016) Clostridium difficile infection: a model for disruption of the gut microbiota equilibrium. Dig Dis 34:217–220
162. Youngster I, Russell GH, Pindar C, Ziv-Baran T et al (2014) Oral, capsulized, frozen fecal microbiota transplantation for relapsing Clostridium difficile infection. JAMA 312:1772–1778
163. Kelly CR, Ihunnah C, Fischer M, Khoruts A et al (2014) Fecal microbiota transplant for treatment of Clostridium difficile infection in immunocompromised patients. Am J Gastroenterol 109:1065–1071
164. Youngster I, Sauk J, Pindar C, Wilson RG et al (2014) Fecal microbiota transplant for relapsing Clostridium difficile infection using a frozen inoculum from unrelated donors: a randomized, open-label, controlled pilot study. Clin Infect Dis 58:1515–1522
165. Bashan A, Gibson TE, Friedman J, Carey VJ et al (2016) Universality of human microbial dynamics. Nature 534:259–262
166. Xu X, Zhang X (2015) Effects of cyclophosphamide on immune system and gut microbiota in mice. Microbiol Res 171:97–106
167. Imhann F, Bonder MJ, Vich Vila A, Fu J et al (2016) Proton pump inhibitors affect the gut microbiome. Gut 65:740–748
168. Devkota S (2016) Microbiome. Prescription drugs obscure microbiome analyses. Science 351:452–453
169. Forslund K, Hildebrand F, Nielsen T, Falony G et al (2015) Disentangling type 2 diabetes and metformin treatment signatures in the human gut microbiota. Nature 528:262–266
170. Kang MJ, Kim HG, Kim JS, Oh DG et al (2013) The effect of gut microbiota on drug metabolism. Expert Opin Drug Metab Toxicol 9:1295–1308
171. Yoo DH, Kim IS, Van Le TK, Jung IH et al (2014) Gut microbiota-mediated drug interactions between lovastatin and antibiotics. Drug Metab Dispos 42:1508–1513
172. Modi SR, Collins JJ, Relman DA (2014) Antibiotics and the gut microbiota. J Clin Invest 124:4212–4218
173. Cho I, Yamanishi S, Cox L, Methe BA et al (2012) Antibiotics in early life alter the murine colonic microbiome and adiposity. Nature 488:621–626

174. Gough EK, Moodie EE, Prendergast AJ, Johnson SM et al (2014) The impact of antibiotics on growth in children in low and middle income countries: systematic review and meta-analysis of randomised controlled trials. BMJ 348:g2267
175. Kozyrskyj AL, Ernst P, Becker AB (2007) Increased risk of childhood asthma from antibiotic use in early life. Chest 131:1753–1759
176. Shaw SY, Blanchard JF, Bernstein CN (2010) Association between the use of antibiotics in the first year of life and pediatric inflammatory bowel disease. Am J Gastroenterol 105:2687–2692
177. Trasande L, Blustein J, Liu M, Corwin E et al (2013) Infant antibiotic exposures and early-life body mass. Int J Obes 37(16–23)
178. Cox LM, Yamanishi S, Sohn J, Alekseyenko AV et al (2014) Altering the intestinal microbiota during a critical developmental window has lasting metabolic consequences. Cell 158:705–721
179. Mahana D, Trent CM, Kurtz ZD, Bokulich NA et al (2016) Antibiotic perturbation of the murine gut microbiome enhances the adiposity, insulin resistance, and liver disease associated with high-fat diet. Genome Med 8:48
180. Fujisaka S, Ussar S, Clish C, Devkota S et al (2016) Antibiotic effects on gut microbiota and metabolism are host dependent. J Clin Invest 126:4430–4443
181. Reijnders D, Goossens GH, Hermes GD, Neis EP et al (2016) Effects of gut microbiota manipulation by antibiotics on host metabolism in obese humans: a randomized double-blind placebo-controlled trial. Cell Metab 24:63–74
182. Membrez M, Blancher F, Jaquet M, Bibiloni R et al (2008) Gut microbiota modulation with norfloxacin and ampicillin enhances glucose tolerance in mice. FASEB J 22:2416–2426
183. Chou CJ, Membrez M, Blancher F (2008) Gut decontamination with norfloxacin and ampicillin enhances insulin sensitivity in mice. Nestle Nutr Workshop Ser Pediatr Program 62:127–137
184. Han J, Lin H, Huang W (2011) Modulating gut microbiota as an antidiabetic mechanism of berberine. Med Sci Monit 17:RA164–RA167
185. Chang W, Chen L, Hatch GM (2015) Berberine as a therapy for type 2 diabetes and its complications: from mechanism of action to clinical studies. Biochem Cell Biol 93:479–486
186. Tailford LE, Owen CD, Walshaw J, Crost EH et al (2015) Discovery of intramolecular trans-sialidases in human gut microbiota suggests novel mechanisms of mucosal adaptation. Nat Commun 6:7624. https://doi.org/10.1038/ncomms8624
187. Arike L, Hansson GC (2016) The densely O-glycosylated MUC2 mucin protects the intestine and provides food for the commensal bacteria. J Mol Biol
188. Ouwerkerk JP, de Vos WM, Belzer B (2013) Glycobiome: bacteria and mucus at the epithelial interface. Best Pract Res Clin Gastroenterol 27:25–38. https://doi.org/10.1016/j.bpg.2013.03.001
189. Johansson MEV, Larsson JMH, Hansson GC (2011) The two mucus layers of colon are organized by the MUC2 mucin, whereas the outer layer is a legislator of host-microbial interactions. Proc Natl Acad Sci U S A 108:4659–4665. https://doi.org/10.1073/pnas.1006451107
190. Gustafsson JK, Ermund A, Johansson MEV, Schutte A et al (2012) An ex vivo method for studying mucus formation, properties, and thickness in human colonic biopsies and mouse small and large intestinal explants. Am J Physiol Gastrointest Liver Physiol 302:G430–G438. https://doi.org/10.1152/ajpgi.00405.2011
191. Johansson ME, Jakobsson HE, Holmén-Larsson J, Schütte A et al (2015) Normalization of host intestinal mucus layers requires long-Term microbial colonization. Cell Host Microbe 18:582–592. https://doi.org/10.1016/j.chom.2015.10.007
192. Juge N (2012) Microbial adhesins to gastrointestinal mucus. Trends Microbiol 20:30–39. https://doi.org/10.1016/j.tim.2011.10.001
193. Tailford LE, Crost EH, Kavanaugh D, Juge N (2015) Mucin glycan foraging in the human gut microbiome. Front Genet 6:131. https://doi.org/10.3389/fgene.2015.00081

194. Rausch P, Rehman A, Kunzel S, Hasler R et al (2011) Colonic mucosa-associated microbiota is influenced by an interaction of Crohn disease and FUT2 (Secretor) genotype. Proc Natl Acad Sci U S A 108:19030–19035. https://doi.org/10.1073/pnas.1106408108
195. Arpaia N, Campbell C, Fan X, Dikiy S et al (2013) Metabolites produced by commensal bacteria promote peripheral regulatory T-cell generation. Nature 504:451–455. https://doi.org/10.1038/nature12726
196. Furusawa Y, Obata Y, Fukuda S, Endo TA et al (2013) Commensal microbe-derived butyrate induces the differentiation of colonic regulatory T cells. Nature 504:446–450. https://doi.org/10.1038/nature12721
197. Zarepour M, Bhullar K, Montero M, Ma C et al (2013) The mucin MUC2 limits pathogen burdens and epithelial barrier dysfunction during Salmonella enterica serovar Typhimurium colitis. Infect Immun 81:3672–3683. https://doi.org/10.1128/IAI.00854-13
198. Earle KA, Billings G, Sigal M, Lichtman JS et al (2015) Quantitative imaging of gut microbiota spatial organization. Cell Host Microbe 18:478–488. https://doi.org/10.1016/j.chom.2015.09.002
199. Desai MS, Seekatz AM, Koropatkin NM, Kamada N et al (2016) A dietary fiber-deprived gut microbiota degrades the colonic mucus barrier and enhances pathogen susceptibility. Cell 167:1339–1353.e21. https://doi.org/10.1016/j.cell.2016.10.043
200. Li J, Lin S, Vanhoutte PM, Woo CW et al (2016) Akkermansia muciniphila protects against atherosclerosis by preventing metabolic endotoxemia-induced inflammation in apoemice. Circulation 133:2434–2446. https://doi.org/10.1161/CIRCULATIONAHA.115.019645
201. Plovier H, Everard A, Druart C, Depommier C et al (2016) A purified membrane protein from Akkermansia muciniphila or the pasteurized bacterium improves metabolism in obese and diabetic mice. Nat Med 23:107–113. https://doi.org/10.1038/nm.4236
202. Zhao S, Liu W, Wang J, Shi J et al (2017) Akkermansia muciniphila improves metabolic profiles by reducing inflammation in chow diet-fed mice. J Mol Endocrinol 58:1–14. https://doi.org/10.1530/JME-16-0054
203. Cockburn DW, Koropatkin NM (2016) Polysaccharide degradation by the intestinal microbiota and its influence on human health and disease. J Mol Biol 428:3230–3252. https://doi.org/10.1016/j.jmb.2016.06.021
204. El Kaoutari A, Armougom F, Gordon JI, Raoult D, Henrissat B (2013) The abundance and variety of carbohydrate-active enzymes in the human gut microbiota. Nat Rev Microbiol 11:497–504. https://doi.org/10.1038/nrmicro3050
205. Cantarel BL, Lombard V, Henrissat B, Appanna VD (2012) Complex carbohydrate utilization by the healthy human microbiome. PLoS One 7:e28742. https://doi.org/10.1371/journal.pone.0028742
206. Larsbrink J, Rogers TE, Hemsworth GR, McKee LS et al (2014) A discrete genetic locus confers xyloglucan metabolism in select human gut bacteroidetes. Nature 506:498–502. https://doi.org/10.1038/nature12907
207. Rogowski A, Briggs JA, Mortimer JC, Tryfona T et al (2015) Glycan complexity dictates microbial resource allocation in the large intestine. Nat Commun 6:7481. https://doi.org/10.1038/ncomms8481
208. Cuskin F, Lowe EC, Temple MJ, Zhu Y et al (2015) Human gut Bacteroidetes can utilize yeast mannan through a selfish mechanism. Nature 517:165–169. https://doi.org/10.1038/nature13995
209. Tauzin AS, Kwiatkowski KJ, Orlovsky NI, Smith CJ et al (2016) Molecular dissection of xyloglucan recognition in a prominent human Gut symbiont. MBio 7:e02134–e02115. https://doi.org/10.1128/mBio.02134-15
210. Foley MH, Cockburn DW, Koropatkin NM (2016) The Sus operon: a model system for starch uptake by the human gut bacteroidetes. Cell Mol Life Sci 73:2603–2617. https://doi.org/10.1007/s00018-016-2242-x

211. Glenwright AJ, Pothula KR, Bhamidimarri SP, Chorev DS et al (2017) Structural basis for nutrient acquisition by dominant members of the human gut microbiota. Nature 541:407–411. https://doi.org/10.1038/nature20828
212. Ze X et al (2015) Unique organization of extracellular amylases into amylosomes in the resistant starch-utilizing human colonic Firmicutes bacterium Ruminococcus bromii. MBio 6:e01058–e01015
213. Bjedov I (2003) Stress-induced mutagenesis in bacteria. Science 300:1404–1409. https://doi.org/10.1126/science.1082240
214. Xu J et al (2007) Evolution of symbiotic bacteria in the distal human intestine. PLoS Biol 5:1574–1586
215. Svanback R, Bolnick DI (2007) Intraspecific competition drives increased resource use diversity within a natural population. Proc R Soc B-Biol Sci 274:839–844. https://doi.org/10.1098/rspb.2006.0198
216. Emerson BC, Kolm N (2005) Species diversity can drive speciation. Nature 434:1015–1017. https://doi.org/10.1038/nature03450
217. Louis P, Flint HJ (2016) Formation of propionate and butyrate by the human colonic microbiota. Environ Microbiol 19(1):29–41
218. Ze X, Duncan SH, Louis P, Flint HJ (2012) Ruminococcus bromii is a keystone species for the degradation of resistant starch in the human colon. ISME J 6:1535–1543. https://doi.org/10.1038/ismej.2012.4
219. Louis P, Scott KP, Duncan SH, Flint HJ (2007) Understanding the effects of diet on bacterial metabolism in the large intestine. J Appl Microbiol 102:1197–1208. https://doi.org/10.1111/j.1365-2672.2007.03322.x
220. Duncan SH, Louis P, Flint HJ (2004) Lactate-utilizing bacteria, isolated from human feces, that produce butyrate as a major fermentation product. Appl Environ Microbiol 70:5810–5817. https://doi.org/10.1128/AEM.70.10.5810-5817.2004
221. Rakoff-Nahoum S, Foster KR, Comstock LE (2016) The evolution of cooperation within the gut microbiota. Nature 533:255–259. https://doi.org/10.1038/nature17626
222. Juge N, Tailford L, Owen CD (2016) Sialidases from gut bacteria: a mini-review. Biochem Soc Trans 44:166–175. https://doi.org/10.1042/BST20150226
223. Crost EH, Tailford LE, Le Gall G, Fons M et al (2013) Utilisation of mucin glycans by the human Gut symbiont Ruminococcus gnavus Is strain-Dependent. PLoS One 8:e76341. https://doi.org/10.1371/journal.pone.0076341
224. Crost EH et al (2016) The mucin-degradation strategy of Ruminococcus gnavus: the importance of intramolecular trans-sialidases. Gut Microbes 7(4):302–312
225. Larsson JMH, Karlsson H, Crespo JG, Johansson MEV et al (2011) Altered o-glycosylation profile of MUC2 mucin occurs in active ulcerative colitis and is associated with increased inflammation. Inflamm Bowel Dis 17:2299–2307. https://doi.org/10.1002/ibd.21625
226. Carbonero F, Benefiel AC, Alizadeh-Ghamsari AH, Gaskins HR (2012) Microbial pathways in colonic sulfur metabolism and links with health and disease. Front Physiol 3:448. https://doi.org/10.3389/fphys.2012.00448
227. Ridlon JM, Kang DJ, Hylemon PB, Bajaj JS (2014) Bile acids and the gut microbiome. Curr Opin Gastroenterol 30:332–338. https://doi.org/10.1097/MOG.0000000000000057
228. Staley C, Weingarden AR, Khoruts A, Sadowsky MJ (2017) Interaction of gut microbiota with bile acid metabolism and its influence on disease states. Appl Microbiol Biotechnol 101:47–64. https://doi.org/10.1007/s00253-016-8006-6
229. Kakiyama G, Pandak WM, Gillevet PM, Hylemon PB et al (2013) Modulation of the fecal bile acid profile by gut microbiota in cirrhosis. J Hepatol 58:949–955. https://doi.org/10.1016/j.jhep.2013.01.003
230. Bäckhed F, Roswall J, Peng Y, Feng Q et al (2015) Dynamics and stabilization of the human gut microbiome during the first year of life. Cell Host Microbe 17:690–703

231. Martin R, Makino H, Yavuz AC, Ben-Amor K et al (2016) Early-Life events, including mode of delivery and type of feeding, siblings and gender, shape the developing gut microbiota. PLoS One 11:e0158498
232. Chu DM, Ma J, Prince AL, Antony KM et al (2017) Maturation of the infant microbiome community structure and function across multiple body sites and in relation to mode of delivery. Nat Med 23:314–326
233. Cox LM, Yamanishi S, Sohn J, Alekseyenko AV et al (2014) The intestinal microbiota during a critical developmental window has lasting metabolic consequences. Cell 158:705–721
234. Dominguez-Bello MG, Costello EK, Contreras M, Magris M et al (2010) Delivery mode shapes the acquisition and structure of the initial microbiota across multiple body habitats in newborns. Proc Natl Acad Sci U S A 107:11971–11975
235. Dominguez-Bello MG, De Jesus-Laboy KM, Shen N, Cox LM et al (2016) Partial restoration of the microbiota of cesarean-born infants via vaginal microbial transfer. Nat Med 22:250–253
236. Negele K, Heinrich J, Borte M, Von Berg A et al (2004) Mode of delivery and development of atopic disease during the first 2 years of life. Pediatr Allergy Immunol 15:48–54
237. Nagpal R, Tsuji H, Takahashi T, Kawashima K et al (2016) Sensitive quantitative analysis of the meconium bacterial microbiota in healthy term infants born vaginally or by cesarean section. Front Microbiol 7:1997
238. Penders J, Thijs C, Vink C, Stelma FF et al (2006) Factors influencing the composition of the intestinal microbiota in early infancy. Pediatrics 118:511–521
239. Wampach L, Heintz-Buschart A, Hogan A, Muller EEL et al (2017) Colonization and succession within the human gut microbiome by archaea, bacteria, and microeukaryotes during the first year of life. Front Microbiol 8:738
240. Fallani M, Young D, Scott J, Norin E et al (2010) Intestinal microbiota of 6-week-old infants across Europe: geographic influence beyond delivery mode, breast-feeding, and antibiotics. J Pediatr Gastroenterol Nutr 51:77–84
241. Biasucci G, Rubini M, Riboni S, Morelli L et al (2010) Mode of delivery affects the bacterial community in the newborn gut. Early Hum Dev 86:13–15
242. Kim KH, Fekety R, Batts DH, Brown D et al (1981) Isolation of clostridium difficile from the environment and contacts of patients with antibiotic-associated colitis. J Infect Dis 143:42–50
243. Rousseau C, Poilane I, De Pontual L, Maherault AC et al (2012) Clostridium difficile carriage in healthy infants in the community: a potential reservoir for pathogenic strains. Clin Infect Dis 55:1209–1215
244. Gabriel I, Olejek A, Stencel-Gabriel K et al (2017) The influence of maternal vaginal flora on the intestinal colonization in newborns and 3-month-old infants[J]. J Matern Fetal Neonatal Med 31(11):1448–1453
245. Al Jumaili IJ, Shibley M, Lishman AH, Record CO (1984) Incidence and origin of Clostridium difficile in neonates. J Clin Microbiol 19:77–78
246. Adlerberth I, Lindberg E, Åberg N, Hesselmar B et al (2006) Reduced enterobacterial and increased staphylococcal colonization of the infantile bowel: an effect of hygienic lifestyle? Pediatr Res 59:96–101
247. Arboleya S, Binetti A, Salazar N, Fernández N et al (2012) Establishment and development of intestinal microbiota in preterm neonates. FEMS Microbiol Ecol 79:763–772
248. Stewart CJ, Embleton ND, Clements E, Luna PN et al (2017) Cesarean or vaginal birth does not impact the longitudinal development of the gut microbiome in a cohort of exclusively preterm infants. Front Microbiol 8:1008
249. Biasucci G, Benenati B, Morelli L, Bessi E et al (2008) Cesarean delivery may affect the early biodiversity of intestinal bacteria. J Nutr 138:1796S–1800S
250. Guaraldi F, Salvatori G (2012) Effect of breast and formula feeding on gut microbiota shaping in newborns. Front Cell Infect Microbiol 2:94
251. Groer MW, Luciano AA, Dishaw LJ, Ashmeade TL et al (2014) Development of the preterm infant gut microbiome: a research priority. Microbiome 2:38

252. Li C, Liu Y, Jiang Y, Xu N et al (2017) Immunomodulatory constituents of human breast milk and immunity from bronchiolitis. Ital J Pediatr 43:8
253. Williams JE, Price WJ, Shafii B, Yahvah KM et al (2017) Relationships among microbial communities, maternal cells, oligosaccharides, and macronutrients in human milk. J Hum Lact 33:540–551
254. Collado MC, Rautava S, Aakko J, Isolauri E et al (2016) Human gut colonisation may be initiated in utero by distinct microbial communities in the placenta and amniotic fluid. Sci Rep Sep 6:23129
255. Perez PF, Doré J, Leclerc M, Levenez F et al (2007) A bacterial imprinting of the neonatal immune system: lessons from maternal cells? Pediatrics 119:e724–e732
256. Pannaraj PS, Li F, Cerini C, Bender JM et al (2017) Association between breast milk bacterial communities and establishment and development of the infant gut microbiome. JAMA Pediatr 171:647
257. Lee SA, Lim JY, Kim BS, Cho SJ et al (2015) Comparison of the gut microbiota profile in breast-fed and formula-fed Korean infants using pyrosequencing. Nutr Res Pract 9:242–248
258. Bezirtzoglou E, Tsiotsias A, Welling GW (2011) Microbiota profile in feces of breast- and formula-fed newborns by using fluorescence in situ hybridization (FISH). Anaerobe 17:478–482
259. Bullen CL, Tearle PV, Stewart MG (1977) The effect of 'humanised' milks and supplemented breast feeding on the faecal flora of infants. J Med Microbiol 10:403–413
260. Le Huërou-Luron I, Blat S, Boudry G (2010) Breast-v. formula-feeding: impacts on the digestive tract and immediate and long-term health effects. Nutr Res Rev 23:23–36
261. Stark PL, Lee A (1982) The microbial ecology of the large bowel of breastfed and formula-fed infants during the first year of life. J Med Microbiol 15:189–203
262. Cresci GA, Bawden E (2015) Gut microbiome: what we do and don't know. Nutr Clin Pract 30:734–746
263. Fanaro S, Chierici R, Guerrini P, Vigi V (2003) Intestinal microflora in early infancy: composition and development. Acta Paediatr Suppl 92:48–55
264. Makino H, Kushiro A, Ishikawa E, Kubota H et al (2013) Mother-to-infant transmission of intestinal bifidobacterial strains has an impact on the early development of vaginally delivered infant's microbiota. PLoS One 8:e78331
265. Brooks B, Firek BA, Miller CS, Sharon I et al (2014) Microbes in the neonatal intensive care unit resemble those found in the gut of premature infants. Microbiome 2:1
266. Touati A, Achour W, Cherif A, Hmida HB et al (2009) Outbreak of Acinetobacter baumannii in a neonatal intensive care unit: antimicrobial susceptibility and genotyping analysis. Ann Epidemiol 19:372–378
267. Costeloe K, Bowler U, Brocklehurst P, Hardy P et al (2016) A randomised controlled trial of the probiotic Bifidobacterium breve BBG-001 in preterm babies to prevent sepsis, necrotising enterocolitis and death: the probiotics in preterm infantS (PiPS) trial. Health Technol Assess 20:1–194
268. Millar M, Seale JJ, Greenland M, Hardy P et al (2017) The microbiome of infants recruited to a randomised placebo-controlled probiotic trial (PiPS Trial). EBioMedicine 20:255–262
269. Biedermann L, Zeitz J, Mwinyi J, Sutter-Minder E et al (2013) Smoking cessation induces profound changes in the composition of the intestinal microbiota in humans. PLoS One 8: e59260. https://doi.org/10.1371/journal.pone.0059260
270. Jiang H, Ling Z, Zhang Y, Mao H et al (2015) Altered fecal microbiota composition in patients with major depressive disorder. Brain Behav Immun 48:186–194. https://doi.org/10.1016/j.bbi.2015.03.016
271. Tyakht AV, Kostryukova ES, Popenko AS, Belenikin MS et al (2013) Human gut microbiota community structures in urban and rural populations in Russia. Nat Commun 4:2469. https://doi.org/10.1038/ncomms3469
272. Francisco G (2003) Malagelada Juan-RGut flora in health and disease[J]. Lancet 361 (9356):512–519

273. Sghir A, Gramet G, Suau A et al (2000) Quantification of bacterial groups within human fecal flora by oligonucleotide probe hybridization[J]. Appl Environ Microbiol 66(5):2263–2266
274. Mm QE (2011) Gut microbiota and the role of probiotics in therapy[J]. Curr Opin Pharmacol 11(6):593–603
275. Sandra C (2018) Intestinal barriers protect against disease[J]. Science 359(6380):1097–1098
276. Sarker S, Gyr K (1992) Non-immunological defence mechanisms of the gut. Gut 33:987–993
277. Szymanowska-Powa owska D, Orczyk D, Leja K (2014) Biotechnological potential of Clostridium butyricum bacteria. Braz J Microbiol 45:892–901
278. Nijiang Y (2012) Research progress on the relationship between intestinal microorganisms and host metabolism[J]. Feed Expo 7:9–12
279. Rooks MG, Garrett WS (2016) Gut microbiota, metabolites and host immunity[J]. Nat Rev Immunol 16(6):341–352
280. Huangxiaoyan W (2012) The molecular mechanism of intestinal microorganisms regulating lipid metabolism[J]. Feed Industry 33(18):59–62
281. Pew G (2017) Zhuweiyun. Advances in the study of the metabolic axis of intestinal microorganisms in animal hosts[J]. J Microbiol 2:161–169
282. Jinlei, W (2018) Advances in the study of the relationship between intestinal microorganisms and host immunity[J]. Modern Animal Husbandry, 358(09):57–63.
283. Peterson DA, McNulty NP, Guruge JL, Gordon JI (2007) IgA response to symbiotic bacteria as a mediator of gut homeostasis. Cell Host Microbe 2:328–339
284. Wu JJ, Lai SM, Pan K et al (2015) Effects of intestinal flora on intestinal development, mucosal morphology and immune organs development of chicks. Chin J Anim Nutr 27 (4):1101–1109
285. Liu L et al (2009) Polyamines regulate E-cadherin transcription through c-Myc modulating intestinal epithelial barrier function. Am J Phys Cell Phys 296:C801–C810
286. Cummings JH, Pomare EW, Branch WJ, Naylor CP, Macfarlane GT (1987) Short chain fatty acids in human large intestine, portal, hepatic and venous blood. Gut 28:1221–1227
287. Lewandowski T et al (2013) Staphylococcus aureus formylmethionyl transferase mutants demonstrate reduced virulence factor production and pathogenicity. Antimicrob Agents Chemother 57:2929–2936
288. Liuruixue LZ (2016) Research progress in microecological balance and human health of intestinal flora[J]. Food Indus Technol 37(06):383–387+391
289. Turnbaugh PJ, Ley RE, Mahowald MA et al (2006) An obesity-associated gut microbiome with increased capacity for energy harvest[J]. Nature 444(7122):1027–1031
290. Chang CJ, Lin CS, Lu CC et al (2015) Ganoderma lucidum reduces obesity in mice by modulating the composition of the gut microbiota[J]. Nat Commun 6(7489):1
291. Barnich N, Carvalho FA, Glasser AL et al (2007) CEACAM6 acts as a receptor for adherent-invasive E. coli, supporting ileal mucosa colonization in Crohn disease[J]. J Clin Invest 117:1566–1574
292. Tang C, Kamiya T, Liu Y et al (2015) Inhibition of dectin-1 signaling ameliorates colitis by inducing Lactobacillus-mediated regulatory T cell expansion in the intestine[J]. Cell Host Microbe 18:183–197
293. Bäumler Andreas J (2016) Sperandio Vanessa. Interactions between the microbiota and pathogenic bacteria in the gut.[J]. Nature 535(7610):85–93
294. Backhed F, Ley RE, Sonnenburg JL, Peterson DA, Gordon JI (2005) Host—bacterial mutualism in the human intestine[J]. Science 307:1915–1920
295. Chenheng, Yesheng, Zuo Yanwen, Pei Xiaofang (2008) Research progress in the relationship between intestinal microorganisms and obesity[J]. Modern Prev Med 035(004):608–609,616
296. Gerba CP, Gramos DM, Nwachuku N (2002) Comparative inactivation of Enteroviruses and adenovirus 2 by UV light[J]. Appl Environ Microbiol 68(10):5167–5169
297. Jialianqun, S, Lvmeijun, S, Chenlijuan, CS, Zhang, L, Yu, YG. Discussion on the relationship between intestinal microbiological homeostasis and glycolipids metabolism based on the theory of "temper astigmatism"[J/OL]. Liaoning J Tradit Chin Med: 1–8[2019-07-15].

298. Tong X, Xu J, Lian F et al (2018) Structural alteration of gut microbiota during the amelioration of human type 2 diabetes with hyperlipidemia by metformin and a traditional Chinese herbal formula: a multicenter, randomized, open label clinical trial. MBio 9(3):e02392–e02317
299. Aleman JO, Bokulich NA, Swann IR et al (2018) Fecal microbiota and bile acid interactions with systemic and adipose tissue metabolism in diet-induced weight loss of obese postmenopausal women[J]. J Transl Med 16(1):244
300. Eckburg PB, Bik EM, Bernstein CN et al (2005) Diversity of the human intestinal microbial flora[J]. Science 308(5728):1635–1638
301. Ewaschuk JB, Diaz H, Meddings L et al (2008) Secreted bioactive factors from Bifidobacterium infantis enhance epithelial cell barrier function[J]. Am J Physiol Gastrointest Liver Physiol 295(5):G1025–G1034
302. Gaorunping LY (2019) Microbial intervention as a new target for advanced treatment of non-alcoholic fatty liver disease[J]. Journal of Clinical Hepatitis (02):35, 333
303. Qian L, Hachengyong Z, Yubin Z (2018) Research progress on treatment of alcoholic liver disease based on liver and intestine axis[J]. Pharm Biotechnol 25(04):368–371
304. Zhangdong (2017) Effects of intestinal flora disorders and probiotics on alcoholic liver disease and its mechanism[D]. Qingdao University
305. Lizhao X, Wangziqian N, Limengtao ZX (2017) The role of intestinal microorganisms in autoimmune diseases[J]. Chin J Clin Immunol Allergy 11(01):61–68
306. Wang Meng, Wang Dengjielin, Mengguannan, Yu Li Radium, Jiang Hong (2018) Advances in the study of intestinal microorganisms and cardiovascular diseases[J]. Medical Overview 24 (18):3543–3547+3553
307. Lederberg J (2000) Infectious history. Science 288:287–293
308. Marchesi JR, Ravel J (2015) The vocabulary of microbiome research: a proposal. Microbiome 3:31
309. Pfeiffer JK, Sonnenburg JL (2011) The intestinal microbiota and viral susceptibility. Front Microbiol 2:1–6
310. Zhao L, Shen J (2010) Whole-body systems approaches for gut microbiota-targeted, preventive healthcare. J Biotechnol 149:183–190
311. Dishaw LJ, Cannon JP, Litman GW, Parker W (2014) Immune-directed support of rich microbial communities in the gut has ancient roots. Dev Comp Immunol 47:36–51
312. Panyutong P, Jun Z (2019) Research progress on the correlation between intestinal flora and kidney disease[J]. Chin J Microecol 31(06):729–733
313. Genglin YC (2019) Research progress on the correlation between intestinal flora and type 2 diabetes[J]. Medical Overview 25(10):2034–2038
314. Sender R, Fuchs S, Milo R (2016) Are we really vastly outnumbered? Revisiting the ratio of bacterial to host cells in humans. Cell 164:337–340
315. Jiménez E, Marín ML, Martín R et al (2008) Is meconium from healthy newborns actually sterile?[J] Res Microbiol 159(3):0–193
316. Horinaka M, Yoshida T, Kishi A et al (2010) Lactobacillus strains induce TRAIL production and facilitate natural killer activity against cancer cells secreted bioactive factors from Bifidobacterium infantis enhance epithelial cell barrier function[J]. FEBS Lett 584(3):577–582
317. Bengmark S , Cocco PD , Clemente K et al (2011) Bio-ecological control of chronic liver disease and encephalopathy[J]. Minerva Med 102(4):309–319
318. Lee KJ, Kim YB, Kim JH et al (2008) The alteration of enterochromaffin cell, mast cell, and lamina propria T lymphocyte numbers in irritable bowel syndrome and its relationship with psychological factors[J]. J Gastroenterol Hepatol 23(11):1689–1694
319. Saleh M, Trinchieri G (2010) Innate immune mechanisms of colitis and colitis-associated colorectal cancer[J]. Nat Rev Immunol 11(1):9–20
320. Progress of Wang Lin, Li Bing, and Zhujian et al (2016) Chin Agri Bull 32(5):10–15
321. Chenxiaolin RH (2014) Relationship between intestinal microbial groups and intestinal immunity[J]. Gastroenterol Hepatol Impur 23(11):1245–1248

Chapter 3
Introduction to Probiotics and Their Potential Health Benefits

Marwa M. El-Dalatony and Xiangkai Li

3.1 Probiotics Definition

The terminology of probiotic is a new word originated from a combination of couple languages Latin and Greek in which "pro" means for and "biotic" stands for life, respectively. Recently, different means to define probiotics have been given on the basis of their mode of action and influence on human health. The term probiotic was coined in 1965, described as the compounds generated by a microbe, that influence the optimum growth conditions for other microbes, and analogous with the term antibiotic [1]. "Organisms and substances that contribute to intestinal balance," was the initial term of probiotic, in use today [2]. The expression of these compounds in Parker's meaning of probiotics provides an extensive association that involved analogy with the antibiotics. However, probiotics and antibiotics differ significantly in terms of their role in humans (Table 3.1).

Fuller in 1989 upgraded Parker's description of probiotic was as "A live microbial feed supplement which beneficially affects the host animal by improving its intestinal microbial balance." The definition was focused on the need of sustainability of probiotics as well as their beneficial impacts on an animal host, as per his descriptions. Fuller's view of probiotics was further extended by The European expert group, they included all mechanisms stimulated by the microflora and defined probiotics as "Probiotics are live microbial food components that have a positive effect on human health." The limitation of the purposed concept was the site of entry of the probiotics and mode of action was not considered in relation with their positive effects. Therefore, various sites such as the intestine, skin, vagina, oral cavity may be used for the administration of the probiotics.

M. M. El-Dalatony (✉) · X. Li
School of Life Sciences, Lanzhou University, Lanzhou, PR China
e-mail: marwa@hanyang.ac.kr

X. Li, P. Liu (eds.), *Gut Remediation of Environmental Pollutants*,
https://doi.org/10.1007/978-981-15-4759-1_3

Table 3.1 Difference between the role of probiotics and antibiotics in relation with human health

Probiotics	Antibiotics
For the life	Against the life
Non-invasive in nature	Emergence of antibiotic resistance microorganisms
Free from undesirable side effects	Associated with several unpleasant side effects
Preventive mode of action in disease conditions	Antibiotics are used against bacteria to treat the disease either by direct killing or inhibition of cell growth
Increase the concentration of *Bifidobacterium* or/and *Lactobacilli*	They are employed to treat various medical concerns, starting from skin infections and urinary tract infections to pneumonia and whooping cough.
Help restore the balance of the gastrointestinal flora when diet is sterile infant formula	Improve the growth enactment and feed conversion effectiveness of the animals.
Increase the population of beneficial microflora by preventing the attachment of pathogens with receptors present on intestinal barriers.	They do not have a specific effect on increasing the gut microbial populace.
Increase the concentration of antibodies in the body such as secretory Immunoglobulin A (IgA).	No specific increase in the antibody concentration due to antibiotic administration has been reported.

Misuse of term "probiotic" both scientifically and commercially was explained by the International Scientific Association for Probiotics and Prebiotics (ISAPP), a non-profit scientific organization in progressing the science of probiotics and prebiotics (Fig. 3.1). Commercially, probiotics is defined as the products that do not authenticate human health benefits, whereas scientifically, it has been explained as the elemental constituents of a bacteria, inactivated or killed bacteria, and bacteria whose health benefits are not well characterized (http://www.isapp.net/Portals/0/docs/ProbioticDefinitionClarification.pdf). No clear explanation and particular definition were given by ISAPP; it only highlights the eminent components that are well characterized in the definition of WHO/FAO. The categorization of probiotics has been done according to the legislation of different countries including: (1) functional foods in far eastern countries such as China, Japan, and Malaysia; (2) dietary additives, biological agent, medical foods, drugs, and live biotherapeutic agents that were applied in USA; (3) supplementary food in Denmark, Finland, and Sweden; (4) Canadian natural health products; and (5) pharmaceuticals with biotherapeutic in Germany and Belgium.

The researchers have claimed the loss of the main aspect of strain and dose specificity of probiotics when categorized improperly by the legislation of different countries. The American-European-Asian legislations need scientific confirmation on efficacy of probiotic species via standardized safety investigations conducted with several clinical trials [3].

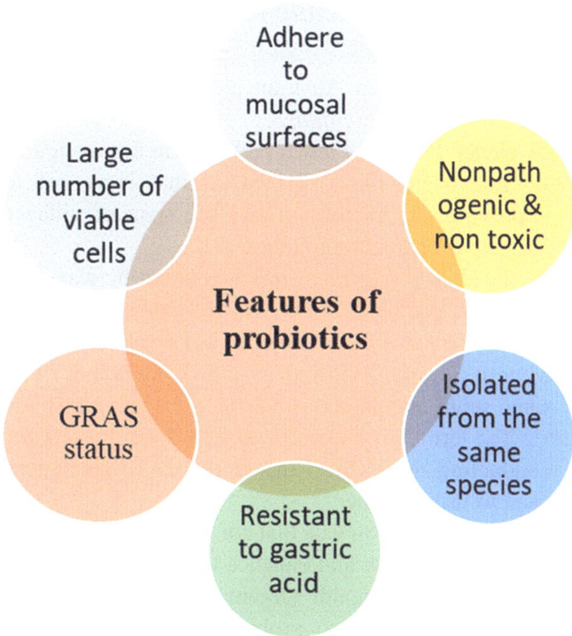

Fig. 3.1 Different features of a bacteria to be defined as a probiotic along with its safety standards

3.2 History of Probiotic

The historical aspects of consuming microbes in diet have various health benefits such as use of bacteria that produce lactic acid to suppress pathogens. The original remark for the explanation of health benefits provided by some specific bacterial strains is credited to a Russian Nobel Prize winner scientist Eli Metchnikoff. In the starting of twentieth century, he proposed "The dependence of the intestinal microbes on the food makes it possible to adopt measures to modify the flora in our bodies and to replace the harmful microbes by useful microbes" at the Pasteur Institute [4]. He also claimed a reduced count of toxin-producing bacteria in the gut followed by the intake of yogurt containing *Lactobacilli* that helped in increasing the lasting power of host [5]. Another observation was given by French pediatrician Henry Tissier, he found a smaller number of Y shaped, peculiar bacteria in the stools of children suffering from diarrhea. However, the count of these "bifid" bacteria were high in healthy children [6]. Based on this observation he suggested that the administration of these bacteria to patients with diarrhea would help in the restoration of normal gut microbes. The first scientific proposal before the invention of the word "probiotics" was given by Metchnikoff and Tissier.

3.2.1 A Brief Timeline for the History of Probiotics

1890: Ernst Moro an Austrian physician discovered an acid producing bacteria *Lactobacillus acidophilus*

1899: Henry Tissier a French pediatrician discovered *Bifidobacteria*

1907: Elie Metchnikoff discovered that intake of fermented yogurt helped in improving gut microbiota and life span of villagers in the Bulgarian regions near the Caucasus mountains.

1923: *Saccharomyces boulardii* as a probiotic was discovered by Henri Boulard in 1923. He also noticed the effect of intake of lychee fruit on human health.

1930: New strain of *L. casei* named as *Lactobacillus casei* Shirota was discovered by Minoru Shirota, followed by the development of a yogurt drink consisting the same strain and named it Yakult, after being inspired by Metchnikoff's work.

1965: The term "probiotics" was first given by RH Stillwell and DM Lilly, while conducting research on the secretions of one microorganism which were then used to form another.

1995: "Prebiotic" term was first coined, and was defined as the foods that encourage the natural growth and enhance the effect of probiotics in the human gut.

2002: The World Health Organization (WHO) and United Nations (UN) gave official recognition to probiotics as microbes that exerts positive effects on humans.

2013: Dr. Ted Dinan coined the term "psychobiotics" after researching the effect of various microorganisms on the mood of the host (psychic effects) [7].

Another definition of probiotics was stated by the association of Joint Food and Agriculture Organization with World Health Organization (WHO) as: "Live microorganisms which, when administered in adequate amounts, confer a health benefit on the host" [8]. According to this description to be probiotic, particular formulations must follow the following criteria: (a) should be administered active; (b) have gone through controlled assessment; (c) have documented beneficial effects on the health of targeted organism; (d) should have a well-defined taxonomic classification from genus to strain level; and (e) characterized as safe (proposed use).

3.3 Relationship among Different Pro-, Pre-, and Symbiotic

Probiotics can be defined as friendly bacteria, recognized to possess valuable influence on humans, and accessible in different formulas such as dairy products (dahi, yogurt, and capsules). The term prebiotic was given by replacing "pro" for "pre," that means "before" or "for" Gibson and Roberfroid [9]. The main properties of prebiotic are the non-digestible components that exerts a positive impact by the mechanism of selective prompt of the useful bacteria in human colon, thus provide health benefits. In a clear way, administration of these additives such as fructose oligosaccharides (FOS) can rise the count of bacteria either in vagina or gut and have a positive effect on human health. Integration of probiotics and prebiotics together form synbiotics. The commonly well-known recently used prebiotic ingredients are non-digestible carbohydrates such as fructo-oligosaccharides (FOS), galacto-oligosaccharides (GOS), inulin, and lactulose. Several non-digestible carbohydrates have been examined to use as potential prebiotics like arabinoxylan, polydextrose, soybean oligosaccharides, xylo-oligosaccharides (XOS), xylo-polysaccharide (XPS), isomalto-oligosaccharides (IMO), and beta glucans. In spite of all the studies the most commonly used prebiotics are inulin and FOS [10].

The chemical nature of the non-digestible oligosaccharides like degree of polymerization, viscosity, glycosidic bonds, and fermentability are responsible for the physiological effects of prebiotic supplements. Therefore, the extent of efficacy of any prebiotic is directly correlated with the end products formed after its complete metabolism by the intestinal microbiota [11]. According to the studies, it is important to completely define the non-digestible carbohydrates used in prebiotics, as these compounds have to be metabolized by different microbes. The final product produced after the metabolism and its consumption by the other microbes defines the therapeutic efficacy. The synbiotic effect of the pre- and pro-biotics formulations is a result of the formation of synergistic activities of these compounds in the food. This reaction in vivo would be influenced by the intake of the prebiotic, that results in a competitive advantage for the microbes in many cases. The synergistic effect of these prebiotics supplements is to improve the growth and survival of the probiotic strains inside the host, thus to increase the composition of beneficial microbes. The selection of particular prebiotic depends on the specific beneficial effect on the probiotics. The synergistic effect of these supplements target different regions of the small and large intestine. Generally, these synbiotic interactions offer huge information to stimulate the potential and efficiency of these types of functional foods [10]. The combination of prebiotic components and probiotic microbes provide not only beneficial health effects to individuals, but also develop the stability of products throughout their storage time era. The addition of ingredients of prebiotic has shown the anti-obesity potential by stimulating the physiological functions responsible for secretion of insulin, by the multiplication of β-pancreatic cells [12].

3.3.1 Need of Pre- and Probiotics

The influence of probiotics on gut microbes have a close relationship with the effects of prebiotics (Fig. 3.2). Prebiotics play various roles in a symbiotic relationship with probiotics and enhance the effects of probiotics in various possible ways, as follows;

1. Not only the gut microflora, prebiotics also play eminent function in the enhancement of calcium and magnesium absorption in the body, thus help in increasing the density of bones.
2. Prebiotics exert a remarkable function in the regulation of appetite by promoting digestion and lipid metabolism thus help in regulating gut flora.

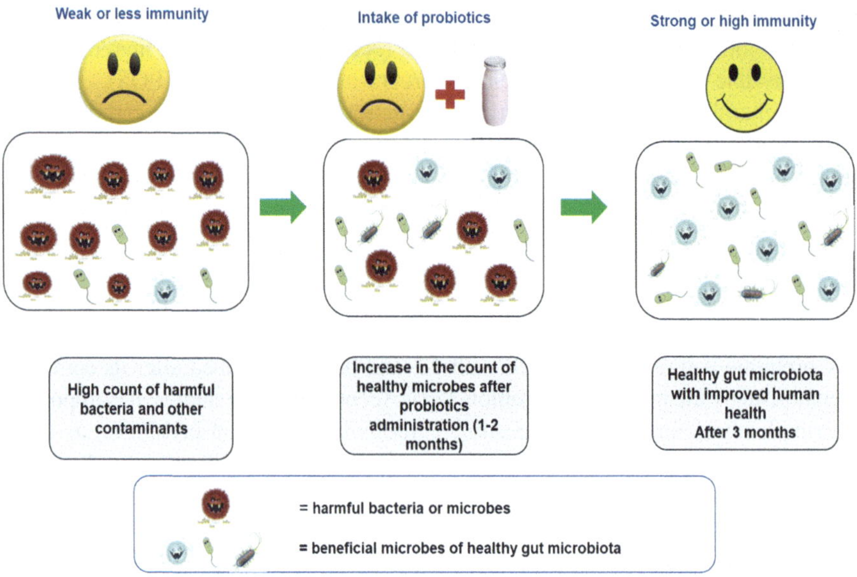

Fig. 3.2 Effects of probiotic strains on both beneficial microflora and pathogen to increase the count of beneficial microbes and development of immunity over specific period of time

3. Helps in the metabolism of carbohydrates, by changing the speed of carbohydrate metabolism by the body during inflammation that further leads to ischemic heart disease if untreated.
4. Helps in reducing insulin resistance.
5. Maintenance of electrolytes such as sodium and potassium, and various minerals to control blood pressure.
6. Hormonal balance

3.4 Where Do Probiotics Produced from

The development of microbiota in neonatal depends on: (a) microbial flora of mother, (b) mode of birth or delivery, (c) birth conditions and surrounding environmental conditions, and (d) genetic components. The primary source of microbes in newborns is the microbes inherited from mother (intestine and vagina), colonizing the intestinal tract of infants. After specific period of time, probiotics (mainly the lactic acid containing yogurt) can be given to the infants, as following;

1. *Immediately after birth*: To establish a healthy gut microbiota and prevent establishment of pathogenic bacteria.

2. *Following antibiotic administration*: To prevent the chances of reinfection by a particular microorganism and the restoration of the healthy gut microbiota which was depleted by the administration of antibiotics.
3. *Treatment of diarrhea*: To reduce the count of pathogens [13]

3.5 Probiotics and Health

Various strains have been identified as potential probiotics in different disease conditions, depending on the results obtained after administration of a particular type of strain. In case of IBD (Inflammatory bowel disease), diarrhea, allergy symptoms, *Bifidobacterium sp.* have shown promising results on the restoration of healthy gut microbiota (Table 3.2).

3.6 Probiotics and Gut Microbiota

Several reports have been studied the probiotics that help in regulation and enhancement of various features of the innate and adaptive immunity in both animals and humans (Table 3.3). Human gut comprises of approximately 100 trillion (10^{14}) microbes which is 10 fold in magnitude than human cells, contributing in total weight of body almost 1.5–2 kg [19]. A regular increase has been observed in the abundance and complexity these microbes in the gut (stomach to colon), levels up to the level of 10^{11} cells/gram of the intestinal composition [20]. Microbiota is defined as the group of microbes residing in human body and their total genetic composition is known as microbiome. The microbes are abundant in the colon, however they are less in respiratory tract, skin, and vagina, these organs also inhabit specific microbes [21].

The eminent roles played by the gut microbiome are maintenance of the functional coalition of gut and intestine, homeostasis of immune response, and metabolic

Table 3.2 Effect of various bacteria used for probiotics on immune system and beneficial gut microflora and their source

Probiotics	Test sps	Observation	References
Bacillus sps	Zebrafish	Antagonistic/inhibitory activity	[14]
Lactobacillus sps	Grouper & Nile tilapia	Growth enhancement, development of resistance against disease, and innate immunity	[15]
Lactobacillus rhamnosus	Rainbow trout	Stimulated respiratory burst	[16]
Bacillus sp.	litopenaeus vannamei	Increase in phagocytic and antibacterial activity due to stimulated immune response	[17]
Bacillus subtilis	Gilthead seabream	Increase in phagocytic activity	[18]

Table 3.3 Effect of *Lactobacilli* sp. and *Bifidobacterium* on innate and adaptive immune response in different animal models

Bacteria strains	Disease model	Disease	Outcomes	References
L. acidophilus	Eight-week-old male C57BL/mice	IBD	↑IL-10, Treg ↓IL-6, IL-1β, IL-17	[31]
L. acidophilus (NCK2025)	Generation of TS4Cre×APC lox468 mice	CRC	↑IL-10, IL-12 ↓Treg	[32]
L. acidophilus	Female BALB/c mice	Crohn's disease	↑IL-17 ↓T17 function, IL-23,	[33]
L. acidophilus	BALB/c mice	Ulcerative colitis	↑*Lactobacilli, Bifdobacteria* ↓*S. aureus*	[34]
L. casei BL23	Female C57BL/6 mice	CRC	↑T17, T 22, IL-10, and IL-22 ↓Treg	[35]
L. fermentum FTDC 812	Eight-week-old BALB/c mice	Hypercholesterolemia	↑Lactobacillus	[24]
L. rhamnosus, B. bifdum	Eight-week C57BL/6 mice	Type 2 diabetes	↑Firmicutes, Actinobacteria ↓Bacteroidetes	[36]
B. breve IPLA20004	Human colon	Inflammatory disease	↑IL-8, IL-10, IL-12	[24]

pathways for energy generation. The microbiota composition constitutes the humungous microbial community that contains at least two-fold more genes compared to the total magnitude of genes in human genome. Another term related to the gut microbiota is "dysbiosis" defined as the changes in the composition of gut microbial population that further leads to amendments in the host-microbiota interactions [22]. The microbial community helps in the regulation of various metabolic and physiological properties and also contribute a vital function in the development of immune response in early life stages, thus maintaining homeostasis of immune system during life. The major dominating bacterial phyla of gut microbiota of humans are: Firmicutes (*Ruminococcus, Clostridium and Eubacteria*), Bacteroidetes *(Porphyromonas and Prevotella*), and Actinobacteria (*Bifidobacterium*). *Escherichia coli, Streptococci, and Lactobacilli*, and also constitute the gut microbiota in smaller number. The specific contribution of the phyla such as Firmicutes (~60 to 65%), Bacteroidetes (~0 to 25%), Actinobacteria (~3%), and Proteobacteria (~5 to 10%), contributes about 97% of the total population of intestinal microbes. A study on the colonization of intestinal microbes has suggested the complex interaction between the different microbes and host-microbes' interactions essential for the establishment of intestinal microbiota. The researchers have also stated the dynamics of bacterial process which not colonize in the intestine [23].

Exposure of metals and other contaminants to gut microbiota through diet leads to various alterations in the composition (dysbiosis), amendments in host-microbial interactions that causes several diseases [24]. Analogous to antibiotics, "probiotic" word was introduced primarily in 1965 by Stillwell and Lilly, means "a microbial substance able to stimulate the growth of another micro-organism." The genus *Bifidobacterium* and *Lactobacillus* are the most often used genera as potential probiotics, whereas *Enterococcus, Leuconostoc,* and *Streptococcus* genera are the less abundantly used. These microorganisms as probiotics are available in fermented dairy products such as Kurut, kefir, milk, and Maasai. Probiotics influence the composition of normal gut microflora by restraining the development of pathogens, stimulate the multiplication of epithelial cell, differentiation, and fortification of the mucosal barriers of intestine by a mechanism known as probiosis. [25].

Studies have clearly demonstrated the perturbances in gut microbial composition in the infections caused by *Clostridium difficile* that leads to dysbiosis. Most of these perturbances in the microbial population of gut were reported to be closely related with various gastrointestinal disorders (colorectal cancer, inflammatory bowel disease, and irritable bowel syndrome). Studies have also reported the dysbiotic microbiota due to some intestinal disorders that indirectly affects the respiratory tract or liver and cause cystic fibrosis, bronchial asthma, and allergy. Any alteration in the composition of *Bifidobacterium* population that constitutes the normal microflora of human gut represents the most common factor in these diseases. *Bifidobacteria* strains as a potential probiotic in preventive medicine to sustain the normal functions of intestine have been well-discussed in various studies. Probiotics have also been reported as therapeutic agents for several gastrointestinal diseases and other related disorders [26].

A recent article published by the World Gastroenterology Organization focused on the administration of several probiotic formulations reported the predominance of *Lactobacilli* sp. in clinical studies. Preparations of *Bifidobacterium* strains either alone or mixed with other bacterial strains, as probiotic for human clinical studies, have been considered in the review. Substantial research and literature are available on probiotics but the magnitude of scientific evidences on the efficacy of probiotics on gut microbiota is inadequate in many cases. Accurate human studies and trials needed to demonstrate the mechanism of probiotics in several intestinal pathologies. In recent decades, development of research area in human health and welfare has grown rapidly due to the vital functions of gut microbiota. These studies have shown that individuals with allergy, obese have an intestinal microbiota varying from a healthy individual. The Hero Child Nutrition Institute and the Department of Biochemistry the University of Granada have developed a draft isolation and characterization of probiotic strains, from feces of infants that were breastfed. These strains have been registered at the Institute Pasteur (France) and patented. Controls have surpassed the toxicology and safety standards in the report of FAO/WHO, and its function as a potential inhibitor of the growth of pathogens. Modulation of the immune response in both in vivo and in vitro assays has been accessed in murine and cellular models [27].

In the field of probiotics extensive research has been conducted with various advances in the characterization and selection parameters for different strains but claims of deterioration of health after consuming probiotics in some individuals is still unclear.

The synonymously used terms, "microbiome" and "microbiota" describe the cumulative genetic composition of all the microbes inhabiting the gut and the microbes themselves, respectively. Both the terms "microflora" and "microbiota" are identical, microflora was more frequently used before but still most of the researches use the term in many articles. Microbiota is defined as "the microscopic living organisms of a region" by the Dorland's Medical Dictionary for Health Consumers (2007) and "the microorganisms of a particular site, habitat, or geological period" by the Oxford Dictionary [13].

Another important aspect that relates microbiota and humans has acquired more recognition is "the human holobiont." According to the theory, humans build as a "superorganism" that involves a complex of microbiota in their evolution not as single species. The evolved "superorganism" is an amalgam of microbial cells present in ten-fold more in number and mammalian cells. The genetic composition of these microbes in humans is hundred times the total genes in humans [28]. The symbiotic relationship between the humans and the inherited microbes provide a mutual benefit. Studies have showed a decrease in the immune system related diseases by modulating the immune response. The mutual benefits of this association include the nutrient availability and environment to grow to the microbes, in return the microbes provide health benefits. The health effects of symbiote include the improvised functioning of digestion and metabolic processes to provide essential nutrients [29].

The colonization of various parts of human body such as skin, mouth, respiratory system, gastrointestinal, and urogenital tracts are evident by numerous microbes, but the functions of these interactions and their mechanism are not clear in most of the cases. Different criteria for the selection of a particular bacterial strain as a probiotic have been provided by researchers. The safety criteria include the source of origin of probiotic strains along with its non-pathogenic nature and it should be categorized under genetically recognized as safe (GRAS) list of microorganisms. The other criteria enlist the various functions of probiotics such as development of resistance against the acids of gastrointestinal tract and bile salts, ability to adhere the surface of intestinal epithelial tissues, ability to modulate the immune response, colonization of gastrointestinal tract to influence the human metabolism. The second criteria involve all the functional aspects of the probiotic strains, whereas the third criteria are the technological criteria. It involves the scale-up production of probiotics and resistance against the various technological processes involved in the scale-up. In conclusion of all these principles, the minimum criteria required for the choice of a probiotic strain includes the following: (i) the specification of probiotic microorganism through genus and species; (ii) it must possess a feasible probiotic species; (iii) should be controlled to possess beneficial effects in appropriate doses till the end of their shelf life (with slight deviations between batches); and (iv) to establish controlled studies in humans that prove its efficiency and safety [30].

3.6.1 Effect of Probiotics on Gut Microflora

The variations in gut microbiota composition are preeminent in the treatment of animal and human diseases due to use of probiotics. The major patented probiotic microbes include the species of lactic acid bacteria (LAB) such as *Lactobacillus* (*acidophilus, paracasei, casei, plantarum, crispatus, reuteri, rhamnosus, gasseri, bulgaricus*), *Bifidobacterium* (*longum, breve, catenulatum, bifidum,* and *animalis*), and *S. boulardii*. The potential use of microbes such as *Bacillus* (*subtilis, coagulans, laterosporus*) and *Enterococcus faecium* has also been studied. The most common commercial strains related to the genera *Lactobacillus* and *Bifidobacteria* [23].

3.6.1.1 *Lactobacillus* sp.

The strains of genus *Lactobacillus* have been reported to directly affect the intestinal microbiota. These strains can intensify the functions of the mucosal barriers in intestine, help in maintaining the immune response, inhibition of passage of pathogens across the mucosal barriers, and in the treatment of diseases such as IBD, gastrointestinal infections, and IBS [24]. The impact of probiotics on the metabolism of normal microbiota has been reported other than the direct effects on the composition.

3.6.1.2 *Bacillus coagulans*

The morphological features of *Bacillus genera* include the spore bearing Gram-positive bacteria which are either strict aerobes or facultative in nature. The spores of *B. coagulans, B. subtilis,* and *B. cereus* can resist the acidic environment of the body when orally up taken thus are suitable for humans as a probiotic. The application of these species in the treatment of *H. pylori* infections and diarrhea in humans can be considered. The use of *Bacillus coagulans* alone or in combination with other microbes have provided successful results in the treatment of diarrhea caused by the antibiotics, other than *Bifidobacteria* [37].

B. coagulans involves various mechanisms to get rid of the intestinal pathogens, one of which includes the formation of acidic and anoxic conditions in the intestine not suitable for pathogenic microbes, thereby hindering their development and supporting the growth of healthy microflora [38]. The mechanism involves the consumption of free oxygen by *B. coagulans* strains in the stomach and intestine due to their facultative nature, and decrease the redox reactions required for the growth of pathogens. This environment is suitable for the growth of beneficial microbes like *Bifidobacterium* and *Lactobacillus* sp. [39].

Another mechanism involves the use of *B. coagulans in therapeutics*, to inhibit the growth of pathogens by produce antimicrobial substances that helps in maintaining balance between the normal microbiota [38]. Reports on the secretion of bacitracin by a few strains of B. *coagulans* are available. The strain I4 of *B. coagulans* was first reported to produce coagulin which is a bacteriocin-like inhibitory substance (BLIS). The bacteriocin is an anionic compound act against Gram-positive bacteria in different ways, involved in various diseases caused by the contaminated food [40]. Bacteriocins have the ability of perforating the surface of pathogens and cause leakage of inorganic salts and amino acids from the cells to prevent the growth of harmful bacteria [37]. *B. coagulans* also secrete acetic acid and lactic acid reported as eminent antimicrobial substances to avoid the growth of harmful bacteria in the gut.

3.6.1.3 *Bifidobacterium*

The genus *Bifidobacterium* constitutes the Gram-positive normal human gut flora, which are non-motile anaerobes, results in the formation of endosymbiotic relationships between the vagina and gastrointestinal tract. The *Bifidobacterium genera* have been identified a potential probiotic as they have the ability to resist bile salts. Most of the functions of probiotics take place in the presence of bile salts thus resistance against bile salt is an important factor. The strain-dependent tolerance of bile salts has already proved, but subculturing of wild type *Lactobacilli* and *Bifidobacteria* strains with subsequent increase in the concentration of bile can lead to development of bile tolerance. The probiotic strains of the *Bifidobacteri*um genera are *B. infantis*, *B. breve B. adolescentis*, *B. longum, B. bifidum, B. animalis* subsp *animalis*, and *B. animalis* subsp lactis. Nowadays, companies are developing names for these compounds that resemble the scientific names and later using them as trademarks [41].

These *Bifidobacterium* species have proved their effectivity as probiotics in the treatment of various diseases such as constipation, diarrhea associated with traveling and use of antibiotics, maintaining remission of inflammation of gut and colon, and moderate ulcers of colon. These strains also involved in the prevention and treatment of food allergies, diarrhea induced by the exposure of radiations, necrotizing enterocolitis in newborns, eczema, and decreasing the cholesterol level [27].

3.6.2 Effect of Probiotics on Removal of Heavy Metals

Heavy metals are defined as the group of compounds occurring naturally and released into the environment by various natural or anthropogenic processes. The rapid growth of human industrial activities like smelting, mining, and formation of synthetic compound has cause to an exponential enhancement in the amounts of heavy metals exposed into the environment such as water sources, and soil. The clear result is that people around the world are exposed and suffering from various diseases, thus new methodologies are required to reduce the accumulation and adverse effects of these compounds [42].

Microorganism removes heavy metals by 3 main mechanisms that involve the attachment of metals to the cell wall of bacteria, as the following:

1. Exchange of ions with the components of cell wall (teichoic acid and peptidoglycan)
2. Formation of precipitates by nucleation reactions
3. Complex formation with oxygen and nitrogen ligands

The presence of high amount of teichoic acid and peptidoglycan and in the cell wall of Gram-positive bacteria, especially *Bacillus spp.*, reflects high metal adsorption capacity. Less metal adsorption capability was shown by Gram-negative bacterial as their cell membranes have low content of these components.

The phylum Firmicutes is the dominating phylum among the colon microbiota, mainly consist Gram-positive bacteria, that includes *Lactobacillus*, *Clostridium,* and *Bacillus* as the dominating groups. Thus, a large population of microbes in the human gut have the ability to prevent the entry of metals in the body. Another mechanism for the removal of heavy metals and toxic chemicals involves detoxification, defined as the ability to remove various harmful agents such as drugs, mutagens from the body, whereas detoxication is the process of prevention of entry of hazardous substances in the body. Detoxication takes place in various organs including the intestine, kidneys, and liver to prevent the spreading of compounds to target sites where damage can occur [43]. The gut microbiota and potential probiotic bacteria play a vital function in heavy metals sequestering, preventing the entry of these compounds in the body to protect the host.

3.6.2.1 Role of Probiotics in Removal of Cadmium

Cadmium is the 7th most toxic heavy metal with that humans and animals may get exposed to at work or in the environment. When this metal gets absorbed by humans, it will gather inside the body during life. Cadmium binds to cysteine-rich protein such as metallothionein to form complexes that can led to hepatotoxicity and then it circulates to the kidney and causes nephrotoxicity [44].

The adverse effect of cadmium on gastrointestinal tract (GIT), especially intestine, through ingestion of cadmium contaminated water and food has been studied. Cadmium uptake causes the inflammation of intestinal tract, disarrangements in tight cellular junctions leading to damage of cells. These disruptions can cause increased permeability of intestinal membrane to cadmium and other pathogens, thus increasing the systemic absorption of cadmium. Probiotics remove the cadmium by binding or sequestering cadmium in intestinal tract, or absorption of cadmium, and protection of intestinal membrane by decreasing the accumulation and systemic absorption of cadmium accumulation in tissues. These activities help in the excretion of cadmium form GIT through fecal route [45].

In Italy commercial formulation of *Lactobacilli* and *Bifidobacterium strains* along with *Streptococcus thermophilus* were given to a pregnant woman for long term, as a part of a random double-blind study to evaluate the effect of cadmium in infants subjected via breast milk. This study did not clarify whether the use of probiotics can help in decreasing the cadmium absorption or not [42].

3.6.2.2 Role of Probiotics in Removal of Arsenic

Arsenic exist in different forms in nature in which the trivalent strains are higher toxic than pentavalent species. The toxicity of arsenic compounds (arsenicals) differs widely based on the chemical content of these compounds and can increase if the compounds are thiolated. It has a semi-metallic property, is highly toxic and carcinogenic, and is mostly available in the form of oxides or sulfides or as a salt of iron, calcium, sodium, and copper. The microbiomes of humans and mice were shown to metabolize arsenic when cultured in vitro. *Faecalibacterium prausnitzii* appears to be a beneficial correlate of microbiome stability throughout arsenic exposure in humans, and also delivers some protection in the GF As3mt-KO murine model [46].

3.6.3 Effect of Probiotic Intake on Immune Health

Human immune response is mediated by innate and adaptive immunity after the exposure of various pathogens or foreign substances (Fig. 3.3). Probiotics have a direct effect on the macrophages, T cells (cell mediated response), B cells (humoral response), and the dendritic cells to regulate immune response [28].

Reports have shown the activation of innate immunity and adaptive immunity when a mixture of *B. bifidum L. casei, L. acidophilus, L. reuteri,* and *Streptococcus thermophilus* probiotics was injected. High expression of indoleamine 2,3-dioxygenase, TGF-β, IL-10, and COX-2 was observed due to the stimulation of the regulatory dendritic cells. CD4+ and CD25 cells enhanced the growth of CD4 +Foxp3+ regulatory T cells (Tregs), which further leads to the enhanced suppression activity of CD4+CD25+Tregs. The administration of the probiotic formulation induced hypo response of both B and T cells, results in the downregulation of T helper Th2, (Th) 1, Th17 cytokines without induction of apoptosis. Suppression of inflammation of intestine induced by 2,4,6-trinitrobenzenesulfonic acid was revealed by the in vivo study, that was linked with high count of CD4+Foxp3+ Tregs in the affected areas. Thus, the probiotic mixtures that help in the enhanced production of the regulatory dendritic cells to further stimulate Tregs can be considered a promising therapeutics for inflammation causing diseases [47].

Reports on the high doses of *L. rhamnosus* Lcr35 probiotics and its effects on dendritic cells (comparing MOI, multiplicity of infection 100 to 0.01), showed huge changes in gene expression of immature dendritic cells derived from human monocytes. Increase in the rate of production of the cytokines pro-Th1/Th17, such as IL-12p40, IL-12p70, IL-1β, IL-23, and TNF, was observed after a high dose administration. Different phenotypic transformations of membrane of dendritic

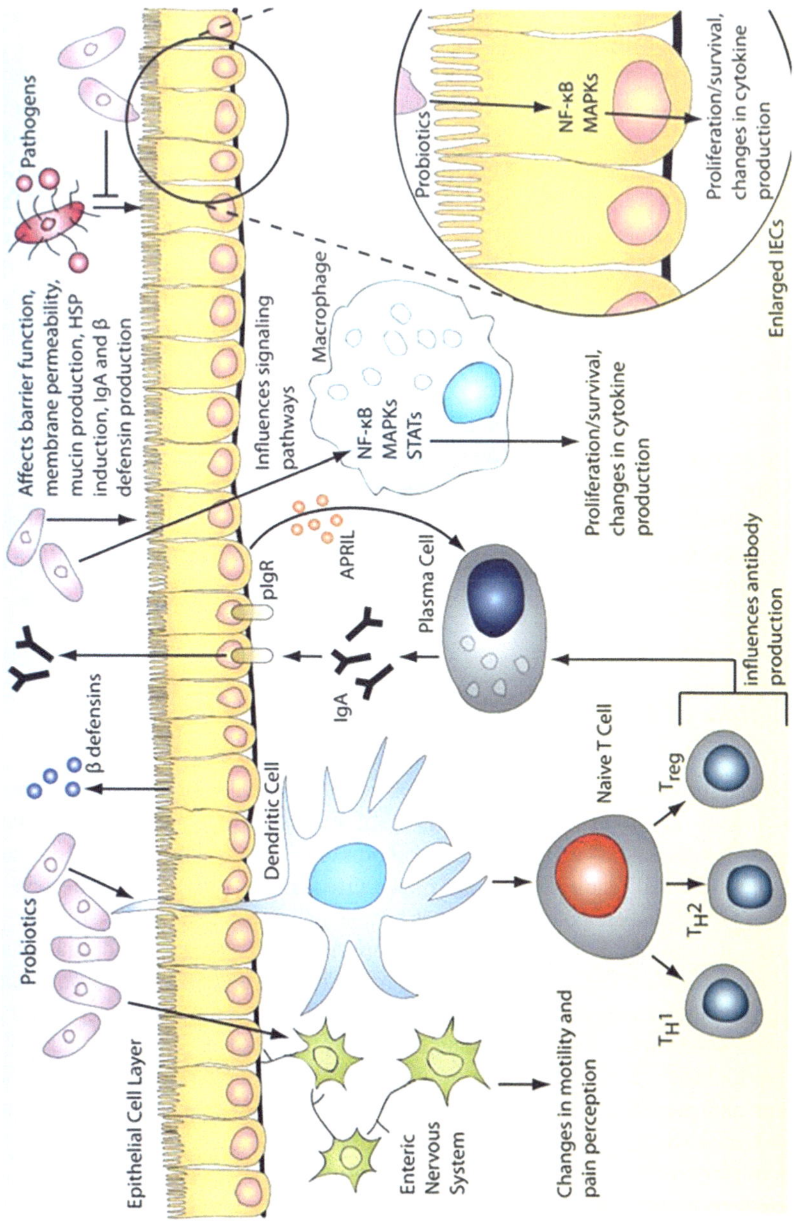

Fig. 3.3 Different roles played by the beneficial microbes in innate and adaptive immune response by suppression and proliferation of immune cells [22]

cells were also observed as a dose-dependent response due the upregulation of, HLA-DR, CD86, TLR4, and CD83 and low activity of CD14, MR, and DC-SIGN cells. A strong pro-inflammatory response due the partial maturation of different cells was observed by the administration of *L. rhamnosus* Lcr35 due the modulations in the immune dendritic cells [48].

Another in vivo study has reported the strain specific effect of *Bifidobacterium longum* AH1206, *Lactobacillus salivarius* AH102, and *B. breve* AH1205, on induction of Foxp3+ Tregs response. These cells are responsible for the protection of ovalbumin against respiratory and allergy associated with intake of Cholera toxin in diet. *B. longum* AH1206 have been reported to protect airway inflammation in infants, adults, and germ-free animals due to these two allergies by increasing the numbers of Foxp3 Tregs. *B. breve* AH1205 showed increase in Foxp3+ Tregs in case of infant mice only and no alteration was observed in the number of Tregs in germ-free animal models after the intake of *L. salivarius* AH102 and *B. breve* AH1205 probiotics [49].

Administration of *L. acidophilus* NCFM and *L. salivarius* Ls-33 showed a complete protection against inflammation of colon (colitis) in mice associated with SCID by decreasing the count of Tregs population that were increased by the disease conditions. Rectum samples of the probiotic administered mice showed similarities with the expression of genes pattern of naïve SCID mice but no similarity was found with the control groups. This revealed that probiotics indirectly affects the Tregs-favorable environment not the Tregs itself [50].

3.6.4 Role of Probiotics in Metabolic Abnormalities

The abnormalities in the human metabolism can lead to various diseases such clinical manifestation include development of resistance against hormones like insulin, dyslipidaemia, elevation in the blood pressure, and obesity in the abdominal region, these clinical factors have a culminate effect in terms of more rapid development of type 2 diabetes mellitus and cardiovascular disease. Several factors have been taken in consideration for the development of these metabolic syndromes that involve changes in lifestyle, development of an individual and the adopted perinatal planning, and the genetic factors. Several therapies have been reported to prevent the metabolic syndromes (MetS), manipulation of lifestyle, and diet intake which undoubtedly have proven the eminent non-pharmacological factors for the treatment and prevention of MetS [25]. The various short-term experiments and cross-sectional studies have explored the relationship between the gut microbiota and parameters influencing the metabolic syndromes (MetS). To investigate these parameters in context to probiotics, several standardized clinical trials and more advanced techniques such as bioinformatics analysis, host genetics, and microbial metabolism need to be explored further [51].

3.6.4.1 Hypercholesteromic Effect

Probiotics administration affect level of lipids in the bloodstream. The mode mechanism of action of reducing the levels of lipids involves the generation of short chain fatty acids (SFCAs) by gut microbiota. These SCFAs are produced by the fermentation of the prebiotic substrate such as inulin, fructo-oligosaccharides (FOS), lactulose, and galacto-oligosaccharides (GOS). The host can absorb these SCFSs and use as a source of energy [11]. In addition to production of SCFAs, probiotics have also been stated to play an eminent role in the regulation of metabolic processes. The production of SCFAs from the fermentation of prebiotic substrates have shown the reduction of hepatic cholesterol synthesis and the redistribution of cholesterol from the plasma to the liver. The effect of probiotics on deconjugation of bile acid salts has been studied. The deconjugation of bile salts involves the activation of bile acid hydrolase that inhibit the reabsorption of bile acid salts, as a result free bile acid excreted in feces. The main mechanism of hypocholesteromic potential of microbial strains comprises the conversion of cholesterol in coprostanol and co-precipitation of cholesterol with deconjugated bile salts [52].

3.6.4.2 Effects on Obesity Related Parameters

Probiotics administration can stimulate the factors responsible for insulin secretion. These functional food supplements with prebiotic ingredients are responsible for the anti-obesity effects by stimulating various physiological functions. The main mechanism of insulin production involves the multiplication of β-pancreatic cells. The stable beneficial flora of gut can potentially alters adiposity in obese individuals and positively affect the peripheral organs to secrete hormones like PYY and GLP-1 and factors responsible for satiety control in the brain [53].

The effect of SCFAs on modulation of numerous specific cellular functions by interacting with exact receptors implanted into the G protein coupled receptors such as GPR41 and GPR43 has been reported. These specific receptors help in the excretion of intestinal hormones like glucagon-like peptide-1 (GLP-1). The secretion of these hormones increases the gastric phase and time of intestinal transit, that further leads to the high rate of nutrient absorption. Another protein YY (PYY) is vital for other functions like inhibition of intestinal motility and stomach emptying, management of caloric intake and appetite, and also enhanced absorption of water, electrolytes, and nutrients in the gastrointestinal tract [30].

3.6.4.3 Anti-hypertensive Effect

The regular administration of *L. casei* and *L. plantarum* strains in hypertensive patients has been reported to show a potential anti-hypertensive response. The study conducted by Lollo et al. [54] with rats reported lowering of blood pressure after probiotics administration. The release of bioactive compounds after degradation of proteins like milk protein possess a hypotensive impact that further affect the renin-angiotensin system (RAAS). Thus, the protein degradation capability of probiotics can help in lowering the blood pressure by stimulating the secretion of ACE inhibitory peptides. The intake of probiotics also help in the induction of multiplication of β-pancreatic cells for insulin secretion to contribute as an anti-obesity compounds [52].

3.6.4.4 Improved Glycemic Control

The administration of probiotics in MetS suffering individuals has led to the formation of a hypothesis by the researchers stating that the modulations of immune response after probiotic administration has contributed for improved glycemic control. A study reported the interaction between key anti-inflammatory and inflammatory cytokines such as resistin, TNF-α, IL-6, and adiponectin have a direct effect on the development of insulin resistance and glycemic control to maintain glucose homeostasis in the host [55].

3.6.4.5 Effect on Low-grade Inflammation

Prebiotics have an effect on the generation of glucagon-like peptide-2 (GLP-2) endogenously. The high production of GLP-2 has an inverse effect of the metabolic endotoxemia that further results in the reduction of low-grade inflammation. In addition, the effect of intestinal microbe count on the disturbances of metabolic processes and obesity has been reported [30]. The management of probiotics has become as a non-pharmacological substitutes to prevent the obesity and metabolic endotoxemia associated with the variations in the microbial gut composition. The high production of Lipopolysaccharides (LPS) by the Gram-negative bacteria can lead to inflammatory processes. Thus, the intake of different probiotic strains along with the prebiotics can affect the MetS by reducing the count of Gram-negative bacteria in the intestine. The decreased count of Gram-negative bacteria can prevent the metabolic endotoxemia by reducing serum LPS levels to decrease the inflammatory process [56].

3.6.4.6 Effect on the Levels of Trimethylamine-N-oxide

Several reports are available on the use of probiotics and the various effects of probiotics intake on the cardiovascular system related diseases. The application of probiotics has been reported to affect the risk factors associated with the cardiovascular disorders, probiotics has shown the manipulation of the count of beneficial intestinal microbiota. Thus, a positive effect of the metabolism of gut microbiota due to probiotics can be considered in metabolic syndromes (MetS). In this context, change in the composition of intestinal microbes due to probiotic administration also causes reduction in the levels of trimethylamine-N-oxide (TMAO) and trimethylamine (TMA) production in the host. This reduction in TMA and TMO levels has grabbed the attention of numerous scientists. The production of TMAO as a co-metabolite by the intestinal microbes has grabbed a lot attention due to its role as a biomarker for CVD risk and as a promoter of atherothrombotic diseases. It has maintained a specific link among the CVD and gut microbiota [57].

3.7 Introduction to Next Generation Probiotics

The development in the use of probiotics for several health benefits has been studied extensively. As the study on microbiota proceed, the use of shotgun metagenomic sequencing, 16S ribosomal RNA (rRNA) gene, and bioinformatics analysis have been explored in human microbiome studies to further investigate new bacterial species to improve disease conditions. The role of commensal microbes in severity of several disease conditions has been reported. The evidences gathered from different reports have stated that several uncharacterized bacterial strains of gut have shown overlapping patterns with the new emerging therapeutics, or live biotherapeutic products (LBP), these strains can be used as next generation probiotics (NGP) [58]. These NGP are as well, known as bacteria established for pharmaceutical use, may or may not be innocently developed as traditional probiotics only in the form of dietary supplements.

3.7.1 *Basic Regulations and Requirements of NGP*

The use of NGP species in large scale set up in fermentation purpose has not been carried out, so the chances of causing disease are still there. Thus, it is eminent to characterize the new NGP strain for safety purpose. The various parameters for the characterization of NGP strain involve genomic analysis, virulence study, antimicrobial resistance genes, and in vitro bacterial physiology, antimicrobial resistant genes. The in vivo study of these NGP strains to check the toxicity levels in both healthy and immunocompromised animal models is also eminent. Other than the safety criteria, the study of targeted disease of NGP strains is also important. The mode of action of the NGP strains in the disease host should be studied specifically. The formation of different patterns of metabolic processes after the intake of NGP strains gave valuable evidences that can further help in the formation of preventive measures against chronic inflammation related diseases. The basic information on the viability of NGP strains under different conditions, the interaction between NGP and host, resistance to the low pH of gastrointestinal tract, tolerance against the bile acids present in the intestine, and their colonization on the mucosal surfaces is also important in this context [59].

3.7.2 Development of NGPs

The effect of NGPs on improvement of specific diseases relies on exact bacterial species and the amount given for treatment. It is not a recommended option to evaluate the functions of NGP by only the terminology of identified bacterial strains [60].

3.7.2.1 *Akkermansia muciniphila*

A. muciniphila shows symbiotic relationship with the mucus layers belonging to the phylum Verrucomicrobia. The bacteria consume mucin as the only energy source, carbon, and nitrogen. *A. muciniphila* as a probiotic have the potential to modulate the immunity and metabolism of host. The effects of *A. muciniphila* on increasing lifespan, enhancement in the efficacy of anticancer immunotherapy such as anti-PD-1 treatment, and improvement of amyotrophic lateral sclerosis [61]. The pasteurized *A. muciniphila* showed an increased potential to avoid the increase of obesity and insulin resistance in mice, as well, purified membrane protein (Amuc_1100) only also presented the valuable impacts of the bacterium. Nevertheless, the abundance of *A. muciniphila* has been reported to cause auto-immune diseases and metabolic syndromes. The *A. muciniphila* is likewise collected from a clinical blood infection and need more clinical trials to be categorized under potential NGPs [62].

3.7.2.2 *Bacteroides* Species

Bacteroides strains are anaerobic bacteria categorized under the phylum Bacteroidetes. The species of this phyla involved in clinical manifestations and can also contribute as NGP. The strains of Bacteroides species may differ completely in physiological and pathological characteristics. The *B. fragilis* strains (ETBF) contain enterotoxin that can act as specific clinical pathogens and cause severe clinical infections. On the other hand, *B. fragilis* strains which do not have the enterotoxin gene are non-toxigenic. *B. fragilis* (NTBF), comprise the genes found in synthesis of a capsular polysaccharides A (PSA) which involved in the control of neuroinflammations and prevention of viral encephalitis. Additionally, the 1st authorization of *Bacteroides xylanisolvens* in food by the European Commission have grabbed the attention of researchers to further investigate the potential applications of this specific genera. Several investigations are currently undergoing to discover other potential *Bacteroide* species (*Bacteroides dorei*, *Bacteroides thetaiotaomicron, Bacteroides ovatus*), and explore their reaction mechanisms with the host, and analysis of the prospective safety risks associated with the strains during commercialization [60].

3.7.2.3 *Bifidobacterium* spp.

Bifidobacterium spp. are mostly analyzed in colitis and cancer therapies as a potential probiotic, categorized under Actinobacteria phylum. These species have been reported to increase the $CD8^{+}T$ cell priming and accumulation, enhanced activation of dendritic cells (DCs) in tumor conditions. Several pathogenic species of *Bifidobacterium* have been reported to cause severe health implications. As well, whether this is partial to precise Bifidobacterium species is still unclear and specific standardized clinical trials are required to gather significant information and use of these strains as NGPs [41].

3.7.2.4 *Christensenella minuta*

C. minuta is an anaerobic Gram-positive bacterium categorized under Firmicutes. The effects of *C. minuta* on reducing metabolic syndromes like obesity have been studied. The microbiota composition of > 1000 fecal trials were attained from the UK twin pairs population, in which the family *Christensenellaceae* was originated to make a cooccurrence network with other heritable bacteria. The latest studies have displayed that *C. minuta* could be a potential pathogen in many cases consequently its use as a potential NGPs have been limited and need more specific trials to check the pathogenic potential [63].

3.7.2.5 *Clostridium* Species

Clostridium sp. categorized under the phylum Firmicutes, which are totally anaerobic spore-forming bacillus. The *Clostridium* bacteria can utilize various substrates like cellulose and sugars. *Clostridia* bacteria such as *Clostridium acetobutylicum, C. ljungdahlii, C. beijerinckii, C. butyricum, C. thermocellum,* and *C. cellulolyticum* that can be used as potential NGPs. However, a lot of non-toxigenic *Clostridium* species are recently used as probiotics, other species such as *Clostridium perfringens* and *Clostridium difficile* are classified among the most pathogenic strains which can led to enteric diseases in humans and animals and also cause the local intestinal mucosa immunity [64]. *C. butyricum* has been found to reduce chemotherapy-induced diarrhea in patients with lung cancer by decreasing the systemic inflammatory response system and influencing the preservation of homeostasis. On the other hand, various toxin genes have been recognized by genome sequencing in pathological strains, and *C. butyricum* was testified more frequent in stool samples from preterm neonates suffering from necrotizing enterocolitis. The factors involved in the development of *C. butyricum* as a potential NGP for human health benefits must be strain dependent and should undergo specific clinical trials [65].

3.7.2.6 *Eggerthellaceae* Family

Eggerthellaceae family is categorized under phylum Actinobacteria. The urolithin generating strains of this family related to *Gordonibacter pamelaeae* and *Gordonibacter urolithinfaciens* obtained from human feces of a non-sick individual. These bacteria have the capability to convert ellagic acid (present in strawberries, pomegranates, and walnuts) into urolithins metabolites. Potential of those urolithin making bacterial species as NGPs depends on the biological activity of isourolithin A, urolithins A and B, that possess cardioprotective, α and β-inflammatory, anti-carcinogenic, and neuroprotective properties that can improve human health [66].

3.7.2.7 *Streptococcus* spp.

Streptococcus spp. are commensals of pathobionts categorized under phylum Firmicutes. The probiotic effect of *Streptococcus thermophilus* has been studied, an eminent modulation of uremic toxins in the patients' gut diagnosed with chronic kidney disease (CKD) was observed after the administration of the strain to prevent the progression of CKD [67]. Another study on *Streptococcus dentisani* 7746 and 7747 showed an oral probiotic potential against tooth decay. This strain had the ability to generate bacteriocins which can hinder the growth of major oral pathogens, and also buffers acidic pH by an arginolytic pathway. The effects of various species of *Streptococcus* genus in gastroenterology has not been extensively explored, thus more clinical trials are required to provide significant data.

3.7.2.8 *Enterococcus* Species

Enterococcus classes are clustered under the phylum Firmicutes. The presence of these species in the human feces and their determination in the environment, these have been considered as an indicator of human fecal pollution in H_2O. *Enterococcus* species are characterized as valuable gut microbiota commensals that sources drug resistant infections. The effect of *Enterococcus hirae* on simplifying cyclophosphamide-induced therapeutic for immunomodulatory response has been studied. The potential use of *Enterococcus mundtii* QAUEM2808 in milk fermentation, and the protection of the model insect *Tribolium castaneum* against *Bacillus thuringiensis* infection by an *E. Mundtii* isolate has been reported. On the other hand, the *Enterococcus* in the human bloodstream, and multidrug-resistant *Enterococcus* species, such as *E. faecium* with *E. faecalis* have emerged as leading causes of nosocomial infections [68]. Thus, the existence of enterococci in the environment and on hands might have significant direct health consequences.

3.7.2.9 *Faecalibacterium prausnitzii*

The family *Ruminococcaceae* of phylum Firmicutes consists Gram-positive *F. prausnitzii* that can ferment glucose to produce short chain fatty acids (SCFAs) including butyrate, formic acid, and lactate. The production of butyrate by the *F. prausnitzii* helps in the conservation of the functional integrity and homeostasis in the intestine to maintain health. The effect of *F. prausnitzii* on immunomodulation has been reported to stimulate the proliferation of regulatory T cells in case of inflammatory bowel disease. The positive influence of genus on the efficacy of immune checkpoint blockade (ICB) therapy have also been reported. The purposed mechanism of immunomodulation by *Faecalibacterium* involves the enhanced $CD8^{+}$T cell infiltration within the tumor environment, as well as the frequency of effector $CD4^{+}$ and $CD8^{+}$ T cells in the peripher. In case of Crohn's disease, no beneficial effects of *F. Prausnitzii* have been reported [69]. On the basis of mode of action and several studies more clinical trials are required to find the potential of *Faecalibacterium* species in patients suffering from cancer.

3.7.2.10 Lactic Acid Bacteria (LAB)

The use of different species of phylum Firmicutes has grabbed attention of the researchers and been extensively studied as probiotics. The *Lactobacillus* species has been used as traditional probiotics, but the recent studies on probiotics has revealed that many other species of *Lactobacillus* genera have the potential to act as NGPs in specific diseases conditions [70]. In case of otitis media (OM) in children due to the colonization of the pathogenic microbes like *Streptococcus pneumoniae*, *Haemophilus influenzae*, and *Moraxella catarrhalis*, studies have shown the effect of other LABs in maintaining a healthy individual. In addition, an antagonistic effect of the use of *Lactobacillus* bacteremia in case of ICU patients treated with probiotics has been studied [56]. The mechanism of these antagonistic effects is still not clear and a standard high quality research need to be incorporated to validate whether this is a public phenomenon or is restricted to definite Lactobacillus strains.

3.7.2.11 *Parabacteroides goldsteinii*

The use of probiotics in the treatment of obesity and various metabolic syndromes has been widely studied. A study has reported the use of *P. Goldsteinii* as a novel potential probiotic for the treatment of obesity. *P. goldsteinii* is a potential NGP due its various effects on MetS such as anti-inflammatory, anti-obesity, and insulin-sensitizing property. In addition to these effects *P. Goldsteinii* strains has also identified to improve the chronic obstructive pulmonary disease (COPD) associated with cigarette smoking and colorectal cancer. Isolation of *P. goldsteinii* from the blood culture of abdominal sepsis patient has been reported [71].

3.7.2.12 *Pediococcus pentosaceus*

The Firmicutes have been reported with several species carrying probiotic potential. *Pediococcus pentosaceus* is an anaerobic Firmicutes, Gram-positive bacteria. The morphological features include cocci shape with a non-spore forming and non-motile features. Based on the various morphological and physiological characteristics it is categorized as a "lactic acid bacteria." As a LAB categorized bacteria its metabolic end product is lactic acid. The ability of *P. pentosaceus* to produce bacteriocins (an antimicrobial agent) has been studied in food preservation [72]. Another study has reported the cholesterol-lowering activity of *P. pentosaceus* strain LAB6 both in the presence and absence of bile salts (14–69% in the presence and 19–59% in the absence of bile salts). An antagonist effect of *P. pentosaceus* has been reported in which endocarditis and some infections have been caused by the use of catheters associated with the bacteria. Thus, high quality research and clinical trials are necessary to prove the potential benefits of *P. pentosaceus* [73].

3.7.2.13 *Prevotella copri*

The potential of *P. copri* as a next generation probiotic has been studied to affect the glucose metabolism. The bacteria of phylum Bacteroidetes play specific role in intestinal gluconeogenesis process to maintain homeostasis in the glucose level. Studies have also reported the role of *P. copri* in the elevation of glucose tolerance. The enhancement in glucose tolerance can further lead to a decrease in insulin resistance which prevent the chances of various metabolic syndromes like the development category 2 diabetes and ischemic cardiovascular disease [73]. Therefore, various *P. Copri* strains can further be analyzed for their different beneficial and deleterious effects in order to use them as probiotic.

3.8 Conclusions

Probiotics have several health benefits such as restoration of normal gut microbiota, modulation of immune response, adsorption and removal of heavy metals from the human body. These benefits can be increased with the simultaneous intake of prebiotics to form a synbiotic association. Probiotics help in the maintenance of optimum growth conditions for the normal flora and prevent the growth and invasion of pathogens on the intestinal barriers. Several specific strains of genus *Lactobacilli* and *Bifidobacterium* were reported to modulate the immune response but the clear mechanism is still unknown. More standardized clinical trials are vital to find the exact mechanism and pathways involved. The use of probiotics for a long duration and the ultimate effects on both the immune health status and gut microbiota need to be emphasis for safety purpose. The study of next generation probiotics (NGPs) is also eminent to find new potential disease specific strains for improved health benefits. A more rigorous safety strategy for the formation GM probiotics is required to prevent their dissemination into the environment.

References

1. Lilly DM, Stillwell RH (1965) Probiotics: growth-promoting factors produced by microorganisms. Science 147(3659):747–748
2. Parker RJANH (1974) Probiotics, the other half of the antibiotic story. Anim Nutr Health 29:4–8
3. Arora M, Baldi A (2015) Regulatory categories of probiotics across the globe: a review representing existing and recommended categorization. Indian J Med Microbiol 33 (Suppl):2–10
4. Metchnikoff E (1907) Lactic acid as inhibiting intestinal putrefaction. In: The prolongation of life, optimistic studies. Mitchell Heinemann, London, UK, pp 161–183
5. Overall C (2004) The prolongation of life: optimistic studies. Gerontologist 44(6):847–851
6. Tissier H (1906) The treatment of intestinal infections by the method of transformation of bacterial intestinal flora. CR Soc Biol 60:359–361
7. Ozen M, Dinleyici EC (2015) The history of probiotics: the untold story. Benefic Microbes 6 (2):159–165
8. Hotel ACP, Cordoba AJP (2001) Health and nutritional properties of probiotics in food including powder milk with live lactic acid bacteria. Prevention 5(1): 1–10.
9. Gibson GR, Roberfroid MB (1995) Dietary modulation of the human colonic microbiota: introducing the concept of prebiotics. J Nutr 125(6):1401–1412
10. Kolida S, Gibson GR (2011) Synbiotics in health and disease. Annu Rev Food Sci Technol 2:373–393
11. Cuello-Garcia C et al (2017) Prebiotics for the prevention of allergies: a systematic review and meta-analysis of randomized controlled trials. Clin Exp Allergy 47(11):1468–1477
12. Raman M, Ambalam P, Doble M (2019) Probiotics, prebiotics, and fibers in nutritive and functional beverages. Nutrients Beverages 12:315–367
13. Gritz EC, Bhandari V (2015) The human neonatal gut microbiome: a brief review. Front Pediatr 3:17
14. Madhavi Rane AM (2013) Effects of probiotic on the growth and survival of Zebra fish (Danio rerio). Int J Sci Res (IJSR) 4(3):1839–1841.
15. Yamashita MM et al (2017) Probiotic dietary supplementation in Nile tilapia as prophylaxis against streptococcosis. Aquac Nutr 23(6):1235–1243
16. Nikoskelainen S et al (2003) Immune enhancement in rainbow trout (Oncorhynchus mykiss) by potential probiotic bacteria (Lactobacillus rhamnosus). Fish Shellfish Immunol 15(5):443–452
17. Balcazar JL et al (2006) The role of probiotics in aquaculture. Vet Microbiol 114(3–4):173–186
18. Salinas I et al (2005) Dietary administration of Lactobacillus delbrueckii and Bacillus subtilis, single or combined, on gilthead seabream cellular innate immune responses. Fish Shellfish Immunol 19(1):67–77
19. Backhed F et al (2005) Host-bacterial mutualism in the human intestine. Science 307 (5717):1915–1920
20. Gill SR et al (2006) Metagenomic analysis of the human distal gut microbiome. Science 312 (5778):1355–1359
21. Kim BS, Jeon YS, Chun J (2013) Current status and future promise of the human microbiome. Pediatr Gastroenterol Hepatol Nutr 16(2):71–79

22. Hemarajata P, Versalovic J (2013) Effects of probiotics on gut microbiota: mechanisms of intestinal immunomodulation and neuromodulation. Ther Adv Gastroenterol 6:39–51
23. Dixit Y, Wagle A, Vakil B (2016) Patents in the field of probiotics, prebiotics, synbiotics: a review. J Food Microbiol Saf Hyg 01(02): 1–3
24. Azad MAK et al (2018) Probiotic species in the modulation of gut microbiota: an overview. Biomed Res Int 2018:9478630
25. He M, Shi B (2017) Gut microbiota as a potential target of metabolic syndrome: the role of probiotics and prebiotics. Cell Biosci 7:54
26. Carlucci C, Petrof EO, Allen-Vercoe E (2016) Fecal microbiota-based therapeutics for recurrent clostridium difficile infection, ulcerative colitis and obesity. EBioMedicine 13:37–45
27. Picard C et al (2005) Review article: bifidobacteria as probiotic agents—physiological effects and clinical benefits. Aliment Pharmacol Ther 22(6):495–512
28. Yan F, Polk DB (2011) Probiotics and immune health. Curr Opin Gastroenterol 27(6):496–501
29. de Moreno de LeBlanc A, LeBlanc JG (2014) Effect of probiotic administration on the intestinal microbiota, current knowledge and potential applications. World J Gastroenterol 20 (44):16518–16528
30. Lee SJ et al (2014) The effects of co-administration of probiotics with herbal medicine on obesity, metabolic endotoxemia and dysbiosis: a randomized double-blind controlled clinical trial. Clin Nutr 33(6):973–981
31. Park JS et al (2018) Lactobacillus acidophilus improves intestinal inflammation in an Acute Colitis Mouse Model by regulation of Th17 and treg cell balance and fibrosis development. J Med Food 21(3):215–224
32. Khazaiea K, Zadeh M, Khana MW, Bereb P, Gounaric F, Dennisa K, Blatnera NR, Owenb JL, Klaenhammerd TR, Mohamadzadeha M (2012) Abating colon cancer polyposis by Lactobacillus acidophilus deficient in lipoteichoic acid. PNAS 109:10462–10467
33. Chen L et al (2015) Lactobacillus acidophilus suppresses colitis-associated activation of the IL-23/Th17 axis. J Immunol Res 2015:909514
34. Chen LL et al (2013) Efficacy profiles for different concentrations of Lactobacillus acidophilus in experimental colitis. World J Gastroenterol 19(32):5347–5356
35. Jacouton E et al (2017) Probiotic strain lactobacillus casei BL23 prevents colitis-associated colorectal cancer. Front Immunol 8:1553
36. Bagarolli RA et al (2017) Probiotics modulate gut microbiota and improve insulin sensitivity in DIO mice. J Nutr Biochem 50:16–25
37. Riazi S, Dover SE, Chikindas ML (2012) Mode of action and safety of lactosporin, a novel antimicrobial protein produced by Bacillus coagulans ATCC 7050. J Appl Microbiol 113 (3):714–722
38. Honda H et al (2011) Use of a continuous culture fermentation system to investigate the effect of GanedenBC30 (Bacillus coagulans GBI-30, 6086) supplementation on pathogen survival in the human gut microbiota. Anaerobe 17(1):36–42
39. Abhari K et al (2016) The effects of orally administered Bacillus coagulans and inulin on prevention and progression of rheumatoid arthritis in rats. Food Nutr Res 60:30876
40. Abdhul K et al (2015) Bacteriocinogenic potential of a probiotic strain Bacillus coagulans [BDU3] from Ngari. Int J Biol Macromol 79:800–806
41. Fukui H et al (2018) Effect of probiotic Bifidobacterium bifidum G9–1 on the relationship between gut microbiota profile and stress sensitivity in maternally separated rats. Sci Rep 8 (1):12384
42. Astolfi ML et al (2019) A prophylactic multi-strain probiotic treatment to reduce the absorption of toxic elements: in-vitro study and biomonitoring of breast milk and infant stools. Environ Int 130:104818
43. Monachese M, Burton JP, Reid G (2012) Bioremediation and tolerance of humans to heavy metals through microbial processes: a potential role for probiotics? Appl Environ Microbiol 78 (18):6397–6404

44. Jaishankar M et al (2014) Toxicity, mechanism and health effects of some heavy metals. Interdiscip Toxicol 7(2):60–72
45. Bhattacharya S (2020) The role of probiotics in the amelioration of cadmium toxicity. Biol Trace Elem Res. https://doi.org/10.1007/s12011-020-02025-x
46. Coryell M et al (2018) The gut microbiome is required for full protection against acute arsenic toxicity in mouse models. Nat Commun 9(1). https://doi.org/10.1038/s41467-018-07803-9
47. Kwon HK et al (2010) Generation of regulatory dendritic cells and CD4+Foxp3+ T cells by probiotics administration suppresses immune disorders. Proc Natl Acad Sci U S A 107 (5):2159–2164
48. Evrard B et al (2011) Dose-dependent immunomodulation of human dendritic cells by the probiotic Lactobacillus rhamnosus Lcr35. PLoS One 6(4):e18735
49. Lyons A et al (2010) Bacterial strain-specific induction of Foxp3+ T regulatory cells is protective in murine allergy models. Clin Exp Allergy 40(5):811–819
50. Petersen ER et al (2012) Consumption of probiotics increases the effect of regulatory T cells in transfer colitis. Inflamm Bowel Dis 18(1):131–142
51. Miremadi F, Sherkat F, Stojanovska L (2016) Hypocholesterolaemic effect and anti-hypertensive properties of probiotics and prebiotics: a review. J Funct Foods 25:49–510
52. Mazidi M et al (2016) Gut microbiome and metabolic syndrome. Diabetes Metab Syndr 10 (2 Suppl 1):S150–S157
53. Remely M, Haslberger AG (2017) The microbial epigenome in metabolic syndrome. Mol Asp Med 54:71–77
54. Lollo PC, Morato PN, Moura CS, Almada CN, Felicio TL, Esmerino EA, Barros ME, Amaya-Farfan J, Sant'Ana AS, Raices RR, Silva MC (2015) Hypertension parameters are attenuated by the continuous consumption of probiotic Minas cheese. Food Res Int 76:611–617
55. Tonucci LB et al (2017) Clinical application of probiotics in type 2 diabetes mellitus: a randomized, double-blind, placebo-controlled study. Clin Nutr 36(1):85–92
56. Thakur BK et al (2016) Live and heat-killed probiotic Lactobacillus casei Lbs2 protects from experimental colitis through Toll-like receptor 2-dependent induction of T-regulatory response. Int Immunopharmacol 36:39–50
57. Tang WH, Hazen SL (2014) The contributory role of gut microbiota in cardiovascular disease. J Clin Invest 124(10):4204–4211
58. Lin T-L et al (2019) Investiture of next generation probiotics on amelioration of diseases—Strains do matter. Med Microecol 1–2:100002
59. Saarela MH (2019) Safety aspects of next generation probiotics. Curr Opin Food Sci 30:8–13
60. Sun F et al (2019) A potential species of next-generation probiotics? The dark and light sides of Bacteroides fragilis in health. Food Res Int 126:108590
61. Zhai Q et al (2019) A next generation probiotic, Akkermansia muciniphila. Crit Rev Food Sci Nutr 59(19):3227–3236
62. Ansaldo E et al (2019) Akkermansia muciniphila induces intestinal adaptive immune responses during homeostasis. Science 364(6446):1179–1184
63. Yang Y et al (2018) Effects of Christensenella minuta lipopolysaccharide on RAW 264.7 macrophages activation. Microb Pathog 125:411–417
64. Dash S et al (2016) Metabolic modeling of clostridia: current developments and applications. FEMS Microbiol Lett 363(4):fnw004
65. Hosny M et al (2019) Multidisciplinary evaluation of Clostridium butyricum clonality isolated from preterm neonates with necrotizing enterocolitis in South France between 2009 and 2017. Sci Rep 9(1):2077
66. Selma MV et al (2017) Isolation of human intestinal bacteria capable of producing the bioactive metabolite isourolithin A from ellagic acid. Front Microbiol 8:1521
67. Vitetta L, Llewellyn H, Oldfield D (2019) Gut dysbiosis and the intestinal microbiome: streptococcus thermophilus a key probiotic for reducing uremia. Microorganisms 7(8):E228
68. Daillere R et al (2016) Enterococcus hirae and Barnesiella intestinihominis facilitate cyclophosphamide-induced therapeutic immunomodulatory effects. Immunity 45(4):931–943

69. Gatti S et al (2017) Effects of the exclusive enteral nutrition on the microbiota profile of patients with Crohn's disease: a systematic review. Nutrients 9(8):832
70. Schnadower D et al (2018) Lactobacillus rhamnosus GG versus Placebo for acute gastroenteritis in children. N Engl J Med 379(21):2002–2014
71. Tsung-ru W, Chuan-Sheng L, Chih-Jung C, Tzu-lung L, Jan M, Yun-Fei K, David MO, Chia-chen L, John DY, Hsin-chih L (2018) Gut commensal Parabacteroides goldsteinii plays a predominant role in the anti-obesity effects of polysaccharides isolated from Hirsutella sinensis gut microbiota. Gut. https://doi.org/10.1136/gutjnl-2017-315458
72. Syakila RN et al (2019) In vitro assessment of pediococci- and lactobacilli-induced cholesterol-lowering effect using digitally enhanced high-performance thin-layer chromatography and confocal microscopy. Anal Bioanal Chem 411(6):1181–1192
73. Pedersen HK et al (2016) Human gut microbes impact host serum metabolome and insulin sensitivity. Nature 535(7612):376–381

Chapter 4
Effects of the Bio-accumulative Environmental Pollutants on the Gut Microbiota

Pengya Feng, Xingpeng Xiao, Tuoyu Zhou, and Xiangkai Li

4.1 Introduction

The gut microbiota (GM) has been currently known as an important "organ" within the human body [1], which consists of approximately 1800 genera and 500–1000 varieties of bacteria species [2, 3]. 90% bacterial species of GM mainly belong to the *Bacteroidetes* and *Firmicutes* phyla, followed by *Proteobacteria*, *Actinobacteria*, and *Fusobacteria* [4]. The overall genome of GM possesses more than 3.3 million genes, approximately 150 times more than human genome [3, 5]. The GM evolves through several transitions in the first few years of life and thereafter remains relatively stable if no obvious disturbance occurs [2]. GM also have substantial inter-individual and intra-individual variation due to different genotypes, lifestyles, ages, and geographic locations. GM differences even exist between monozygotic twins [6, 7]. Furthermore, the GM are also highly dynamic, which can rapidly respond to altered diet within a few hours [8].

It is well known that GM plays a crucial role in regulating host metabolism, including mucus layer shaping [9], food digest, vitamins, and amino acids synthesize [10]; storage and energy metabolism [11], immune system modulation [12], neurodevelopment, and even behavior regulation [13, 14]. Therefore, the effects of GM on host metabolism extend far beyond local effects in the intestine, and extend

Pengya Feng and Xingpeng Xiao are contributed equally to this work.

P. Feng · T. Zhou · X. Li (✉)
Ministry of Education Key Laboratory of Cell Activities and Stress Adaptations, School of Life Science, Lanzhou University, Lanzhou, People's Republic of China
e-mail: xkli@lzu.edu.cn

X. Xiao
Medical Molecular Imaging Research Group, Vall d'Hebron Research Institute (VHIR), Universitat Autònoma de Barcelona (UAB), Barcelona, Spain

X. Li, P. Liu (eds.), *Gut Remediation of Environmental Pollutants*,
https://doi.org/10.1007/978-981-15-4759-1_4

to various remote organ systems, including adipose, liver, muscle, brain, and so forth [13, 15]. Qualitative, quantitative, or metabolic perturbation of the GM (called gut dysbiosis) may mediate or link with the development of numerous diseases [16, 17], such as obesity [18], diabetes [19], hypertension [20], allergy [21], cardiovascular diseases [22], metabolic disorders [23], and others. For example, a metagenomic study revealed that *Bacteroidetes* populations are less and *Firmicutes* populations are more abundance in obese individuals than in normal individuals [24]. Likewise, individuals with diabetes exhibit a remarkable increase in the opportunistic pathogens and a relative reduction in the beneficial butyrate-producing bacteria compared with healthy individuals [25]. The GM are also highly sensitive to external factors such as drug, bacterial infection, and even environmental pollutants.

Tremendous xenobiotic compounds were released into the environment resulting from fast urbanization and industrialization [26], especially in developing countries [27]. EPs are generally classified as non-biodegradable and biodegradable. Non-degradable EP refers to those that are not degraded by microorganisms, e.g., some heavy metals (HMs), antibiotics, pesticides, persistent organic pollutants (POPs), and several biological contaminants. Human populations' exposure to these EPs are mainly from a route of food ingestion, which directly interact with GM and might be contributing to the development of health disorders. For example, arsenic chemicals exposure has been linked with diabetes [28, 29] and an increased incidence of perturbation of GM [30]. Besides, it has also been reported that metabolic disturbance associated with alterations of the GM composition contribute in a significant way to develop a variety of diseases [22, 31]. However, it is unclear that the interaction between EPs and GM and whether the altered GM by EPs could be relevant for (or a cause of) disease. Here, we aimed to comprehensively discuss the GM-related metabolic alterations associated with the EPs-perturbed GM community, as well as the possibility that exposure to EPs lead to GM changes (dysbiosis) as a mechanism by which environmental chemicals exert their detrimental effects on host health (Fig. 4.1).

4.2 The Gut Microbiota is Disturbed by a Variety of Bio-accumulative Environmental Pollutants

4.2.1 Heavy Metals

HMs are commonly considered an omnipresent toxic environmental pollutant associated with health concerns. As the no-degradable nature of HMs, about 40–60% of ingested metals are absorbed through intestinal barrier [32, 33], ultimately be absorbed by host tissues [34]. Oxidative stress is one of the features of heavy metals included redox-active (Cd, Pb, and Hg) and redox-inactive metals (Cr, Fe, Cu), which result in oxidative tissue damage. Under oxidative stress, host usually displays a deplete in antioxidative enzyme and an increase in lipid peroxidation. HMs

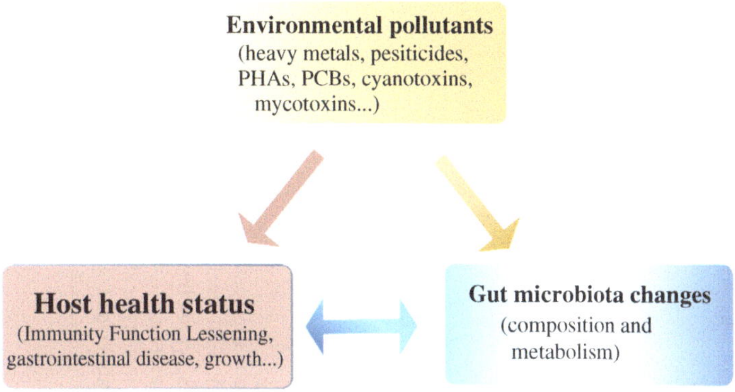

Fig. 4.1 Relationship between environmental pollutants exposure, host health status, and GM. Environmental pollutants-induced host health status changes may result in GM changes. Environmental pollutants might also directly result in gut dysbiosis that could in turn affect the host health

can also be accumulated and provide a consistent selective pressure on GM [35], including alterations in the composition and function of the GM along with various health hazards (Table 4.1).

4.2.1.1 Cadmium (Cd)

Cd was one of the most widely studied heavy metals, involved in a series of environmental pollution, such as batteries, paint, electroplating, fertilizers, and plastics [44]. Dietary intake of Cd is the common route to contact with populations of developing nations such as Nigeria, Bangladesh, and China [45]. Cd exposure causes oxidative stress, hepatic toxicity, renal dysfunction, cardiovascular diseases, osteoporosis, hypoimmunity, and tumorigenesis [46–48]. Low-dose cadmium exposed to mouse was found to fat accumulation in adult male mice rather than female mice by profiling hepatic genes.

Fazeli et al. [49] observed a sharp decrease in all bacterial species in biopsy samples and fluid contents derived from all intestinal regions of C57BL/6 mice by using Cd chloride exposure (23–50 mg/kg) for 45 days. The toxic profile in the small intestine was greater than in the colorectum under the same Cd stress, suggesting a regional preference in the gut. *Bacilluscereus* and *Enterococcus* spp. that belonging to gram (+) bacteria were more sensitive to *Escherichiacoloi* and *Klebsiella* spp. that belonging to gram (−) bacteria, which can be possibly explained by the varied metal ion uptake ability [50].

In a previous study [49], oral gavage of Cd (20 and 100 mg/kg) was administered to BALB/c mice for 3 weeks. Cd exposure has resulted in a significantly decrease in both the overall growth rate and abundance of GM. The ratio of *Bacteroidetes* and *Firmicute* significantly increased, and *Bifidobacteria* and *Lactobacilli* members

Table 4.1 The effect of the GM composition and related metabolic exposed to heavy metal

Type	Model	Exposure	Composition of GM	Metabolism of GM	Host effects	References
Cd	Carp (*Cyprinus carpio L.*)	500 ug/L Cd in drinking water for 4 weeks	*Cetobacterium* (−); *Akkermansia* (−); norank_f_*Erysipelotrichaceae* (+); A. *muciniphila* (−); *Aeromonas* (−); *Unclassified_f_Comamonadaceae* (+); norank_c_*Cyanobacteria* (+); *Acidovorax* (+); *Candidatus_Odyssella* (+); *Pseudomonas alcaligenes* (+);	N.S.	N.S.	[36]
Cd	Female mice (Balb/C)	20 or 100 ppm-mg/L in drinking water for 8 weeks	*Lachnospiraceae* (−); *Lactobacillaceae* (+); *Erysipelotrichaceae* (+); *Turicibacter* (+); *coprococci* (+); *streptococci* (+); *Blautia* (+); *Barneselia* (+); *Allistipes* (+)	N.S.	N.S.	[37]
Cd	Male mice (C57BL/6J)	100 nM	Alpha and beta diversity (−); Bacteroidetes (+); Firmicutes (−); Bifidobacterium (−); Prevotella (−); Sphingomonas (+);	N.S.	Plasma triglycerides, total cholesterol level, free fatty acids, leptin, high-density lipoprotein (+); triglycerides levels, fatty acid and lipid metabolism of liver (+); lipid deposits and fat accumulation	[38]
As	Female mice (C57BL/6)	10 ppm As for 4 weeks in drinking water	o_*Bacillales*_Other (+); *Clostridiales* Family XIII Incertae Sedis (+); *Streptophyta*_unassigned (−); *Clostridia*_Other_Other (−);	Indolelactic acid (−)	N.S.	[30]

			*Clostridiales*_unassigned (−); f_*Catabacteriaceae* (−); f_*Clostridiaceae* (−); f_*Erysipelotrichaceae* (−)			
As	ICR mice	3 mg/L drinking water for 90 days	*Firmicutes* (+); *Tenericutes* (+); *Proteobacteria* (+); *Bacteroidetes* (−); TM7 (−); *Acidobacteria* (+); *Cyanobacteria/Chloroplast* (+); *Lactobacillus* spp. (−); *Barnesiella* (−); *Bacteroides* (−)	Carbohydrate transport and metabolism (+); secondary metabolites biosynthesis (−); inorganic ion transport and metabolism (−); tetracycline resistance genes (+); transport and catabolism (−)	Small-bowel mucosal edema of intestine (+)	[39]
As	Tac male mice (C57Bl/6)	0, 10, or 250 ppb As(III) for 2, 5, or 10 weeks	*Bacteroidetes* (+); *Firmicutes* (−); *Bacteroidia_Porphyromonadaceae* (−); *Clostridia* (−); *Lachnospiraceae* (−); *Ruminococcaceae* (−); *Bacteroidaceae* (−)	The gene expression of nitrite reductase (nrfA) (+)	plasma arginine and monomethylarginine levels (−); dimethylated arginine metabolite levels (+)	[40]
Pb	Female mice (Balb/C)	100 or 500 ppm- mg/L in drinking water for 8 weeks	*Lachnospiraceae* (−); *Lactobacillaceae* (+); *Erysipelotrichaceae* (+); *Turicibacter* (+); *coprococci* (+), *streptococci* (+), *Blautia* (+), *Barneselia* (+); *Allistipes* (+)	N.S.	N.S.	[37]
Pb	Female mice (C57BL/6)	10 ppm Pb for 13 weeks	*Clostridiaceae* (+); *Lachnospiraceae_other* (−); *Blautia* (−); *Coprococcus* (−); *Ruminococcus* (−); *Ruminococcaceae_other* (−)	Ursodeoxycholic acid (UDCA) (−); vitamin E (-α-tocopherol and γ-tocopherol) (−);the primary bile acids cholic acid (CA) (−);the secondary bile acid deoxycholic acid (DCA) and cholesterol and its derivative coprostanol (−)	N.S.	[41]

(continued)

Table 4.1 (continued)

Type	Model	Exposure	Composition of GM	Metabolism of GM	Host effects	References
Cr	Bufo gargarizans	416 μg Cr6+ L−1	The bacterial community diversity (−); *Firmicutes* (−); *Aeromonas* (−); *Norank_f_ Peptostreptococcaceae* (−); *Norank_f_ Hados. Sed. Eubac.3* (−); *Saccharibacteria* (+); *TM6_Dependentiae* (+)	carbohydrate metabolism (+); energy metabolism (+); metabolism of cofactors and vitamins (+); amino acid metabolism (+); nucleotide metabolism (+); metabolic diseases, (+); xenobiotics lipid metabolism (+); biodegradation metabolism (+); immune system diseases (+); infectious diseases (+); neurodegenerative diseases (+)	Total body length (−); body wet weight (−); intestinal length (−); wet weight (−); mucosal injury and inflammation (+)	[42]
Ni	Broilers	0–900 mg/kg NiCl2 for 42 days	The bacterial diversity (−); *Bifidobacterium* spp. (−); *Lactobacillus* (−); *Escherichia* coli (+); *Enterococcus* spp. (+)	N.S.	N.S.	[43]

Note: N.S. in this study represents the item not shown in the published studies. "−" and "+" means decrease and increase, respectively were suppressed in response to Cd exposure. In addition, alterations in gut barrier disturbance, TNF-α, SCFAs were also noted in the colon, possibly due to Cd-mediated gut dysbiosis leading to the suppression of commensal bacteria.

In a previous study, the effect of C57BL / 6 mice on GM composition after 10 weeks of exposure to subchronic low-dose Cd (10 mg/L) was examined [51]. 16S rRNA gene amplicon sequencing analysis and quantitative PCR of cecum and feces revealed that Cd treatment mediated the composition of gut microbiota, with the abundance of *Bacteroidetes* (*Bacteroidaceae* and *Paraprevotellaceae*) elevated, and with the abundance of *Firmicutes* (*Ruminococcaceae, Lachnospiraceae*, *Streptococcaceae*, and *Clotridiaceae*), and *γ-Proteobacteria* diminished. Additionally, *Bifidobacterium longum* was detected to dramatically increase in response to Cd treatment. The above of the changes in microbial composition could be correlated with an increased LPS production, which may result in developing chronic liver disease, such as cirrhosis [52].

In another study, C56BL/6J mice received 100 nM of Cd (low dose) in early life showed the long-term effects on GM and host metabolism [38]. Gender differences of early Cd treatment (12–16 weeks) on adiposity enhancement and hepatic lipid metabolism dysfunction (increase in levels of free fatty acids, hepatic TG, and serum TG) were observed, which were more obvious in male than female mice. After 8 weeks of exposure Cd to mice, a decrease in *Firmicutes populations*, *Bifidobacterium,* and *Prevotella* was detected. Together with the results above, the male intestinal microbiome is more sensitive to early cadmium exposure, which may be related to fat accumulation and metabolic disorders later in life [53, 54]. Fecal microflora transplant in germ-free mice has also confirmed gender differences in GM when exposed to chromium

4.2.1.2 Lead (Pb)

Pb is a ubiquitous toxic pollutant in the environment. Humans are primarily exposed to lead by ingesting it from marine products (e.g., fish). Pb exposure is linked with obesity, inflammation, liver toxicity, nervous system disorders, and gut microbiome dysbiosis.

Orally exposure Pb to Avy-mice for 40 weeks has been reported to significantly shift the relative abundance of Firmicutes and Bacteroidetes during gestation and lactation periods [55]. The increase in populations of *Desulfovibrionaceae, Barnesiella,* and *Clostridium* cluster XIVb, and the decrease in populations of *Enterorhabdus*, *Lactococcus,* and *Caulobacterales* were detected in maternal Pb exposure. At the genus level, *Desulfovibrio* members that produce trimethylamine and leads to influence obesity, cardiovascular diseases, inflammation, and even colorectal cancer were elevated, whereas *Akkermansia* spp. members that hold the homeostasis of intestinal mucus layer were diminished [56–58]. Additionally, the alterations in GM composition were also associated with sex-dependent shifts in body weights [55].

In another study, Gao et al. [41] analyzed C57BL/6 mice fecal microbiota after 10 ppm Pb exposure for 4 and 13 weeks [30]. Compared to controls, Pb treatment lead to the decrease in phylogenetic diversity of GM, with diminished abundance of *Ruminococcaceae, Clostridiales, Ruminococcus* spp., *Blautia* spp., *Oscillospira* spp., and *Lachnospiraceae*. This resulted in changes in the GM metabolism, with the synthesis of cholesterol, vitamin E, and bile acids reduced, whereas with nitric oxidative stress induction, oxide generation, activation of defensive microbial mechanisms, and energy deprivation enhanced.

Xia et al. [59] treated ICR mice to Pb by orally treatment, consisting of 0.01, 0.03, or 0.1 mg/L for 15 weeks, and investigated the effects on the structure, abundance, and diversity of cecal and fecal microbiota. The results demonstrated an increase in *Proteobacteria* and *Bacteroidetes*, whereas a decrease in *Firmicutes* under Pb stress. Parabacteroides considered as known opportunistic pathogens were also increased. On the other hand, *Bacteroides*, *Oscillospira*, and *Ruminococcus* that belonging to SCFA-producing bacteria were also detected decreased. Elevated triglyceride and pyruvate levels in the liver, and changes in isobutyrate, glutamate, glycine, alanine, and among others in the intestine, demonstrate a disorder of hepatic and gut metabolic as a result of Pb exposure.

4.2.1.3 Hexavalent Chromium (Cr(VI))

Cr(VI) is another inorganic heavy metal, and mainly exists in hexavalent (VI) and trivalent (III) forms [60]. Cr (VI) is more toxic than Cr(III). The World Health Organization (WHO) has considered the concentration of 50 μg/L in drinking water as the minimum hazardous level of Cr(VI) [61]. Cr(VI) exposure causes liver toxicity, kidney damage, inflammation, and gut microbiome dysbiosis [62]. So far, the impact of Cr(VI) on GM has been very limited to study.

Yao et al. [42] exposed *Bufo gargarizans* to chronic Cr(VI) (0, 13, 104, and 416 μg Cr(VI) L^{-1}) Cr(VI) for a total of 70 days. The Shannon's index was found to be dramatically declined in Cr(VI) exposed groups, suggesting a decrease in the diversity of GM. The loss of GM diversity has been reported to have potential risk factors for animal diseases [16, 63]. Significant changes in the *Bacteroidetes* and *Firmicutes* phyla were also observed. Both of their abundant in Cr(VI) group lower than that in the control group, while *Proteobacteria* was significantly increased in 13 μg Cr(VI) L^{-1} exposure groups and *Saccharibacteria* and TM6_Dependentiae were also dramatically enhanced under 416 μg Cr(VI) L^{-1} exposure groups relative to untreated groups. *Saccharibacteria* in previous studies have been considered as an opportunistic pathogen and associated with inflammatory bowel disease [64] and mucosal carcinoma [65] in humans. Moreover, since *Saccharibacteria* as well as TM_*Dependentiae* lack of all essential amino acid and vitamins biosynthesis pathways [65, 66], leading to their dependence on symbionts to synthesize essential nutrients such as host tissues and other symbionts [67, 68]. At the genera level, Cr (VI) exposure was associated with diminished populations of other 13 genera. Moreover, based on 16S rRNA analysis, the alterations in the structure of GM caused by Cr(VI) ultimately affected the metabolism of GM, with an increase in cancers, infectious diseases, and immune system diseases. Together, exposure to Cr (VI) induced dysbiosis of GM with both composition and metabolism, therefore affected the host's health.

In a recent study by Wu and Xiao et al. [69], female Kunming mice were treated with potassium dichromate (1 mM) through drinking water and fecal samples were collected for analysis after 4 weeks' post-exposure. Although the diversity of GM was found no changes in mice treated with Cr(VI) compared to controls, it was showed that Cr(VI) exposure dramatically decreased the abundances of Firmicutes and increased the abundances of Bacteroidetes. The alterations in GM also observed at family levels, with an increase in *Paraprevotellaceae* (*Prevotella*, *Clostridiales*) and S24-7, whereas with a decrease in *Lachnospiraceae*, *Prevotella,* and *Clostridiales*, which have been reported to involve in various function disturbance including the transferable and colitogenic activity, hepatic metabolic activity, immune function, and short chain fatty acids production [70–73]. *Lachnospiraceae* as a butyrate-producing bacteria plays an important role in intestinal health by supplying energy and promoting intestinal epithelial cells development [74]. All of the above results suggested that oral exposure of Cr(VI) have potential to weaken the health of hosts by regulating GM community.

4.2.1.4 Mercury (Hg)

Hg is considered one of the most toxic heavy metals. It is discharged in the atmosphere during the progressive processing of industrial chemicals or waste electrical products. All forms of mercury can have adverse effects on health at high doses, such as gingivitis, gastrointestinal dysfunctions, and acute hepatotoxicity [75, 76]. Minamata disease is the typical consequence of consuming Hg-contaminated fish [77].

Ruan et al. [78] reported mice exposed to Hg (2 mg/kg body weight) through dietary for 90 days, resulting in significant bodyweight loss, cecal tissues damage, and GM changes. For the GM, only phylum *Tenericutes* were changed after Hg treatment, with significant increased. At the genus level, *Butyricimonas*, *Bilophila*, *Coprococcus*, *Dehalobacterium*, and *Oscillospira* were significantly elevated, whereas *Acinetobacter*, *Jeotgalicoccus*, *Sporosarcina*, and *Staphylococcus* were markedly diminished in the Hg exposure mice compared to the untreated mice. The result also revealed that the presence of Hg decreased the bacteria relevant to oxygen and increased anaerobic bacteria.

4.2.1.5 Nickel (Ni)

A previous study [43] that broilers exposed to NiCl2 (from 300 to 900 mg/kg) through diet for 42 days. Plate counting and PCR-DGGE analysis of ileal and cecal content substantiated that $NiCl_2$ exposure and reduced the abundance and diversity of the intestinal microbiome, with *Bifidobacterium* spp. and *Lactobacillus* decreased and *Escherichia coli and Enterococcus spp.* increased. It has been reported that *Bifidobacterium* spp. and *Lactobacillus* spp. as commensal resident of the GM that contribute to maintain GM balance, and increase mucin synthesis and secretion [79]. *E. coli* and *Enterococcus* spp. are both a part of normal bacteria of the GM but an important pathogen in humans and animals [80] that possibly harmful to the animals.

4.2.1.6 Mixtures

Humans are more exposure to the combined heavy metals than the single one in daily life. Nevertheless, the possible impacts of heavy metals mixture on the GM have not been tested. Despite that, a number of studies have focused on the differences in the effects of various heavy metals alone on the GM during the same period of experiment.

In a previous study, BALB/c mice were orally supplemented with Cd (20 or 100 ppm) or Pb (100 or 500 ppm) for 8 weeks [37]. The heavy metals significantly diminished *Lachnospiraceae* members, whereas elevated *Erysipelotrichaceacae* (especially *Turicibacter* spp.) and *Lactobacillaceae* when compared with control. *Lachnospiraceae* reduction and high abundance of *Turicibacter* have been correlated with gut inflammation [81, 82]. The finding suggests that heavy metals exposure promotes gut inflammation by bidirectional GM response.

Richardson et al. [83] exposure rats with different kinds and doses of As ($NaAsO_2$, 15, 22, 31 mg/kg/day), Cd ($CdCl_2$, 35, 54, 85 mg/kg/day), Co ($CoCl_2$, 27, 47, 82 mg/kg/day), Ni ($NiCl_2$, 177, 232, 300 mg/kg/day), and Cr ($Na_2Cr_2O_7$, 44, 62, 88 mg/kg/day) for 5 days. The study showed the alterations in GM community using 16S rRNA analysis, with significant changes observed in Ni, As, and Cd post-exposure samples, while with no significant changes in Co and Cr post-exposure samples. This study emphasized the specific changes in the composition of GM is not always similar with other studies previous reported. Specifically, the relative abundance of phylum *Bacteroidetes* relative to *Firmicutes*, which generally reported to increase after Cd exposure [74], but no difference in their study. In addition, the phylum TM7 nearly absent in base line of control samples, whereas exhibit a significant decrease post-exposure to Cd.

In a more recent study, either Cu (1 g/L), or Al (1.8 g/L), or Pb (1.83 g/L), or Cd (100 mg/L) were orally treated with C57BL/6 mice for 8 weeks disturbed the GM in a metal-specific manner [84]. No changes in gut microbial diversity were observed in different heavy metals exposed groups except Cu, which displayed a significant decline in that of mice. Exposure to any metal significantly caused alterations in GM with specificity at phylum, family, or genus levels.

4.2.1.7 Heavy Metals Induce the Generation of Antibiotic Resistance Genes (ARGs)

HMs exposure can induce the co-selection of ARGs in GM. HMs such as Zn, Cd, and Cr alone remarkably increased the diversity and abundance of ARGs in collembolan guts. As the no-degradable nature of HMs, they can be accumulated and produce a consistent selective pressure on the GM [35]. The gastrointestinal tract provides a unique habitat for a variety of microorganisms.

In summary, exposure to any heavy metals disturbed the GM, which leads to: (1) decline in diversity, (2) decrease in F/B ratio, (3) disturb the GM in a metal-specific manner at either phylum, family, or genus levels, (4) change the trends of the GM structure with not always similarity either exposed to different type heavy metals exposure and even the same heavy metal, (5) change the GM populations not only at the abundance and composition level, but also concurrently disturb its metabolism and function level, (6) promote gut inflammation by bidirectional GM response. Together, the perturbations of the GM and its function have been regarded as a potential mechanism that heavy metals lead to or exacerbate host diseases, although many studies are still underway to determine the relationship between heavy metal exposure-induced GM changes and relevant physical health.

4.3 Antibiotics

Antibiotics are widely applied in veterinary medicine, animal husbandry, and human medicine [85, 86]. Some antibiotics in animals and humans can enter the environment through feces or urine [87]. A large number of antibiotics have been detected in the ecosystem [88–90]. Therefore, humans are also vulnerable to passive exposure to antibiotic contamination. The overuse of antibiotics negatively affects the human organs, potentially resulting in metabolic deficiences such as diarrhea, allergy, and the gut microbiota dysbiosis [91].

Over the past decades, ABs has become a solid cornerstone of public health knowledge in the past few decades, but they also had a profound impact on the GM including the alterations in the taxonomic, genomic as well as functional capacity. The effects of ABs on the GM are profound with rapid and sometimes persistent. Moreover, these effects have two sides. Some cases demonstrated the intervention of antibiotics on GM can improve the insulin and glucose tolerance, which was beneficial for disease therapy. But on the other hand, the destruction of GM by antibiotics in healthy hosts is often associated with higher occurrence rates of diseases (Table 4.2). Here, we review recent studies that investigated the effects of ABs on the animals and human GM, highlighting the profound implications for lateral transfer of resistance genes.

Table 4.2 The effect of the GM composition and metabolic exposed to antibiotics

Type	Dosage	Composition of GM	Metabolism of GM	Effect on host	References
β-lactam	Intravenous cefazolin alone for 14 days.	The diversity of total bacteria (−); *Bacteroides* (+); *Parabacteroides* (+)	Five putative glycerophospholipids and fatty acid carnitines (+); GTPases (−); aerobic CobN cobaltochelatases (−); antimicrobial peptide transporters (+); multidrug efflux pumps (+	N.S.	[92]
Mixture	Bacitracin (108.0 mg), ampicillin (43.2 mg), neomycin (108 mg), vancomycin and meropenem (21.6 mg) (6.48 mg) dissolved in 4.5 mL for 11 days by oral gavage (10 mL/kg)	The diversity of microbiota in colon (+)	Acetate (−); n-butyrate (−); propionate (−);	Corticosterone (+); sphingomyelin (+); phosphatidylinositol (+); phosphatidylcholine (+); lysophosphatidylcholine (−); p-cresyl sulfate (−); circulating trimethylamine-N-oxide (−); deoxycholic acid/ chenodeoxycholic acid (−); memory index (−);	[93]
Mixture	norfloxacin and ampicillin (1 g/L each) for 2 weeks	Cecal aerobic bacteria (−); anaerobic bacteria (−)	N.S.	blood glucose (−); liver triglycerides (−); liver glycogen (+); oral glucose tolerance (+); fat storage in the liver (−); hepatic steatosis (−); plasma LPS levels (−);	[94]
Mixture	penicillin VK, vancomycin for 7 weeks	*Lachnospiraceae* (+)	Acetate (+); propionate (+); butyrate (+)	total fat mass (+); Per cent body fat (+); Bone mineral density in early life (3 weeks) (+); glucose-dependent insulinotropic polypeptide (+); hepatic fatty acids and lipids metabolism (+)	[95]

Tylosin	0.333 mg mL−1 at days 10–15, 28–31 and 37–40 of life	richness and Shannon evenness (−); *Ruminococcaceae* (−); *Erysipelotricaceae* (−); unclassified *Clostridiales* (+); *Streptococcaceae* (+); *Rikenellaceae* (+); *Firmicutes* other (+); *Prevotellaceae* (−); *Bacteroidetes* other (+); *Bacteroidales* other (+); *Lactobacillus* (−); *Enterococcus* (+); delay microbiota maturation	Glycolysis (−); gluconeogenesis (−); tRNA biosynthesis (−); nucleoside (inosine) and amino acid (leucine) biosynthesis (+); the citric acid cycle (+); the alternate Entner–Doudoroff pathway (+); the classic Embden–Meyerhoff pathway (−); isoprenoid biosynthesis (−); ribosomes (−); proline and vitamin biosynthesis (+); LPS synthesis, (+); pyruvate oxidation and molecular transport (+)	Total and lean +mass (+); micro- and overall hepatic steatosis (+); bones growth (+); ghrelin (−); lipid metabolism and cellular movement and assembly (+); ghrelin (−)	[96]
Amoxicillin	0.167 mg mL−1 at days 10–15, 28–31 and 37–40 of life	Richness (−); Shannon evenness (−); *Ruminococcaceae* (+); *Erysipelotricaceae* (+); unclassified *Clostridiales* (+); *Streptococcaceae* (+); *Rikenellaceae* (−); Firmicutes other (+); *Bacteroidales* other (−); *Prevotellaceae* (−); *Bacteroidetes* other (−); delay microbiota maturation	Glycolysis (−); isoprenoid biosynthesis (−); tRNA biosynthesis (−); ribosomes (−); proline and vitamin biosynthesis (+); LPS synthesis, (+); oxalate-degrading capacity (−);pyruvate oxidation and molecular transport (+)	Total and lean mass (+); bones growth (+); hepatic microsteatosis (−)	[96]

(continued)

Table 4.2 (continued)

Type	Dosage	Composition of GM	Metabolism of GM	Effect on host	References
Mixture	Neomycin sulfate (1 g/L), vancomycin (500 mg/L) and ampicillin (1 g/L), and metronidazole (1 g/L) each day by gavage for 2 weeks	Shannon index (–)	N.S.	The proliferation capacity of corneal cells (–); the thickness of cornea (–); the spatial range of the cornea (–); impaired corneal neurogenesis; impaired the angiogenesis in limbal blood vessels; macrophage distribution in the postnatal mouse cornea (–)	[97]

4.3.1 Antibiotics Improve Disease by Affecting the Composition and Metabolism of Gut Microbiota

Some studies had showed antibiotics treatment could be beneficial for improving diseases by modulating GM, such as the improvement of insulin and glucose tolerance. A previous study reported that exposed high-fat fed mice to 8 weeks treatment with antibiotics (ampicillin, neomycin, and metronidazole) improved insulin signaling by modulation of GM [98]. Based on the analysis of metagenomic sequencing from feces, the total bacterial count was found to be reduced, and concurrent with decreased *Bacteroidetes* and *Firmicutes* following antibiotics treatment. Alteration of GM by antibiotics treatment decreased the levels of circulating LPS, consequently downregulating the TLR4 signaling pathway in HFD-fed mice. In the gut, LPS from gram-negative bacteria is a ligand for TLR4, and the activation of TLR4 leads to the expression TNF-α. LPS has been commonly considered to play a significant role in developing the insulin resistance [99]. Moreover, antibiotics treatment also found to increase insulin-induced insulin receptor, Akt, and IRS-1phosphorylation and reduce inflammation, which thus improves insulin signaling. The inhibition of macrophage infiltration in adipose and liver tissue using histology and immunohistochemistry is another obvious evidence to corroborate the beneficial for antibiotics usage. Membrez et al. has reported [94] that modulation of GM by a combination of ampicillin and norfloxacin after 2 weeks significantly improved fasting oral and glycemia glucose tolerance in male *ob/ob* mice. The total bacterial in cecum including both the aerobic and anaerobic bacteria were observed to be significantly suppressed at the end of antibiotics treatment period using culture-based microbial analysis. Mice treated with antibiotics had lower liver triglycerides and higher liver glycogen compared with the untreated mice, which are correlated with enhanced glucose tolerance. The effect of antibiotics treatment was further supported by the reduction in plasma LPS and the increase in adiponectin.

4.3.2 Antibiotics Regulated Changes in Gut Microbiota are Linked to Various Diseases

Antibiotics treatment more often caused side effects at GM level, however, it is considerable limited to understand that at GM level. Oral intake of antibiotics leads to dysbiosis of composition and more importantly the metabolism of GM, which might be closely correlated with a multitude of diseases [100]. Antibiotics associated *Clostridium difficile* infections and diarrhea can be the most common symptom following antibiotics treatment [101, 102]. A previous study shown that mice exposed to ampicillin, streptomycin, and clindamycin dramatically reduced the microbial diversity in the large and cecal intestine contents of mice. The Bacteroidetes was drastically decreased and the two dominant genus, *Xanthomonas* and *Stenotrophomonas* were significant elevated [103]. The genus

Stenotrophomonas is known to be as potential emerging opportunistic pathogen and highly antibiotic resistant bacterium and [104]. The patients undergoing clindamycin and ampicillin therapy were susceptible to *Clostridium difficile* infection and lead to a decrease in the count of *Clostridium scindens*, which as a secondary modulator during the processing of bile acid metabolism [105]. Numerous studies recently revealed that the phylum *Proteobacteria* in GM was remarkably increased as a result of antibiotic treatment [106–108]. *Proteobacteria* encompass various pathogens, such as *Yersinia, Escherichia*, *Helicobacter*, *Salmonella*, and others [109]. Among some species of genus *Escherichia* has been known to be implicated in human diseases, for example, *E. coli* belonging to *Escherichia* is responsible for most of the *Escherichia*-related pathogenesis [110, 111]. *Salmonella* is a known intracellular pathogen and certain serotypes cause disease [112]. Another study in piglets treated with a combination of metronidazole, metronidazole, and gentamicin also altered the GM composition, with notable decreased *Lactobacillus* and *Bifidobacterium* and increased in *Escherichia* both in ileum and feces, which was associated with decreased SCFAs metabolism, neurotransmitter expression in hypothalamus, and increased aromatic amino acids metabolism [113].

Fröhlich et al. [93] investigated the influence of a mixture of antibiotics (bacitracin, neomycin, ampicillin, meropenem, and vancomycin) exposure to adult male C57BL/6N mice by oral gavage for 11 days on GM. The result suggested both the microbial composition and metabolism in the colon are strongly disrupted following the antibiotics treatment, with the diversity of bacteria and the levels of SCFAs (acetate, butyrate, propionate), uracil, adenine, and trimethylamine significantly decreased. Furthermore, compared to vehicle-treated mice, antibiotics treatment also impacted the production of circulating plasma metabolites. Specifically, the levels of corticosterone, sphingomyelin, phosphatidylinositol, and phosphatidylcholine were significantly increased, whereas the levels of p-cresyl sulfate and lysophosphatidylcholine were decreased. As a result of antibiotics treatment, it is observed that the recognition memory was impaired and the expression of tight junction protein (CLDN5, TJP1, OCLN) was differentially altered in the amygdala and hippocampus as a result of antibiotics treatment. The compositional and metabolic effect of β-lactam on GM has been investigated with multi-omics approaches usage, including 16s rRNA, 16S rDNA, metagenome, metatranscriptome, metametabolome, and metaproteome [92]. The biodiversity of both the total and active GM was decreased during antibiotic treatment. At the phylum level, it was observed that the abundance of Firmicutes was significantly elevated and the abundance of Bacteriodetes was reduced. Moreover, the protein and gene expression of GM appeared to corroborate the metabolic disturbance of GM during antibiotic treatment. It is noted that the genes belonging to the "mobile and extrachromosomal element functions" category were declined as a consequence of antibiotic therapy and were correlated with clustered regulatory interspaced short palindromic repeats, which functioned as a type of bacterial adaptive "immune" response [114]. Concretely speaking, the CRISPR / Cas system are involved in protecting cells from foreign DNA (viruses and plasmids) through a process similar to RNA interference [115]. Therefore, reduced expression of these genes may make it easier for bacteria to obtain foreign DNA. This may be advantageous in an antibiotic-containing environment because it increases the possibility of obtaining resistance genes through horizontal gene transfer. Together with these findings, the altered structure of GM is correlated with the pathogenesis of diseases.

4.3.3 Antibiotics Cause the Loss of Colonization Resistance

Antibiotics can cause the loss of colonization resistance in the gut, which may be one of the majority features as a result of antibiotics [116, 117]. This conclusion has been verified by several studies, including cefoperazone [118], tigecycline [101], clindamycin [102]. These results are basically consistent with those of human studies [119, 120]. The opportunistic pathogens *Salmonella* can be easily colonized in the gut following antibiotics treatment, and therefore result in relevant diseases through multiple routes including direct interference competition and indirect resource competition. Another recent study in mice has been reported that antibiotics cause an increase in host-derived free sialic acid level in the gut, which can be used by opportunistic pathogens such as *Clostridium difficile* and *Salmonella typhimurium* to enhance their growth [121].

4.3.4 Antibiotic Cause ARGs Generation

As modern abuse of antibiotics with high dose, the coevolution of antibiotic resistant bacteria or genes is often a common concern in recent years. It is particularly paid attention to human GM from a host ecological perspective. In the laboratory, bacterial populations treated with antibiotics selected and enriched for resistant strains and species [122, 123], similar findings have been observed in vivo as a result of using antibiotics. Jakobsson et al. [124] studied patients received a mixture of clarithromycin-containing antibiotic for H. pylori-associated peptic ulcers, the ermB resistance gene immediately increased 1000-fold following the course of antibiotics treatment, which encodes the macrolide target–modifying RNA methylase. Four years later, the antibiotics resistance genes still exist in the GM with comparable levels, although patients without additional antibiotics therapy in this course of study. Subtherapeutic antibiotics doses also appear to develop resistance genes. For example, in swine feed a cocktail of chlortetracycline, sulfamethazine, and penicillin following only 3 days of treatment, multiple resistance genes were also significantly enriched [125]. It is noted that genes resistant to drugs, such as aminoglycoside, that not present in the feed source were also enriched, which provide evidence for antibiotics to promote the enrichment of resistance genes to unused drugs in commensal GM.

4.3.5 Antibiotics Cause a Persistent Effect on the Gut Microbiota

Antibiotics can affect GM with persistent term. Various researches based on 16S rDNA analysis have demonstrated that important differences in baseline bacterial composition recovery after AB treatment depend on the individual and ABs (type and dose) used [124, 126, 127]. In general, the effects of antibiotics on gut microbial composition have found to be disturbed by varying factors, leading to a decrease in microbial diversity, with different degree diminishes and increases in certain taxa populations. There was some degree of recovery in most individuals but were persistent effects in others, like individual host-specific and antibiotic effects. The distal gut bacterial communities before and after treatment with ciprofloxacin (500 mg twice a day) for 5 days has been investigated using deep 16S rRNA sequencing, showing a declined diversity of GM, with about 30% of the bacterial taxa influenced. Despite this pervasive disturbance of GM has been largely recovered to the before treatment state within 4 weeks, several taxa failed to return within 6 months [126]. Seven days of metronidazole, omeprazole, and clarithromycin exposure on the fecal and pharyngeal taxonomic composition discovered broad taxonomic compositional effects with rapid but only partial recovery in some cases and persistent effects at least 4 years after exposure [124]. Fouhy et al. [106] studied the short-term recovery of the GM after parenteral gentamicin and ampicillin exposure for infant within 48 h of birth and found the number of *Bifidobacterium* species was reduced and the *Proteobacteria* abundance remained significantly high and in the infants after 8 weeks of exposure with ABs. Therefore, it is clear that the certain ABs can dramatically affect the evolution of infant GM. Another study investigates short- and long-term effects of macrolides on 2–7 years old children and found depletion of abundance of *Actinobacteria*, increased macrolide resistance and populations of *Proteobacteria as well as Bacteroidetes* [107]. A study in mice showed that *Bacteroidetes* was significantly reduced following treatment with the antibiotic mixture of clindamycin, ampicillin, and streptomycin and never fully recovered after cessation of ABs exposure [103]. In addition, there are still a numerous studies suggested that treatment with clindamycin have significantly caused the long-term effects on GM composition [128–130].

4.3.6 Antibiotic Increase the Risk of Being Overweight

Antibiotics exposure are associated with later risk of overweight by altered GM. For example, during early exposed to children (<6 months of age) of normal weight mothers with antibiotics, the risk of overweight was increased at age 7 years (95% confidence interval: 0.95–1.47). This result indicates that the usage of antibiotics, rather than the genetic background induced the risk of overweight among children [131]. In a previous study [96], metagenomic analysis in mice treated with early life therapeutic-dose pulsed tylosin indicated that the cumulative weight gain was achieved, with increased total and lean mass, as well as bones growth, which might be correlated with the decrease in tRNA biosynthesis, gluconeogenesis, and hepatic glycolysis and the increase in amino acid biosynthesis, nucleoside, and hepatic citric acid cycle. Meanwhile, the diversity of GM including richness and Shannon evenness were observed to be decreased. The bloom of *Lachnospiraceae* in antibiotics exposed mice altered regulation of hepatic metabolism of fatty acids and lipids, as well as increased the copies of critical genes involved in SCFAs synthesis in cecal and fecal samples. Together with the results discussed, the study revealed the mechanisms of the effect of low-dose antibiotics on adiposity in detail. The tendency for weight gain following antibiotics treatment was also substantiated among children who received macrolides in their early life [107].

4.4 Pesticides

Pesticides were easily found in our ordinary life, no matter where you live. According to the World Health Organization (WHO) (2017), pesticides defined as chemical compounds that are used to kill pests, including rodents, fungi, insects, and unwanted plants (weeds) [132]. The use of pesticides significantly increased the yield of crops and improved the quality of people's life. Therefore, pesticide use increased dramatically between the 1960s and 1990s and the speed goes slowly this century. From statistics of Environmental Protection Agency of U.S., nearly 6 billion pounds pesticides were produced annually in the world in both 2011 and 2012 [133]. After these decades, pesticides have aroused wide attention that they have caused serious environmental problems due to their abuse, high toxicity, and low degradation, which have affected people's life and health. The term "pesticides" is a general name which includes a variety of substances. According to target pests, pesticides can be classified as herbicides, insecticides, rodenticides, fungicides, bactericides, miticides, and so on [134]. According to chemical structure, they can also be categorized into four main groups: organophosphorus, organochlorines, carbamates, and pyrethrin and pyrethroids [135].

Nowadays, a large number of pesticides have been detected in food ingredients, soil, and water, which are the main exposure sources for us. Another one important exposure pathway is indoor pesticides for vector control and elimination of nuisance pests, such as mosquitos, black beetles, acarids, and rodents.

Between 2008 and 2012, herbicides accounted for the largest portion of global usage (approximately 50% annually in all years), followed by insecticides, fungicides, and fumigants, respectively [133]. There are some common herbicides such as paraquat, diquat, and 2,4-dichlorophenoxyacetic acid (2,4-D), especially the first two chemical agents or their mixtures. Paraquat and diquat are bipyridyl herbicides because they both have the structure of two pyridine rings. The former is usually manufactured as a salt with chloride ion, while the latter with bromide. Paraquat and diquat are highly poisonous to crops and weeds. Paraquat is easily absorbed through the respiratory tracts gastrointestinal (GI) and skin, so it has high risks for farmer when they spray in the farmland. On the contrary, diquat is less toxic and poorly absorbed through intact skin. Most cases of toxicity result from ingestion [136]. They would induce free radicals in the body and then result in a serious damage. 2,4-D is chlorophenoxy derivatives and is also very toxic for skin, eyes, and respiratory and GI tracts.

Organophosphates (OPs) are a kind of insecticides containing phosphorous derived from phosphoric acid, which were used predominately last century because of the most toxic of all pesticides to vertebrate animals. OPs and carbamates inhibit the function of carboxylic ester hydrolases, such as chymotrypsin, plasma or butyrilcholinesterase (BuChE), plasma and hepatic carboxylesterases (aliesterases), paraoxonases (asterases), acetylcholinesterase (AChE), and other nonspecific esterases within the body. It is reported that OPs can be detected in sewage sludge, river water, waste water, and even in rain and snow with high concentrations. Some experts said that they have evidence to be global occurrence in the atmosphere, because they were also found in soils that had no history of sewage sludge application or irrigation [137]. Chlorpyrifos (CPF) is an organophosphate insecticide usually applied to treat vegetable crops, vineyards, and fruits [138]. CPF can be metabolized by cytochrome P450 enzymes in the liver and gut [139]. Perinatal CPF exposure can reduce the weight and length of rat pups and inhibit their intestinal development. In addition, CPF induced an increase of *Clostridium*, *Clostridium, Enterococcus* and a decrease of *Bifidobacterium* spp. and *Lactobacillus* spp. in rat intestines. CPF-induced microbiome malnutrition damages the mucosal barrier, increases bacterial translocation, and stimulates the innate immune system [140]. CPF was also reported that it can result in significantly decreased Firmicutes and increased Bacteroidetes and bring imbalance of gut ecosystem [141].

Organochlorines are insecticides containing element hydrogen, chlorine, and carbon. They are divided into distinct groups, including DDT (dichlorodiphenyltrichloroethane) and related analogs, cyclodienes, hexachlorocyclohexane, and related compounds. They are lipid soluble and stored in fatty tissues, so a long-time exposure with even low concentrations also can result in accumulation and eventual clinical toxicity. Pentachlorophenol (PCP) is a widely used pesticide worldwide [142]. PCP exposure resulted in genetic and reproductive toxicity in aquatic animals even at very low concentrations [143]. PCP can accumulate in the intestinal tract and liver of fish and inhibit fish growth and cause histopathological damage and hepatic oxidative. Moreover, in goldfish, PCP exposure (100 mg/L) for 28 days also altered the composition of gut microbiome by decreasing the ratio of Bacteroidetes/Firmicutes. At the genus level, relative abundance of *Bacteroides* increased and relative abundance of *Microbacterium*, *Arthrobacter*, *Chryseobacterium*, and *Legionella* decreased [144].

Pyrethroids are popular insecticides in public areas because they have high toxicity to a wide range of insects and low toxicity to birds and mammals, and rapid biodegradability. Pyrethroids exert their effects through delaying closure of the inward sodium channel of the nerve membrane. Permethrin (PEM), one of the most representative pyrethroid compounds, can decrease the proportion of

Porphyromonas Bacteroides and *Prevotella* and increase the abundance of *Lactobacillus* and *Enterobacteriaceae* at low-doses exposure in rats [145]. These shifts of gut microbiota may contribute to its neurotoxicity [146].

Propamocarb (PM) is a systemic carbamated fungicide, which widely used in the growth and yield of fruit trees, because it can control fungal diseases caused by Oomycetes in roots, leaves, and soil. PM residues can accumulate on fruits and subsequently people ingest them containing its residues. Wu et al. have studied the acute and chronic effect of PM exposure on mice, respectively [147, 148]. 300 mg/mL PM exposure-induced acute inflammatory reactions such as obvious changes of overall microbial structure and fecal metabolites. They found that up to 32.2% of OTUs were changed and the proportion of Proteobacteria and Firmicutes decreased while the proportion of Acidobacteria, Bacteroidetes, and Chloroflexi increased in the highest dose group. In the study of chronic exposure, they found that hepatic bile acids (BAs) were significantly increased in the PM treated group, in addition, atherosclerosis-promoting molecule trimethylamine was markedly increased in feces. Genes related to BA synthesis and transportation and hepatic energy metabolism were also significantly altered. And in gut microbiota, they obtained similar results to the acute study in the alteration of Firmicutes and Bacteroidetes in feces and cecal contents. Apart from PM, there are many other broad-spectrum fungicides, such as imazalil (IMZ), epoxiconazole (EPO), and carbendazim (CBZ). They have been assessed their toxicity on rats or mice and they all can cause inflammations and damages and structural alterations of gut microbes [149–151].

Multiple pesticides are usually combined to use in the farmland, warehouse, and house, as it is, so we are actually faced with a high-risk and complicated problem in our daily life. Several single chemicals have already been studied about their toxic effects on gut microbiota and body organs, and they are summarized in Table 4.3. However, studies on multiple pesticides or pesticides mixtures are very rare and very necessary.

4.5 Persistent Organic Pollutants

Persistent organic pollutants (POPs) are a class of synthetic chemical compounds that can persist in the environment for long periods of time and are difficult to degrade, so they can accumulate in organisms, including humans, through the food chain. They include polychlorinated biphenyls (PCBs), polycyclic aromatic hydrocarbons (PAHs), organochlorine pesticides, and polybrominated diphenyl ethers. They are gained much attention in recent years because they are closely associated with the rising global rates of diabetes, autoimmune diseases, obesity, and some developmental disorders [154, 155]. Ingestion of water and food are considered as the main POP exposure routes, and the gut is the largest immune organ via forming a physical barrier against potentially harmful substances. Thus, it is necessary for us to comprehend the effects on animals or human GM.

Table 4.3 Effects of pesticides on gut microbiota and host health

Pesticides	Chemicals	Animal model and dosage	Impacts on gut microbiota	Outcomes	References
Herbicides	Dichlorodiphenyl-dichloroethylene (DDE)	Male mouse, 1 mg/kg body weight per day via oral gavage	(+): Firmicutes, Proteobacteria (−): Bacteroidetes, Verrucomicrobia, Actinobacteria	Disorders of entero-hepatic bile acid metabolism	[152]
	Glyphosate (Roundup)	Male and female Sprague-Dawley rat, 50 ng/L, 0.1 g/L and 2.25 g/L in drinking water	Both genders (+): Bacteroidetes (−): Firmicutes Female rats (+): *S24-7* (−): *Lactobacillaceae*	Intestinal disorders	[153]
Insecticides	Chlorpyrifos	Male mouse (Mus musculus KM), 1 mg/kg body weight per day by oral gavage	(+): Bacteroidetes (−): Firmicutes (*Lactobacillaceae*)	Intestinal injury, abnormal intestinal permeability and perturbed energy metabolism	[141]
	Pentachlorophenol	Goldfish (Carassius auratus), 0–100 μg/L in water	(+): Bacteroidetes (*Bacteroides*) (−):*Microbacterium*, *Chryseobacterium*, *Legionella,* and *Arthrobacter*	hepatic oxidative, Growth inhibition and histopathological damage	[144]
	Permethrin	Rat, 34 mg/4 mL/kg body weight per day by oral gavage	(+): *Enterobacteriaceae* and *Lactobacillus* (−): *Bacteroides*, *Prevotella* and *Porphyromonas*	Not studied	[145]

Fungicides	Propamocarb	Male mouse, 3, 30, and 300 mg/L in drinking water	At the dose of 300 mg/L (+): At phylum level, Bacteroidetes, Acidobacteria, Chloroflexi and Planctomycetes; at genus level, *Bacteroides*, *Dehalobacterium* and *Butyricimonas* (−): At phylum level, Firmicutes, Proteobacteria, TM7, Actinobacteria and Tenericutes; at genus level, *Oscillospira*, *Parabacteroides*, *Desulfovibrio* and *Ruminococcus*	Bile acid metabolic disorder	[147, 148]
	Epoxiconazole	Female Sprague-Dawley rat, 4 and 100 mg/kg/day in food	(+): At phylum level, Proteobacteria and Bacteroidetes; At family level, *Enterobacteriaceae* and *Lachnospiraceae* (−): At phylum level, Firmicutes; At family level, *Lactobacillaceae*	Liver toxicity	[150]
	Carbendazim	Male mouse, 100 and 500 mg/kg/bw in food	(+):Proteobacteria, Firmicutes and Actinobacteria (−): Bacteroidetes	Hepatic lipid metabolism disorder	[151]
	Imazalil	Male mouse, 25, 50 and 100 mg/kg body weight in food	(+): Firmicutes, Proteobacteria and Actinobacteria; *Deltaproteobacteria* and *Desulfovibrio* (−): Bacteroidetes; *Lactobacillus* and *Bifidobacterium*	Colonic inflammation	[149]

PCBs, a large group of chemicals with unique electrochemical properties, are easily found in many industrial products, e.g., capacitors, transformers, hydraulic fluid, cooling liquids, and lubricants and its exposure is mainly due to its improper disposal. PCBs are initially metabolized in the liver and then mostly bind to the aryl hydrocarbon receptor (AhR) to elicit their toxicity. There is some evidence that gut microbiome involve in PCB metabolism [156]. A study showed that PCBs can markely alter the composition of GM in mice within 2 days [157]. Low-dose PCB exposure significantly increased the abundance of *Enterococcus, Bacteroides, Bifidobacterium, Akkermansia muciniphila*, and *Clostridium scidens* in C57BL/6 mice. These changes in microbiota composition were associated with variation of bile acid metabolism [158].

PAHs, a ubiquitous group of several hundred related chemical compounds with various structures, are generated by the incomplete combustion of carbon-containing fuels of vehicles and smoke of tobacco and charcoal. Some PAHs have oestrogenic or carcinogenic properties in humans and the main exposure route is inhalation and food or water ingestion. Benzo[a]pyrene (B[a]P), one compound of the PAH group which is well described, ranks class 1 of human carcinogen. Administration of B[a]P with a dosage 50 mg/kg BW on murine model dramatically perturbed the populations of the mucosal and fecal gut microbiome, increasing pro-inflammatory bacteria of the families *Turicibacter, Bacteroidaceae, Alcaligenaceae, Erysipelotrichaceae, Porphyromonadaceae,* and *Alcaligenaceae*. Contrarily, abundance of beneficial bacteria including *Verrucomicrobiaceae, Ruminococcaceae Mucispirillum, Lactobacillaceae*, and *Lachnospiraceae* were reduced [159].

A summary of studies on effects of POPs on gut microbiota and host health is given in Table 4.4.

4.6 Biological Contaminants

Heavy metals, antibiotics, pesticides, and persistent organic pollutants can bring out chemical pollution if we abuse them without any restrictions or in an improper way, which are closely associated with people's life and then constitute the main potential threats. Other than these chemicals, there are also some biological contaminants, and they come from organisms such as plants, microorganisms, and algae.

Cyanotoxins are toxins generated by bacteria Cyanobacteria which are also blue-green algae. This type of algae is very common in many ponds and lakes, and in the ocean. Under high concentration of phosphorus conditions, they grow exponentially to form blooms and produce cyanotoxins in a certain concentration which can poison and even kill other organisms. Cyanotoxins are bio-active secondary metabolites of Cyanobacteria and can remain and bioaccumulate into the environment. Ingestion of contaminated water is the main route of exposure to cyanotoxins for animals and humans. Sébastien Duperron et al. have studied the effect of pure microcystins which belongs to cyanotoxins, and crude extracts of metabolites from *Microcystis aeruginosa* on gut microbiota of Medaka fishes (*Oryzias latipes*) [163]. They found

Table 4.4 Impacts of persistent organic pollutants (POPs) and mixtures on host health and gut microbiota

Chemicals	Animal model and dosage	Impacts on gut microbiota	Outcomes	References
Fox River mixture (PCBs mixture)	Female C57BL/6J mouse, 6 or 30 mg/kg once daily via oral gavage	At low dose, (+): *Clostridium scindens*, *Akkermansia muciniphila*, *Bifidobacterium*, *Enterococcus and Bacteroides*	Altered BA homeostasis	[158]
PCB congeners	Female C57BL/6J mouse, 5 mg/kg (P77 and P153) and 50 mg/kg (P126) per week via oral gavage then with HF diet	(+): Firmicutes (−): Bacteroidetes	Abdominal fat accumulation, Abdominal greater size of subcutaneous adipocytes, increased expression of pro-inflammatory cytokines	[160]
Benzo[a]pyrene	Male C57BL/6 mouse, 50 mg/kg body weight per day via oral gavage	(+): At phylum level, Bacteroidetes; at family level, *Bacteroidaceae*, *Porphyromonadaceae*, *Paraprevotellaceae* and *Alcaligenaceae*; at genus level, *Bacteroides*, *Parabacteroides*, *Paraprevotella* and *Allobaculum* (−): At phylum level, Verrucomicrobia; at family level, *Lactobacillaceae*, *Verrucomicrobiaceae* and *Ruminococcaceae*; at genus level, *Lactobacillus*, *Oscillospira* and *Mucispirillum*	Ileal and colonic inflammation	[159]
2,3,7,8-tetrachlorodibenzofuran (TCDF)	Male wild-type and $Ahr^{-/-}$ C57BL/6J mouse, 24 μg/kg per day in the diet	(+): Bacteroidetes, *Flavobacteria*, *Butyrivibrio* (−): Firmicutes, *Clostridia*, *Oscillobacter*	Intestinal inflammation, bacterial fermentation, alterations of bile acid metabolism, glucose metabolism disorder and hepatic lipid.	[161]

(continued)

Table 4.4 (continued)

Chemicals	Animal model and dosage	Impacts on gut microbiota	Outcomes	References
2,3,7,8-tetrachlorodibenzo-p-dioxin (TCDD)	Male CD-1 mice, 6 μg/kg bodyweight biweekly by gavage	(+): Firmicutes, *Lactobacillaceae*, *Desulfovibrionaceae* and *Lactobacillus* (−): Bacteroidetes, *Prevotellaceae*, *Ruminococcus*, *Prevotella* and *Anaerostipes*	polydipsia and polyphagia, Hepatic and immune toxicity	[162]

that crude extracts, rather than pure microcystins, have an important influence on gut microbiota composition of Medaka fish. Relative abundances of pathogen-related bacteria *Mycobacterium* spp. and *Nocardia* spp. increased markedly after exposure to one extract. Two bacterial orders *Sphingomonadales* and *Saprospirales* displayed significantly higher abundances in another extract treatment group. These alterations of gut microbiota would have severe influences on associated functions of fish.

Mycotoxins are also a secondary metabolite produced by organisms fungus, mainly *Aspergillus*, *Fusarium,* and *Penicillium.* They are very toxic for humans and animals and usually cause disease and death. These metabolites frequently occur in natural food contaminants and misidentified mushrooms. The major mycotoxins include deoxynivalenol, zearalenone, ochratoxin, fumonisins, aflatoxin, and patulin. Hervé Robert et al. have summarized these major mycotoxins, predominant contaminated crops, major producing fungi, adverse effects, and health-based guidance values [164]. These adverse effects includes gastrointestinal alteration, immunotoxicity, and neurotoxicity. According to the United Nations Food and Agriculture Organization and the World Health Organization, it has been reported that 25% of the world's crops are contaminated by molds and fungi [165]. Due to frequent food poisoning cases, mycotoxins have been drawing more and more people's attention. Gut microbiota are very important for the host and are the first to be exposed to contaminated food, so it is very essential to know their interactions.

Deoxynivalenol (DON) is one of the most prevalent mycotoxins present in cereal crops worldwide which is produced by *Fusarium* species. It is very common, toxic, and stable, so it induces a health risk for both animals and humans. Manuel J. Saint-Cyr and his colleagues investigated the effect of a chronic exposure of DON on rats and analyzed the changes of the composition of gut microbiota [166]. They found a significant increase of species from *Prevotella/Bacteroides* group in rats during the first 3 weeks of administration at the dose of 100 μg/kg DON. The relative level of *Escherichia coli* decreased at day 27 and this decrease remains stable to the end.

Mingzhang Guo et al. evaluated the interactions between gut microbiota and ochratoxin A (OTA) by traditional and metagenomic methods [167]. Male F344 rats were treated by 0, 70, and 210 μg/kg body weight of ochratoxin A via oral gavage, respectively. OTA exposure decreased the diversity of GM and increased the abundance of genus *Lactobacillus* considerably. Furthermore, they isolated *Lactobacillus* species from fecal samples and researched the effects of *Lactobacillus* species on OTA *in vitro*. They found OTA could be absorbed by the strains rather than be degraded by them.

Similar to DON, aflatoxin is also a common contaminant of foods and is considered as an unavoidable food contaminant by the US Food and Drug Administration (FDA). Food and Agriculture Organization of the United Nations (FAO) indicated that approximately 4.5 billion people from developing countries lived in a zone at risk of chronic aflatoxin exposure [168]. Aflatoxin is produced by a fungus species *Aspergillus* and can occur in all periods of crops production, harvest, storage, and food processing. Wang et al. designed an experiment to know the impacts of Aflatoxin B1 (AFB1) on the gut microbiota in a rat model [169]. Their discovery declared that AFB1 could shift the GM in a dose-dependent manner. Accompanied with increasing dose of AFB1, the diversity of microbial community decreased, while evenness of community composition increased. Clustered analysis showed some lactic acid bacteria were significantly reduced by AFB1.

However, current studies about the interactions between xenobiotic toxins and gut microbiota and their *in vivo* metabolisms in humans and animals are very few and unclear. More and more studies are very necessary and important for people to have further insights into the effects of these toxins and to propose some alternative protective strategies. At the same time, particularly, multiple mycotoxins frequently exist in the contaminated food. Target to these multiple contaminations including chemical sources and biological sources, we have a long way to go.

4.7 Conclusions and Future Perspectives

In summary, EPs exposure induces a significant structural perturbation in the GM, which in turn substantially alters the GM metabolism, as evidenced by alterations of diverse GM-related metabolites. The changes in GM can possibly result in immune diseases and several inflammatory. Although the alterations trends of GM sometimes observed similar among different studies, many discrepancies still clearly exist. Of note, GM disturbance may be a secondary consequence of EPs on body health systems, which may be not a causative of EPs on health disorders but further exacerbate the disease conditions. (1) Various environmental pollutants are ubiquitous, and humans are more vulnerable to comprehensive pollution than a single chemical substance. Thus, combined effects of mixtures should also be studied. (2) Most of the researches conventionally focused only the impacts of EP on the composition or metabolism of GM rather than the comprehensive analysis. (3) Since the 16S rRNA sequencing is not capable of reflecting the true conditions of GM, future research should make more use of a combination of sequencing techniques such as metabonomics, metatranscriptomics, and metagenomics and, to a better understanding of host-GM interactions and the causal relationship between GM-associated symptoms and GM. (4) Almost all current studies on the GM and contaminants solely rely on stool microbiota, which is part of the GM and may yield limited conclusions [170]. Hence gut mucosal sampling should also be considered in future studies. (5) Although most studies were performed with causing prominent toxic effects, they are not representative of real-life exposures scenario.

Further studies should focus on explaining the chronic toxicity of EPs on human health by exposing laboratory animals to human relevant doses for a long term.

Conflict of interest statement No conflict of interest has been reported by the authors.

References

1. Forsythe P, Kunze WA (2013) Voices from within: gut microbes and the CNS. Cell Mol Life Sci 70(1):55–69
2. Clemente JC et al (2012) The impact of the gut microbiota on human health: an integrative view. Cell 148(6):1258–1270
3. Qin J et al (2010) A human gut microbial gene catalogue established by metagenomic sequencing. Nature 464(7285):59–65
4. Young VB (2012) The intestinal microbiota in health and disease. Curr Opin Gastroenterol 28 (1):63
5. Ley RE, Peterson DA, Gordon JI (2006) Ecological and evolutionary forces shaping microbial diversity in the human intestine. Cell 124(4):837–848
6. Yatsunenko T et al (2012) Human gut microbiome viewed across age and geography. Nature 486(7402):222–227
7. Goodrich JK et al (2014) Human genetics shape the gut microbiome. Cell 159(4):789–799
8. David LA et al (2013) Diet rapidly and reproducibly alters the human gut microbiome. Nature 505:559
9. Jakobsson HE et al (2015) The composition of the gut microbiota shapes the colon mucus barrier. EMBO Rep 16(2):164–177
10. Spanogiannopoulos P et al (2016) The microbial pharmacists within us: a metagenomic view of xenobiotic metabolism. Nat Rev Microbiol 14:273
11. Charbonneau MR et al (2016) Sialylated milk oligosaccharides promote microbiota-dependent growth in models of infant undernutrition. Cell 164(5):859–871
12. Round JL, Mazmanian SK (2010) Inducible Foxp3$^+$ regulatory T-cell development by a commensal bacterium of the intestinal microbiota. Proc Natl Acad Sci 107(27):12204
13. Heijtz RD et al (2011) Normal gut microbiota modulates brain development and behavior. Proc Natl Acad Sci 108(7):3047
14. Hsiao EY et al (2013) Microbiota modulate behavioral and physiological abnormalities associated with neurodevelopmental disorders. Cell 155(7):1451–1463
15. Claus SP et al (2011) Colonization-induced host-gut microbial metabolic interaction. MBio 2 (2):e00271–e00210
16. Marchesi JR et al (2016) The gut microbiota and host health: a new clinical frontier. Gut 65 (2):330
17. Scott KP et al (2015) Manipulating the gut microbiota to maintain health and treat disease. Microb Ecol Health Dis 26(1):25877
18. Ley RE et al (2005) Obesity alters gut microbial ecology. Proc Natl Acad Sci U S A 102 (31):11070
19. Tilg H, Moschen AR (2014) Microbiota and diabetes: an evolving relationship. Gut 63 (9):1513
20. Yang T et al (2015) Gut dysbiosis is linked to hypertension. Hypertension 65(6):1331–1340
21. Ling Z et al (2014) Altered fecal microbiota composition associated with food allergy in infants. Appl Environ Microbiol 80(8):2546
22. Wang Z et al (2011) Gut flora metabolism of phosphatidylcholine promotes cardiovascular disease. Nature 472(7341):57–63
23. Lippert K et al (2017) Gut microbiota dysbiosis associated with glucose metabolism disorders and the metabolic syndrome in older adults. Benefic Microbes 8(4):545–556
24. Turnbaugh PJ et al (2006) An obesity-associated gut microbiome with increased capacity for energy harvest. Nature 444(7122):1027–1031
25. Qin J et al (2012) A metagenome-wide association study of gut microbiota in type 2 diabetes. Nature 490(7418):55–60
26. Tabrez S, Ahmad M (2011) Oxidative stress-mediated genotoxicity of wastewaters collected from two different stations in Northern India. Mutat Res-Gen Tox En 726(1):15–20

27. Alam MZ, Ahmad S, Malik A (2011) Prevalence of heavy metal resistance in bacteria isolated from tannery effluents and affected soil. Environ Monit Assess 178(1):281–291
28. Hughes MF et al (2011) Arsenic exposure and toxicology: a historical perspective. Toxicol Sci 123(2):305–332
29. Wiele TV et al (2010) Arsenic metabolism by human gut microbiota upon *in vitro* digestion of contaminated soils. Environ Health Perspect 118(7):1004–1009
30. Lu K et al (2014) Arsenic exposure perturbs the gut microbiome and its metabolic profile in mice: an integrated metagenomics and metabolomics analysis. Environ Health Perspect 122 (3):284–291
31. Jones BV et al (2008) Functional and comparative metagenomic analysis of bile salt hydrolase activity in the human gut microbiome. Proc Natl Acad Sci 105(36):13580
32. Valentine JL, Kang HK, Spivey G (1979) Arsenic levels in human blood, urine, and hair in response to exposure via drinking water. Environ Res 20(1):24–32
33. Wester RC et al (1992) In Vitro percutaneous absorption of cadmium from water and soil into human skin. Toxicol Sci 19(1):1–5
34. Wang B, Hu L, Siahaan TJ (2016) Drug delivery: principles and applications. John Wiley & Sons, Hoboken
35. Ding J et al (2019) Heavy metal-induced co-selection of antibiotic resistance genes in the gut microbiota of collembolans. Sci Total Environ 683:210–215
36. Chang X et al (2019) Effects of cadmium exposure on the composition and diversity of the intestinal microbial community of common carp (Cyprinus carpio L.). Ecotoxicol Environ Saf 171:92–98
37. Breton J et al (2013) Ecotoxicology inside the gut: impact of heavy metals on the mouse microbiome. BMC Pharmacol Toxicol 14:62–62
38. Ba Q et al (2017) Sex-dependent effects of cadmium exposure in early life on gut microbiota and fat accumulation in mice. Environ Health Perspect 125(3):437–446
39. Guo X et al (2014) Metagenomic profiles and antibiotic resistance genes in gut microbiota of mice exposed to arsenic and iron. Chemosphere 112:1–8
40. Dheer R et al (2015) Arsenic induces structural and compositional colonic microbiome change and promotes host nitrogen and amino acid metabolism. Toxicol Appl Pharmacol 289 (3):397–408
41. Gao B et al (2017) Multi-omics reveals that lead exposure disturbs gut microbiome development, key metabolites, and metabolic pathways. Chem Res Toxicol 30(4):996–1005
42. Yao Q et al (2019) Effects of hexavalent chromium on intestinal histology and microbiota in Bufo gargarizans tadpoles. Chemosphere 216:313–323
43. Wu B et al (2014) Toxicological effects of dietary nickel chloride on intestinal microbiota. Ecotoxicol Environ Saf 109:70–76
44. Alghasham A, Salem TA, Meki A-RM (2013) Effect of cadmium-polluted water on plasma levels of tumor necrosis factor-α, interleukin-6 and oxidative status biomarkers in rats: protective effect of curcumin. Food Chem Toxicol 59:160–164
45. Al-Rmalli SW, Jenkins RO, Haris PI (2012) Dietary intake of cadmium from Bangladeshi foods. J Food Sci 77(1):T26–T33
46. Jin Y et al (2016) Cadmium exposure to murine macrophages decreases their inflammatory responses and increases their oxidative stress. Chemosphere 144:168–175
47. Ke S et al (2015) Benchmark dose estimation for cadmium-induced renal effects based on a large sample population from five Chinese provinces. Biomed Environ Sci 28(5):383–387
48. Solenkova NV et al (2014) Metal pollutants and cardiovascular disease: mechanisms and consequences of exposure. Am Heart J 168(6):812–822
49. Fazeli M, Hassanzadeh P, Alaei S (2010) Cadmium chloride exhibits a profound toxic effect on bacterial microflora of the mice gastrointestinal tract. Hum Exp Toxicol 30(2):152–159
50. Morozzi G et al (1986) Cadmium uptake by growing cells of gram-positive and gram-negative bacteria. Microbios 48(194):27–35

51. Zhang S et al (2015) Subchronic exposure of mice to cadmium perturbs their hepatic energy metabolism and gut microbiome. Chem Res Toxicol 28(10):2000–2009
52. Giannelli V et al (2014) Microbiota and the gut-liver axis: bacterial translocation, inflammation and infection in cirrhosis. World J Gastroenterol 20(45):16795–16810
53. Cox LM et al (2014) Altering the intestinal microbiota during a critical developmental window has lasting metabolic consequences. Cell 158(4):705–721
54. RodrÍguez JM et al (2015) The composition of the gut microbiota throughout life, with an emphasis on early life. Microb Ecol Health Dis 26(1):26050
55. Wu J et al (2016) Perinatal lead exposure alters gut microbiota composition and results in sex-specific bodyweight increases in adult mice. Toxicol Sci 151(2):324–333
56. Bae S et al (2014) Plasma choline metabolites and colorectal cancer risk in the Women's Health Initiative Observational Study. Cancer Res 74(24):7442–7452
57. Lukovac S et al (2014) Differential modulation by Akkermansia muciniphila and Faecalibacterium prausnitzii of host peripheral lipid metabolism and histone acetylation in mouse gut organoids. MBio 5(4):e01438–e01414
58. Cani PD, Everard A (2014) Akkermansia muciniphila: a novel target controlling obesity, type 2 diabetes and inflammation? Med Sci 30(2):125
59. Xia J et al (2018) Chronic exposure to low concentrations of lead induces metabolic disorder and dysbiosis of the gut microbiota in mice. Sci Total Environ 631–632:439–448
60. Djane N-K et al (1999) Chromium speciation in natural waters using serially connected supported liquid membranes. Talanta 48(5):1121–1132
61. Cotruvo JA (2017) 2017 WHO guidelines for drinking water quality: first addendum to the fourth edition. J—Am Water Work Assoc 109(7):44–51
62. Younan S et al (2016) Chromium(VI) bioremediation by probiotics. J Sci Food Agric 96 (12):3977–3982
63. Serra D, Almeida LM, Dinis TCP (2018) Dietary polyphenols: a novel strategy to modulate microbiota-gut-brain axis. Trends Food Sci Technol 78:224–233
64. Kuehbacher T et al (2008) Intestinal TM7 bacterial phylogenies in active inflammatory bowel disease. J Med Microbiol 57(12):1569–1576
65. Bor B et al (2016) Phenotypic and physiological characterization of the epibiotic interaction between TM7x and its Basibiont Actinomyces. Microb Ecol 71(1):243–255
66. McLean JS et al (2013) Candidate phylum TM6 genome recovered from a hospital sink biofilm provides genomic insights into this uncultivated phylum. Proc Natl Acad Sci 110 (26):E2390
67. Delafont V et al (2015) Shedding light on microbial dark matter: a TM6 bacterium as natural endosymbiont of a free-living amoeba. Environ Microbiol Rep 7(6):970–978
68. He X et al (2015) Cultivation of a human-associated TM7 phylotype reveals a reduced genome and epibiotic parasitic lifestyle. Proc Natl Acad Sci 112(1):244
69. Wu G, Xiao X, Feng P, Xie F, Yu Z, Yuan W, Liu P, Li X (2017) Gut remediation: a potential approach to reducing chromium accumulation using Lactobacillus plantarum TW1-1. Sci Rep 7(1):1–2
70. Brinkman BM et al (2013) Gut microbiota affects sensitivity to acute DSS-induced colitis independently of host genotype. Inflamm Bowel Dis 19(12):2560–2567
71. De Filippis F et al (2016) High-level adherence to a Mediterranean diet beneficially impacts the gut microbiota and associated metabolome. Gut 65(11):1812
72. Delzenne NM, Cani PD (2011) Interaction between obesity and the gut microbiota: relevance in nutrition. Annu Rev Nutr 31(1):15–31
73. Imhann F et al (2018) Interplay of host genetics and gut microbiota underlying the onset and clinical presentation of inflammatory bowel disease. Gut 67(1):108
74. Liu Y et al (2014) Exposing to cadmium stress cause profound toxic effect on microbiota of the mice intestinal tract. PLoS One 9(2):e85323
75. Duruibe JO, Ogwuegbu M, Egwurugwu J (2007) Heavy metal pollution and human biotoxic effects. Int J Phy Sci 2(5):112–118

76. Wang X et al (2017) Toxicity of mineral Chinese medicines containing mercury element. Zhongguo Zhong Yao Za Zhi 42(7):1258–1264
77. Funabashi H (2006) Minamata disease and environmental governance. Int J Jpn Sociol 15 (1):7–25
78. Ruan Y et al (2019) High doses of copper and mercury changed cecal microbiota in female mice. Biol Trace Elem Res 189(1):134–144
79. Smirnov A et al (2005) Mucin dynamics and microbial populations in chicken small intestine are changed by dietary probiotic and antibiotic growth promoter supplementation. J Nutr 135 (2):187–192
80. Dozois CM, Daigle F, Curtiss R (2003) Identification of pathogen-specific and conserved genes expressed *in vivo* by an avian pathogenic Escherichia coli strain. Proc Natl Acad Sci 100 (1):247
81. Lepage P et al (2011) Twin study indicates loss of interaction between microbiota and mucosa of patients with ulcerative colitis. Gastroenterology 141(1):227–236
82. Presley LL et al (2010) Bacteria associated with immunoregulatory cells in mice. Appl Environ Microbiol 76(3):936
83. Richardson JB et al (2018) Exposure to toxic metals triggers unique responses from the rat gut microbiota. Sci Rep 8(1):6578
84. Zhai Q et al (2017) Effects of subchronic oral toxic metal exposure on the intestinal microbiota of mice. Sci Bull 62(12):831–840
85. Brüssow H (2015) Growth promotion and gut microbiota: insights from antibiotic use. Environ Microbiol 17:2216–2227
86. Du L, Liu W (2012) Occurrence, fate, and ecotoxicity of antibiotics in agro-ecosystems: a review. Agron Sustain Dev 32:309–327
87. Sarmah AK, Meyer MT, Boxall AB (2006) A global perspective on the use, sales, exposure pathways, occurrence, fate and effects of veterinary antibiotics (VAs) in the environment. Chemosphere 65(5):725–759
88. Dong H et al (2016) Occurrence and removal of antibiotics in ecological and conventional wastewater treatment processes: a field study. J Environ Manag 178:11–19
89. Ferro G et al (2016) Antibiotic resistance spread potential in urban wastewater effluents disinfected by UV/H2O2 process. Sci Total Environ 560–561:29–35
90. Qian M et al (2016) Occurrence of trace elements and antibiotics in manure-based fertilizers from the Zhejiang Province of China. Sci Total Environ 559:174–181
91. Demoly P et al (2000) Allergy to macrolide antibiotics. Review of the literature. Presse Med 29:321–326
92. Pérez-Cobas AE et al (2013) Gut microbiota disturbance during antibiotic therapy: a multi-omic approach. Gut 62(11):1591
93. Fröhlich EE et al (2016) Cognitive impairment by antibiotic-induced gut dysbiosis: Analysis of gut microbiota-brain communication. Brain Behav Immun 56:140–155
94. Membrez M et al (2008) Gut microbiota modulation with norfloxacin and ampicillin enhances glucose tolerance in mice. FASEB J 22(7):2416–2426
95. Cho I et al (2012) Antibiotics in early life alter the murine colonic microbiome and adiposity. Nature 488:621
96. Nobel YR et al (2015) Metabolic and metagenomic outcomes from early-life pulsed antibiotic treatment. Nat Commun 6:7486
97. Wu M et al (2020) Antibiotic-induced dysbiosis of gut microbiota impairs corneal development in postnatal mice by affecting CCR2 negative macrophage distribution. Mucosal Immunol 13(1):47–63
98. Carvalho BM et al (2012) Modulation of gut microbiota by antibiotics improves insulin signalling in high-fat fed mice. Diabetologia 55(10):2823–2834
99. Cani PD et al (2007) Metabolic endotoxemia initiates obesity and insulin resistance. Diabetes 56(7):1761

100. Wang J, MacNeil JD, Kay JF (2011) Chemical analysis of antibiotic residues in food, vol 38. John Wiley & Sons, Hoboken
101. Schubert AM, Sinani H, Schloss PD (2015) Antibiotic-induced alterations of the murine gut microbiota and subsequent effects on colonization resistance against Clostridium difficile. MBio 6(4):e00974
102. Buffie CG et al (2012) Profound alterations of intestinal microbiota following a single dose of clindamycin results in sustained susceptibility to Clostridium difficile-induced colitis. Infect Immun 80(1):62–73
103. Grazul H, Kanda LL, Gondek D (2016) Impact of probiotic supplements on microbiome diversity following antibiotic treatment of mice. Gut Microbes 7(2):101–114
104. Brooke JS (2012) Stenotrophomonas maltophilia: an emerging global opportunistic pathogen. Clin Microbiol Rev 25(1):2–41
105. Buffie CG et al (2015) Precision microbiome reconstitution restores bile acid mediated resistance to Clostridium difficile. Nature 517(7533):205
106. Fouhy F et al (2012) High-throughput sequencing reveals the incomplete, short-term, recovery of the infant gut microbiota following parenteral antibiotic treatment with ampicillin and gentamycin. Antimicrob Agents Chemother 56:5811–5820
107. Korpela K et al (2016) Intestinal microbiome is related to lifetime antibiotic use in Finnish pre-school children. Nat Commun 7:10410
108. Rea MC et al (2011) Effect of broad-and narrow-spectrum antimicrobials on Clostridium difficile and microbial diversity in a model of the distal colon. Proc Natl Acad Sci U S A 108:4639–4644
109. Madigan M-T, Martinko J (2005) Brock biology of microorganisms. Prentice Hall, Upper Saddle River, NJ. isbn:0-13-144329-1
110. Chaudhury A et al (1999) Enteropathogenicity and antimicrobial susceptibility of new *Escherichia* spp. J Diar Dis Res 17:85–87
111. Pien FD et al (1985) Colonization of human wounds by Escherichia vulneris and Escherichia hermannii. J Clin Microbiol 22:283
112. Jantsch J, Chikkaballi D, Hensel M (2011) Cellular aspects of immunity to intracellular Salmonella enterica. Immunol Rev 240:185–195
113. Gao K et al (2018) Antibiotics-induced modulation of large intestinal microbiota altered aromatic amino acid profile and expression of neurotransmitters in the hypothalamus of piglets. J Neurochem 146(3):219–234
114. Barrangou R et al (2007) CRISPR provides acquired resistance against viruses in prokaryotes. Science 315(5819):1709
115. Makarova KS et al (2006) A putative RNA-interference-based immune system in prokaryotes: computational analysis of the predicted enzymatic machinery, functional analogies with eukaryotic RNAi, and hypothetical mechanisms of action. Biol Direct 1(1):7
116. Bohnhoff M, Miller CP (1962) Enhanced susceptibility to salmonella infection in streptomycin-treated mice. J Infect Dis 111(2):117–127
117. Miller CP, Bohnhoff M, Rifkind D (1956) The effect of an antibiotic on the susceptibility of the mouse's intestinal tract to Salmonella infection. Trans Am Clin Climatol Assoc 68:51–58
118. Reeves AE et al (2014) The interplay between microbiome dynamics and pathogen dynamics in a murine model of Clostridium difficile infection. Gut Microbes 2(3):145–158
119. Schubert AM et al (2014) Microbiome data distinguish patients with Clostridium difficile infection and non-C. difficile-associated diarrhea from healthy controls. MBio 5(3):e01021–e01014
120. Vincent C et al (2013) Reductions in intestinal Clostridiales precede the development of nosocomial Clostridium difficile infection. Microbiome 1(1):18
121. Ng KM et al (2013) Microbiota-liberated host sugars facilitate post-antibiotic expansion of enteric pathogens. Nature 502:96
122. Lee HH et al (2010) Bacterial charity work leads to population-wide resistance. Nature 467:82

123. Toprak E et al (2011) Evolutionary paths to antibiotic resistance under dynamically sustained drug selection. Nat Genet 44:101
124. Jakobsson HE et al (2010) Short-term antibiotic treatment has differing long-term impacts on the human throat and gut microbiome. PLoS One 5(3):e9836
125. Looft T et al (2012) In-feed antibiotic effects on the swine intestinal microbiome. Proc Natl Acad Sci 109(5):1691
126. Dethlefsen L et al (2008) The pervasive effects of an antibiotic on the human gut microbiota, as revealed by deep 16S rRNA sequencing. PLoS Biol 6(11):e280
127. Antonopoulos DA et al (2009) Reproducible community dynamics of the gastrointestinal microbiota following antibiotic perturbation. Infect Immun 77(6):2367
128. Jernberg C et al (2007) Long-term ecological impacts of antibiotic administration on the human intestinal microbiota. ISME J 1(1):56–66
129. Jernberg C et al (2010) Long-term impacts of antibiotic exposure on the human intestinal microbiota. Microbiology 156(11):3216–3223
130. Jernberg C et al (2005) Monitoring of antibiotic-induced alterations in the human intestinal microflora and detection of probiotic strains by use of terminal restriction fragment length polymorphism. Appl Environ Microbiol 71(1):501
131. Ajslev TA et al (2011) Childhood overweight after establishment of the gut microbiota: the role of delivery mode, pre-pregnancy weight and early administration of antibiotics. Int J Obes 35:522
132. World Health Organization (2017) Health topics: pesticides.
133. Atwood D, Paisley-Jones C (2017) Pesticides industry sales and usage 2008–2012 market estimates.
134. Kim KH, Kabir E, Jahan SA (2017) Exposure to pesticides and the associated human health effects. Sci Total Environ 575:525–535
135. Li Z, Jennings A (2017) Worldwide regulations of standard values of pesticides for human health risk control: a review. Int J Environ Res Public Health 14(7):826
136. Abdollahi M et al (2004) Pesticides and oxidative stress: a review. Med Sci Monit 10(6): RA141–RA147
137. Trapp S, Eggen T (2013) Simulation of the plant uptake of organophosphates and other emerging; pollutants for greenhouse experiments and field conditions. Environ Sci Pollut Res Int 20(6):4018–4029
138. Joly C et al (2013) Impact of chronic exposure to low doses of chlorpyrifos on the intestinal microbiota in the simulator of the Human Intestinal Microbial Ecosystem (SHIME®) and in the rat. Environ Sci Pollut Res 20(5):2726–2734
139. Poet TS et al (2003) In vitro rat hepatic and intestinal metabolism of the organophosphate pesticides chlorpyrifos and diazinon. Toxicol Sci 72(2):193–200
140. Condette CJ et al (2015) Chlorpyrifos exposure during perinatal period affects intestinal microbiota associated with delay of maturation of digestive tract in rats. J Pediatr Gastroenterol Nutr 61(1):30–40
141. Zhao Y et al (2016) Effects of chlorpyrifos on the gut microbiome and urine metabolome in mouse (Mus musculus). Chemosphere 153:287–293
142. Dong Y-L et al (2009) Induction of oxidative stress and apoptosis by pentachlorophenol in primary cultures of Carassius carassius hepatocytes. Comp Biochem Phys C 150(2):179–185
143. Luo Y et al (2009) EPR detection of hydroxyl radical generation and its interaction with antioxidant system in Carassius auratus exposed to pentachlorophenol. J Hazard Mater 171 (1):1096–1102
144. Kan H et al (2015) Correlations of gut microbial community shift with hepatic damage and growth inhibition of carassius auratus induced by pentachlorophenol exposure. Environ Sci Technol 49(19):11894–11902
145. Nasuti C et al (2016) Changes on fecal microbiota in rats exposed to permethrin during postnatal development. Environ Sci Pollut Res 23(11):10930–10937
146. Nasuti C et al (2014) Neonatal exposure to permethrin pesticide causes lifelong fear and spatial learning deficits and alters hippocampal morphology of synapses. J Neurodev Disord 6(1):7

147. Wu S et al (2018) Exposure to the fungicide propamocarb causes gut microbiota dysbiosis and metabolic disorder in mice. Environ Pollut 237:775–783
148. Wu S et al (2018) Chronic exposure to fungicide propamocarb induces bile acid metabolic disorder and increases trimethylamine in C57BL/6J mice. Sci Total Environ 642:341–348
149. Jin C et al (2016) Oral imazalil exposure induces gut microbiota dysbiosis and colonic inflammation in mice. Chemosphere 160:349–358
150. Xu C et al (2014) Changes in gut microbiota may be early signs of liver toxicity induced by epoxiconazole in rats. Chemotherapy 60(2):135–142
151. Jin Y et al (2015) Oral exposure of mice to carbendazim induces hepatic lipid metabolism disorder and gut microbiota dysbiosis. Toxicol Sci 147(1):116–126
152. Liu Q et al (2017) Organochloride pesticides modulated gut microbiota and influenced bile acid metabolism in mice. Environ Pollut 226:268–276
153. Lozano VL et al (2018) Sex-dependent impact of roundup on the rat gut microbiome. Toxicol Rep 5:96–107
154. Neel BA, Sargis RM (2011) The paradox of progress: environmental disruption of metabolism and the diabetes epidemic. Diabetes 60(7):1838
155. Jin Y et al (2014) Sub-chronically exposing mice to a polycyclic aromatic hydrocarbon increases lipid accumulation in their livers. Environ Toxicol Pharmacol 38(2):353–363
156. Brandt I et al (1982) Metabolism of 2,4′,5-trichlorobiphenyl: tissue concentrations of methylsulphonyl-2,4′,5-trichlorobiphenyl in germfree and conventional mice. Toxicol Lett 12(4):273–280
157. Choi Jeong J et al (2013) Exercise attenuates PCB-induced changes in the mouse gut microbiome. Environ Health Perspect 121(6):725–730
158. Cheng SL et al (2018) Gut microbiota modulates interactions between polychlorinated biphenyls and bile acid homeostasis. Toxicol Sci 166(2):269–287
159. Ribiere C et al (2016) Oral exposure to environmental pollutant benzo[a]pyrene impacts the intestinal epithelium and induces gut microbial shifts in murine model. Sci Rep 6:31027
160. Chi Y et al (2018) PCBs-high-fat diet interactions as mediators of gut microbiota dysbiosis and abdominal fat accumulation in female mice. Environ Pollut 239:332–341
161. Zhang L et al (2015) Persistent organic pollutants modify gut microbiota-host metabolic homeostasis in mice through aryl hydrocarbon receptor activation. Environ Health Perspect 123(7):679–688
162. Lefever DE et al (2016) TCDD modulation of gut microbiome correlated with liver and immune toxicity in streptozotocin (STZ)-induced hyperglycemic mice. Toxicol Appl Pharmacol 304:48–58
163. Duperron S et al (2019) Response of fish gut microbiota to toxin-containing cyanobacterial extracts: a microcosm study on the medaka (oryzias latipes). Environ Sci Technol Let 6 (6):341–347
164. Robert H et al (2017) Impact of mycotoxins on the intestine: are mucus and microbiota new targets? J Toxicol Environ Health B Crit Rev 20(5):249–275
165. Reddy KRN et al (2010) An overview of mycotoxin contamination in foods and its implications for human health. Toxin Rev 29(1):3–26
166. Saint-Cyr MJ et al (2013) Evaluation of an oral subchronic exposure of deoxynivalenol on the composition of human gut microbiota in a model of human microbiota-associated rats. PLoS One 8(11):e80578
167. Guo M et al (2014) Combination of metagenomics and culture-based methods to study the interaction between ochratoxin A and gut microbiota. Toxicol Sci 141(1):314–323
168. Williams JH et al (2004) Human aflatoxicosis in developing countries: a review of toxicology, exposure, potential health consequences, and interventions. Am J Clin Nutr 80(5):1106–1122
169. Wang J et al (2016) Aflatoxin B1 induced compositional changes in gut microbial communities of male F344 rats. Toxicol Sci 150(1):54–63
170. Zmora N et al (2018) Personalized gut mucosal colonization resistance to empiric probiotics is associated with unique host and microbiome features. Cell 174(6):1388–1405. e21

Chapter 5
Environmental Pollutants That Can Be Metabolized by the Host (Gut Microbiota)

Marwa M. El-Dalatony and Xiangkai Li

5.1 Introduction

The rise in the level of environmental pollutants is a major concern that possess hazardous effects on human health. Several studies have demonstrated an intense relationship between the human health and gut microbial composition [1]. Human gut microbes are the living microorganisms residing the human GI tract and play vital role in human health. In humans, the gut flora is established in the first one to two years of life. It develops in conjunction with the intestinal epithelium in a manner that also provides a barrier to pathogenic microorganism [2]. Most of the pathologies are linked with alteration of the structural composition of gut microbiota, which is also known as dysbiosis, like IBD [3], and other digestive disorders, which include autoimmune disorders, diabetes, obesity, and neurological disturbances [4]. Since decades, human gut microbiota is reported in biotransformation of xenobiotics [5]. Number of drugs substrates have been reported for the gut microbes, showing the potential of gut microbes to execute diverse chemical conversions on drugs and other environmental chemicals [6]. Xenobiotic compounds mainly enter the human body through GI tract and influenced the metabolism of microbiota by the number of compounds reaching the distal gut. After ingestion various environmental chemicals poorly absorbed in the body and then further transported and metabolized in the distal part of small intestine, caecum, and large intestine [5]. Probiotics are used for gut remediation, may treat diet pollutants such as PAHs, pesticides, and nitrotoluenes.

The microbes which are utilized as probiotics are of various categories including bacteria, yeast or mold and the most common species of these types are 1—Bacteria: (1) *Lactobacillus: acidophilus, brevis sporogenes, fermentum rhamnosum, lactus,*

M. M. El-Dalatony · X. Li (✉)
School of Life Science, Lanzhou University, Lanzhou, Gansu, People's Republic of China
e-mail: xkli@lzu.edu.cn

X. Li, P. Liu (eds.), *Gut Remediation of Environmental Pollutants*,
https://doi.org/10.1007/978-981-15-4759-1_5

plantarum, reuteri, delbrueck cellobiosus, casei, gasseri, farciminis, paracasei, and *crispatus*; (2) *Bifidobacterium: adolescentis, animalis, bifidum, breve, infantis, lactis, longum,* and *thermophilum*; (3) *Streptococcus: alivarius, cremoris lactis, diacetylactis, intermedius,* and *thermophilis*; (4) *Leuconostoc mesenteroides*; (5) *Bacillus*; (6) *Propionibacterium*; (7) *Pediococcus*; (8) *Enterococcus faecium* [7] 2—Molds and yeast like: *Saccharomyces cerevisiae, S. bourlardii, Aspergillus niger, A.s oryzue,* and *Candida pintolopesii* [8]. The use of these types of the microorganisms as probiotics has enlarged owing to the rise in the study related with the discovery of new potential strains [9]. The questions that are most commonly asked by the public is why must we use probiotics? The suitable answer for this question is the exposure of humans with different microbes during its life, these microbes can deteriorate human health. Antibiotic treatment can be a real tactic to prevent the growth of pathogens but it can also destroy the normal healthy flora of gut [10]. In order to prevent these normal flora of gut it is essential to take some supplements that can beneficially affect the gut microbes. Probiotic supplements in the diet is the best possible cheapest way to improve human health by recovering the composition of gut microbes. The evidences of use of probiotics in food by olden civilization are also reported [11].

5.1.1 Heavy Metals

Heavy metals are high density metallic elements [12] that can be considered as a potential contaminant if present in high concentration and possess negative impact on human health. These are comprised of various transition metals, lanthanides, metalloids, and actinides [13]. The sources of these heavy metals are rapid industrialization, agricultural residue, geogenic sources, atmosphere and domestic effluents. The essential heavy metals are required by the body for diverse biochemical and physiological functions, the deficiency of these metals can cause various short-term or chronic disorders. These metals include copper, chromium, iron, molybdenum, magnesium, cobalt, nickel, zinc, and selenium [14]. The natural distribution of these metals in the environment include spring waters, volcanic eruptions, erosion, bacterial activities. Heavy metals can also spread by anthropogenic activities such as combustion of fossil fuels, industrialization, and agricultural activities [15]. The accumulation of these heavy metals in living organisms can cause negative effects on human health (Table 5.1). After transportation of these heavy metals inside the human body, they are stored in different compartment of cells and tissues of body. Binding of these compounds to cellular proteins and nucleic acid can destroy the macromolecules and cell functioning. Heavy metals can exert various complications on the human body. These complications include effects on the central nervous function (mental disorders), damage to the blood constituents, lungs, kidneys, liver, and other important organs [21]. The toxicity and carcinogenic activity of heavy metals employ various mechanisms.

Table 5.1 Negative effect related with heavy metals (HMs) exposure and toxicity

Metal	Acute toxicity	Chronic toxicity	References
Arsenic	Blood in urine, discomfort in GI, vomiting, headaches, convulsions, coma, diarrhea, and death	Skin lesions, Blackfoot disease; organ destruction; blisters or failure; Mutagenic characteristics; risk of cancer with diabetes	[16]
Cadmium	Hepatic toxicity, pulmonary injury, and testicular injury	Renal injury osteoporosis; Carcinoma associated with kidney and prostrate; Toxic effect on other organs	[17]
Chromium	Vomiting with diarrhea; hemorrhage with blood loss in GI tract	Necrosis (renal or hepatic); skin and nasal ulcers, "chrome Holes," puncture of the nasal septum; nasal, Pharyngeal, and gastrointestinal Carcinomas, irrigative dermatitis	[18]
Lead	Mild fatigue, headache, vomiting, nausea, neurobehavioral problems like short attention span and impulsivity distractibility,	Antisocial actions; impaired hemoglobin Synthesis; impaired renal function; Deafness, blindness, retardation; reduced IQ, memory loss; decreased libido, fatigue	[19]
Mercury	Weakened neurodevelopment; injury of IQ; reduction in memory, care, language, and visual-spatial perception tests; associations with autism plus amyotrophic lateral sclerosis	Impaired synthesis of hemoglobin; neurodevelopment; loss of IQ; decrease in memory, attention, language, and visual-spatial perception tests; associations with autism besides amyotrophic lateral sclerosis	[20]

Chromium (Cr) occurs naturally in the earth's crust, with a variation in its oxidation states from chromium (II) to chromium (VI) [22]. The main sources of chromium are chromium steel, fertilizers, petroleum oil well drilling and coal,, pigment oxidants, catalyst, and metal plating tanneries. It has wide use in industries such as electroplating, wood preservation, making of paints and pigments, metallurgy, chemicals industry, tanning, and pulp and paper industry. These industries are the major source of chromium pollution in the environment that can adversely affect the ecological species [21]. In addition, the various anthropogenic activities like repeated use of fertilizers and improper disposal of sewage can also contribute in chromium pollution [21]. The most toxic from of chromium is its hexavalent form, whereas less toxicity of Chromium (III) compounds has been studied. Chromium

(VI) can cause various allergic reactions like irritation and ulcers in the lining of the nose, anemia ulcers in the GI tract. The low concentrations of chromium (VI) compounds can cause respiratory, cardiovascular, hematological, renal, neurological and hepatic effects in humans which may lead to death [23]. DNA damage has been reported under various in vivo and in vitro studies, that caused chromosomal aberrations and effect the DNA transcription [24].

5.1.1.1 The Effect of Cr (VI) on Gut Microbiota

Heavy metals can cause remarkable changes in the framework of gut microbes. A reduction in the count of gut microbes has been observed after the exposure of heavy metals [25]. Exposure of HMs has increased the ratio of *Bacteroidetes* to *Firmicutes*, linked with weight loss [26]. Another study on Cr(VI)-treated mice has resulted in increased count of *Tenericutes* and *Bacteroidetes*, whereas a decline in the count of Firmicutes [27]. A decrease in the relative abundance of *Lachnospiraceae* was also observed in Cr(VI) exposed mice [28].

Gut microbes play an eminent function in maintaining important physiological function in the host [29] and influence the immunity and metabolism of host [30]. A study on earthworm exposed with chromium revealed the increase of *Enterobacteriaceae* of *Burkholderiaceae* (13.1%), and *Microscillaceae* sp., whereas a decrease in the *Aeromonadaceae* (5.6%). The study provided the understanding of toxic effects of chromium in soil and the effect of chromium on the organisms present in soil. This data can be used to further investigate the pathways involved in toxic effects of chromium in earthworm and relate the information on the basis of the genomics and proteomics [31].

Alterations in the structure of intestinal tissues have been reported when 0.416 mg Cr^{6+} L^{-1} is given to the tadpoles. A significant reduction was observed in the body wet weight, total body length, besides length of intestine length and wet of *B. gargarizans* tadpoles. The information of 16S rRNA gene sequencing showed alteration in the structural composition of intestinal microbes after exposure to chromium in tadpoles. A significant change was observed in the composition of gut microbes as the Fusobacteria phylum changes in all chromium cured tadpole groups. The groups exposed to a high concentration of chromium showed the presence of TM6_*Dependentiae* and *Saccharibacteria* which were not reported in other groups. *Aeromonas* genus followed a decline trend in Cr(VI) treated groups. The exposure of Cr(VI) induced metabolic syndromes associated with the alterations of structural composition of gut microbes. On a cumulative account the study has demonstrated various effects of Cr(VI) applied on *B. gargarizans* tadpoles, causing fluctuations in the intestinal histology and microbiota composition [32]. Although, the clear mechanism of heavy metal toxicity and changes in gut microbial composition are still not clear.

5.1.1.2 Potential Roles of Probiotics and Gut Microbiota in Cr (VI) Remediation

The study of heavy metal remediation has been in practice from last several years, including various chemical, physical, and biological techniques (microbial and phytoremediation). The use of living organism in removal of heavy metals is one of the most powerful approaches [33]. The use of microbes serves various advantages like low cost, minimal site disruption, and high acceptance by the public. The binding of *lactobacilli* sp. with heavy metals for their removal is reported [20]. Research carried out in mice has established that the intestinal microbiota is the first line of defense in the body and alters extra toxic. The conversion of Cr(VI) to slight toxic Cr(III) by intestinal microbes working as first line of defense in rats was reported. This study indicates the resistance of bacteria against chromium Cr (VI) [34]. The main route of exposure of gut microbes to chromium is the consumption of contaminated food and water. In potable water the concentration of chromium from 1 to 10 ppm is generally safe because the microbes of the gut can strongly reduce Cr(VI) to Cr(III) [35].

Streptomyces werraensis LD22 isolated from feces can tolerate heavy metals like $NiCl_2$, $K_2Cr_2O_7$, $ZnCl_2$, $CuSO_4$, and $PbNO_3$ in humans and animals [36]. A long exposure of *Lactobacillus* to chromium can generate resistance in the strains [37]. Shrivastava et al. [35] has reported the transformation of highly toxic forms of chromium to less toxic forms, changes in immune response by *Lactobacillus* and other gut associated microbes, and the presence of sequestering elements in the human fecal material. The use of *Lactobacillus* spp. in food items and as probiotics is safe as it has a long history. The use of *Lactobacillus* in food industry is associated with its presence as a member of normal gut flora, and it can be used as an adjunct to remove the toxicity of heavy metals. The resistance mechanisms of microbes are effective in preventing cell damage and are related with the removal of heavy metals by binding or sequestering them and their further release through defecation [38]. These microbes can also reduce the oxidative stress initiated via heavy metals toxicity and other food toxins, reported by in vitro and in vivo investigations [39]. The antioxidant properties of *Lactobacillus casei* strain 17 can prevent the liver and kidney from oxidative stress induced by potassium dichromate (Fig. 5.1). The subcutaneous administration of Cr(VI) in the formula of potassium dichromate showed important alterations in kidney and liver, alanine transaminase, functional enzyme markers in the serum, total protein content, creatinine, and urea.

The preeminent histopathological modifications were congestion of vascular region of liver, several degenerative changes, the deterioration of tubular epithelial cells of kidneys showed by the infiltration of mononuclear cells and dilation of sinusoids, and the cystic dilation of tubules plus hyaline. *Lb. casei* strain 17 has overturned the alteration caused by Cr(VI) when co-administered in the body [40]. The effect of oral intake of increasing doses of potassium dichromate more than 3 months was studied (0, 0.12, 0.24, and 0.36 g kg^{-1} diet), which showed a reduction in the performance and changes in the glycemic, renal, lipidic, and hepatic

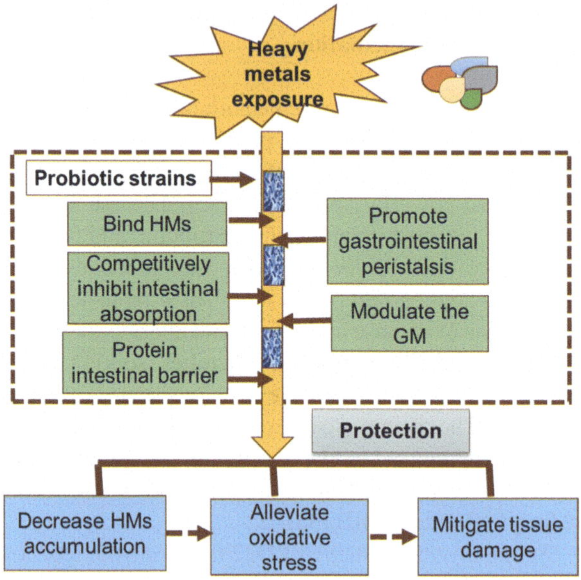

Fig. 5.1 Planned defensive mechanisms of probiotics on heavy metals (HMs) removal in vivo

profiles. The incorporation of 0.2% of probiotic (*Lactobacillus acidophilus*, 2.22 × 109 colony-forming units (CFU); *Streptococcus faecium*, 2.22 × 109 CFU; *Bifidobacterium thermophilum*, 2.22 × 109 CFU; *Bifidobacterium longum*, 2.22 × 109 CFU) in the diet can remarkably reduce the toxic and histopathological effects of chromium on above mentioned organ profiles [41]. A study reported the defensive effects of *L. plantarum* TW1–1, against Cr toxicity in mice revealed that TW1–1 has the potential to diminish Cr toxicity. The effectiveness of "gut remediation," was analyzed, which involves both direct and indirect remediation of heavy metal pollution by *L. plantarum.* Almost 60% removal of 0.5 mM Cr(VI) was observed by Strain TW1–1 within 48 h of incubation. Oral administration of *Lactobacillus plantarum* TW1–1 to Kunming rat for 49 days with 1 mM $K_2Cr_2O_7$ in drinking water was done to find out the mechanism and remediation potential of the strain. A reduction in Cr gathering in tissues and high excretion of Cr in feces was observed. Significant changes in the histopathology and oxidative stress and histopathological were also observed. A double effect on the activity of fecal bacteria to remove chromium was observed after administration of TW1–1. The altered microbial gut population after chromium exposure was restored by TW1–1 and revealed by the MiSeq sequencing of fecal bacterial 16S rRNA genes TW1–1 which has reversed the count of 49 of the 79 operational taxonomic units that were changed by Cr. Based on these observations, TW1–1 was proposed as a working model against Cr as it can effectively eliminate Cr from the host and also regulate the gut microbiota, which helps in chromate decrease and offer defense against Cr [42].

5.1.2 *Pesticides*

Pesticides are defined as the natural or chemically synthesized compounds used to prevent the growth of undesired plants and kill animal pests. Pesticides are amalgamation of compounds predominantly utilized in agriculture related public health protection plans to provide protection against pests, weeds or diseases, and vector-borne diseases infecting humans like schistosomiasis, malaria, and dengue fever. The typical examples of pesticides contain insecticides, herbicides, rodenticides and insecticides, fungicides, and some plant growth promoters classified on the basis of their specific roles [43, 44].

The contribution of pesticides in health deuteration and environmental problems is a major concern [45]. The main routes of exposure include skin contact, incidental oral consumption, and inhalation. The prominent health determining factors associated with the exposure of pesticides are the route of pesticides, duration, and type of pesticides. Pesticides have been reported to be metabolized, and accumulated inside body fat in humans [46]. The diverse negative health effects associated with the use of chemical pesticides such as gastrointestinal effects dermatological, neurological, respiratory, carcinogenic, endocrine, and reproductive effects [47, 48]. The exposure of pesticides to the crops is the primary source of transportation of pesticidal residues in the GI tract of humans. The use of highly toxic pesticides in low concentration can lead to prolonged inauspicious effects on endocrine system, skin, and nervous system by stimulating the production of free radicals that could cause destruction of DNA, lipid peroxidation, carcinogenic effects, and cell death [49, 50].

Pesticide residues are persistent in the environment and cause significant environmental issues. These residues are present in in different water sources, soil, foods, beverages, instance prepared meals, animal feedstuffs, fruit juices [51]. Recently, tremendous development in the field of health concerns in context to the effects of pesticides exposure on animals have done [52]. Several pesticides carry antimicrobial properties that can influence the composition of microbes present in the gut of animals and humans, leading to symptomatic changes [53]. Traces of pesticides have been found in human breast milk samples, thus more concern is required about prenatal exposure and health of children [46].

5.1.2.1 Carbamate

Carbamate pesticides are the chemical pesticides associated with endocrine-disrupting activity in insects, [48]. The commercial forms of carbamates are carbofuran, ziram, and aldicarb, responsible for reproductive disorders, effects on mitochondrial function, and cellular metabolism in insects [54]. Aldicarb is a potential oxime carbamate insecticide against early season nematodes and insects to protect both nonfood crops and food crops [55]. Aldicarb is highly toxic which has been categorized and labeled in the United States under "restricted use pesticide" due to its high toxicity. The use of aldicarb has been reported to cause various food poisonings associated outbreaks [56]. The in vitro study conducted on hamster ovarian cells have shown the genotoxic and cytotoxic effects of carbamates, whereas effects on induction of necrosis and apoptosis in natural killer cells, apoptosis in *T lymphocytes* are also reported [57, 58]. A study on carbaryl has reported the function of carbaryl in dioxin toxicity by acting as a ligand for the hepatic aryl hydrocarbon receptor (a transcription factor) [59]. The neurobehavioral effects of carbamates [60], non-Hodgkin's lymphoma and increased risk for dementia have been discussed by researchers [61].

5.1.2.2 Pyrethroids

Pyrethroids are extensively used broad-spectrum insecticides that differentiates among each other due to the presence of chiral carbon. The natural source of pyrethroids is *Chrysanthemum cinerariaefolium* flower, the first synthetic pyrethroids was formed in 1949 and named as allethrin [62, 63]. These can be classified into two categories, category I pyrethroids (with cyclopropane carboxylic group) and category II pyrethroids (with cyano group) [64]. The category II pyrethroids are more effective against insects as compared to category I pyrethroids due to the presence of cyano group, pyrethroids have been reported to account for minimum four stereoisomeric forms assigned with different biological activities. These pyrethroids are marketed either as single chemical isomer or a racemic mixture of stereoisomers. The use of Piperonyl butoxide as synergist has been given in commercial formulation of pyrethroids to inhibit the degradation of active compounds metabolically [65].

The use of Deltamethrin to control the spread of malaria-spreading mosquitoes in different countries has been reported. Pyrethroids target the chloride and sodium channels in insects and are 2250 times comparatively more toxic to insect than mammals [66]. The inhibition of gamma amino butyric acid (GABA) gated chloride ion channel is reported due to high use of pyrethroids [62]. The application of pyrethroids is mainly against insect pests of horticulture and agriculture, and household insects. The use of pyrethroids is comparatively safe but the extensive use of these pesticides after a certain concentration has been considered hazardous to both animal and humans [67]. The deleterious effects of pyrethroids on non-target species such as aquatic animals have been well reported [68]. The toxicity biomarkers of pyrethroid in fish are also reported [69]. The repeated use of pyrethroids in agriculture can also affect human health and cause symptoms such as antiandrogenic activity contaminated urine and low serum quality. A study has provided information on the bioabsorption of pyrethroids by analyzing the urine sample of outdoor workers in California [70]. In rats bifenthrin neurotoxicity was reported as mixed type (category I/II) [62], whereas in zebrafish neurotoxicity has been detected [71].

5.1.2.3 The Effect of Pesticides on Gut Microbiota

Researchers have illustrated the crucial function performed by gut microbes in the metabolism and removal of pesticidal residues in host. The enzymes produced by GM have also reported to metabolize some pesticides. Biological conversion of chlorpyrifos into 3,5,6-trichloro-2-pyridinol a more toxic form via GM has been reported and consequently forming toxic effects on host health (Fig. 5.2). On the other hand, *L. lactis, Pseudomonas* spp. (ATCC700113), *E. coli*, and L. *fermentum* present in GIT, have been reported to use 3,5,6-trichloro-2-pyridinol as their sole energy source [85, 86]. Thus, the prolonged exposure of pesticides can also affect the count and function of GM and cause various immune and metabolic diseases [87]. The use of PEM in agriculture, residential pest control, and public health objectives and agriculture [88, 89] has reported the exposure of gut microbes to PEM possibly by contaminated food [90]. Previous studies provided data on the reduction of *Bacteroides, Porphyromonas*, and *Prevotella* sp., and increment in the count of *Lactobacillus* and *Enterobacteriaceae*, after exposure of PEM for 4 months [76]. *Bacteroides* species are capable of producing SCFAs to prevent inflammation of gut [91]. Inhibition of *Bifidobacterium* and *Lactobacillus paracasei*, *Staphylococcus aureus,* and *Escherichia coli* by PEM is reported in an in vitro study [76]. In rats' low dose postnatal exposure of PEM reduces the count of microbes, these alterations can be considered as a significant factor determining the development of diseases, so additional studies to find crucial evidences are required (Table 5.2).

PM is a universal fungicide utilized to prevent Oomycetes associated diseases in roots, leaves, and soil [92]. Accumulation of PM residues at high concentrations in fruits can affect humans [93]. The exposure of the PM can potentially disturb metabolism by altering the composition of gut microbial metabolites. A study reported the effect of exposure of 0.3 g L^{-1} PM for 28 days on gut microbiota.

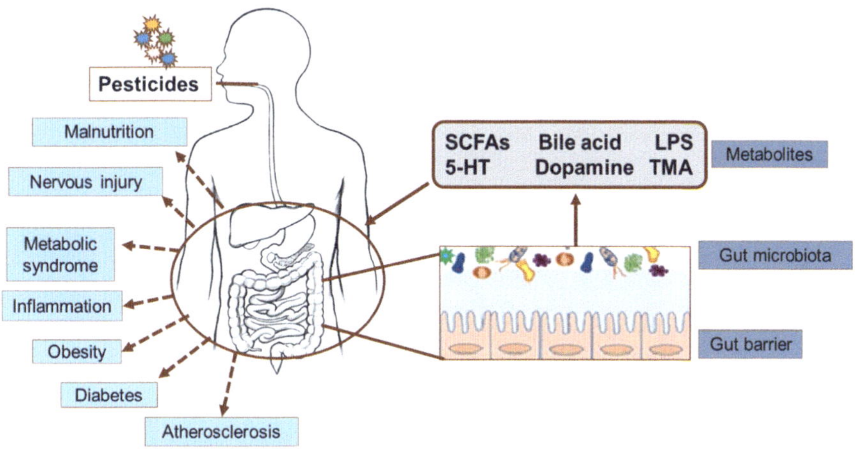

Fig. 5.2 Influences of various types of environmental contaminants on gut microbiota and their consequent results

Table 5.2 Impact of a variety of pesticides on gut microbiota

Pesticide	Classes	Model	Gut microbiota dysbiosis	References
Insecticides	Organophosphates	Mice	Unknown	[72]
		Mice	Firmicutes ↓ Bacteroidetes ↓ Lactobacillaceae↓ Bacteroidaceae ↑	[73]
		Mice	*Lachnospiraceae* ↓	[74]
	Organochlorines	Mice	*Lactobacillus*↑	[75]
	Permethrin (PEM)	Rat pups	*Bacteroides* ↓ *Prevotella* ↓ *Enterobacteriaceae*↑ *Lactobacillus*↑	[76]
	Neonicotinoids	*D. melanogaster*	*Acetobacter* ↑ *Lactobacillus*↑	[77]
Herbicides	Glyphosate	Green turtles *Apis mellifera*	*Pantoea*↓ *Proteus*↓ *Shigella*↓ *Staphylococcus*↓ Species diversity plus richness of gut microbiota were changed	[78] [79]
	Pentachlorophenol (PCP)	Goldfish	*Bacteroidetes* ↑	[52, 80]
Fungicides	Carbendazim (CBZ)	Mice	*Bacteroidetes*↓ *Firmicutes* ↑ *Proteobacteria* ↑ *Actinobacteria* ↑	[52]
	Imazalil (IMZ)	Mice Zebrafish	*Bifidobacterium*↓ *Lactobacillus*↓ *Deltaproteobacteria* ↑ *Desulfovibrio* ↑ *Proteobacteria*↓ *Bacteroidetes* ↓ *Fusobacteria*↑ *Firmicutes* ↑	[81] [82]
	Propamocarb (PM)	Mice	*Oscillospira*↓ *Parabacteroides* ↓ *Desulfovibrio* ↓ *Ruminococcus* ↓ *Bacteroides* ↑ *Dehalobacterium* ↑ *Butyricimonas* ↑	[83]
	Epoxiconazole	Rat	*Firmicutes*↓ *Bacteroidetes*↑ *Proteobacteria*↑ *Lachnospiraceae*↑ *Enterobacteriaceae*↑	[84]

The results showed changes in the composition of microbes and fecal metabolites (20 different kinds of metabolites) such as SCFAs, bile acids, succinate, and trimethylamine (TMA), that can deteriorate human health [83]. TMA production after metabolism of dietary fiber was further metabolized to TMAO (involving the oxidation of flavin mono-oxygenase 3 (FMO3) mediated by the foresaid X receptor). The enhanced level of TMAO can further lead to atherosclerosis [94, 95]. The compositional changes of the gut microbiome were observed after exposing rats to small dosages of PM. Significant alterations in the metabolites of fecal matter and energy metabolism were observed. An increase in the concentration of TMA in feces associated with atherosclerosis was observed. Significant disturbances in the cardiac NO/NOS pathway and improvement of the NF-kB transcriptional levels were also reported. The long-term exposure of PM can induce disorders in enterohepatic metabolism and potentially increase the risk of CVD [83].

5.1.2.4 Potential Roles of Probiotics and Gut Microbiota in Pesticides Remediation

To prevent the damages caused by the exposure of pesticides several drugs and therapies have been employed [96]. To suppress the negative impact of pesticides more economically feasible practices are required. Effect of *Lactobacilli* on downstream cellular damage and oxidative stress induced by pesticides has been studied. *L. plantarum* BJ0021 has shown reduction in MDA concentration and oxidative stress level in liver and kidney, when exposed to endosulfan [97]. *L. casei* ATCC334 administration in rats exposed to carcinogen 1,2-dimethylhydrazine have shown reduction in DNA damage [98]. The role of probiotics in maintaining the integrity of intestinal barrier and reduced absorption of pesticides has been reported. The positive effect of t *L. plantarum* MB452 on the expression of tight junction proteins occludin, ZO-2, ZO-1, and cingulin in the Caco-2 intestinal cell-line was reported [99].

Reduced absorption of parathion or CP in a Caco-2 transwell model after administration of *L. rhamnosus* strain GG (LGG) and LGR-1 was studied [100]. Recently, effects of *Lactobacillus* isolated from dairy products and wheat on degradation of OCP enzymatically with phosphohydrolase was studied [101]. *Lactobacilli* administration motivate host's immunity and detoxification mechanisms to prevent invasion of pesticides and pathogens. Stimulation of phase-II detoxification system was observed in insects administered with *L. casei* and physiological improvements in *Caenorhabditis elegans* in malathion stimulated effects [101]. A positive effect on immunity, reduction in the count of pathogens (*Serratia marcescens)* was observed after consumption of *L. plantarum* ATCC14917 in fruit flies subjected to imidacloprid [77].

5.1.3 Polycyclic Aromatic Hydrocarbons and Nitrated Polycyclic Aromatic Hydrocarbons (PAHs & Nitro-PAHs)

These are the groups of toxic organic compounds which have been made during by the partial combustion of organic matter. USEPA (United States Environmental Protection Agency) and EEA (European Environment Agency) have been declared the PAHs as pre-dominant environmental pollutants [102]. In urban and metropolitan cities, the highest and major source of PAHs comes from the combustion of fossil fuels [103]. For the development of industries and their higher energy demand the assessment of coal and its residues should be done in eco-friendly manner [104]. PAHs characterization is important and necessary by the management coal and its combustion residues because their emission are very high and main sources included processing of coal and by the partial combustion of organic substances [105]. Polycyclic hydrocarbons comprise of fused aromatic ring compounds and these are formed by the partial combustion of petroleum and fossil fuels. PAHs have relationship to non-cancerous (neurobehavioral effects, adverse birth outcomes, decreased fertility) and cancerous effects (lung, breast, and colon cancers) [106]. 16 of these compounds are known as main pollutants because of their toxic, mutagenic carcinogenic properties, 7 of them are potentially human carcinogens [102].

Rapid industrialization progressively leads to the worsening and contamination of ecosystem due to the environmental pollution and it will be reach to its higher disturbing levels in the coming years. PAHs have been the major concern for public due to their deleterious effects on health as PAHs are capable of inducing toxicity, teratogenicity, and carcinogenicity [107]. Benzo[a]pyrene (BaP) has the highest carcinogenic potential among all the various PAHs [108] and it is variously spread in the ecosystem with potential toxicity, mutagenicity, and carcinogenicity [109]. There are seven PAHs compounds which are human carcinogen: Benz[a] anthracene, Benzo[a]pyrene, Benzo[k]fluoranthene, indeno(1,2,3-cd) pyrene, Benzo [b]fluoranthene, Dibenz[a,h]anthracene, and chrysene [110]. Through sewage sludge PAHs may enter in agricultural soil and produce an environmental risk to the organisms present in soil, as well, the crops grown in that soil and eaten by humans may cause various kind of disease in humans [111]. Pharmaceuticals, petrochemical, fertilizers, and other disinfectants generate organic and inorganic compounds and mainly composed of PAHs [112].

Lung cancer is a disease mainly caused by smoking and so many other factors but high exposure of PAHs also cause lung cancer and it is the highest mortality disease in the USA [113]. PAHs have so many negative effects on environment and human health which include organ system effect, carcinogenesis respiratory problems, neurotoxic, and developmental effect [114]. Exposure of PAHs can directly affect the liver and may be responsible for lung cancer and any other carcinogens [115]. Childhood hospitalization in the developing countries is increased day by day because PAHs in the air cause asthma in children [116]. PAHs are considered as the toxic, dangerous, and important pollutant [117], and it may cause the serious health problems also affect the habitat of flora and fauna [118].

It is reported that due the excess emissions of polycyclic aromatic hydrocarbons over 27,000 tons year^{-1} have a direct effect on the environment of China [119]. Ultimately, human health is affected due the uptake of PAHs contaminated water and food, polluted air, and through direct skin contact. Furthermore, the serious risk level of urban residents is higher than rural residents, which may be the result of developed industries [120].

Nitro-PAHs are PAHs derivatives with minimum one nitro-functional group linked with the aromatic benzene ring [121]. PAHs and nitro-PAHs release in the environment in the same way or some nitro-PAHs often release after the transformation of PAHs. These compounds have been categorized as potential carcinogens by International Agency for Research on Cancer [122]. Toxicological studies have identified the nitro-PAHs as mutagens like 1,6-dinitropyrene, 3,9-dinitrofluoranthene 3,6-dinitrobenzo[e]pyrene, 1,3-dinitropypene, 1,8-dinitropyrene, 1,3,6-trinitropyrene, and their toxic properties are much higher than their related PAHs [121]. Nitro-PAHs adversely affect the DNA which include DNA damage, adduction in DNA, changes in protein and gene level expression, aryl hydrocarbon receptor activation, pro inflammation, alter cell cycles, and increased levels of ROS [123].

5.1.3.1 The Effect of PAHs on Gut Microbiota

It is reported that the estrogenicity of 4-PAHs (phenanthrene, pyrene, Benzo[a] pyrene, and naphthalene) before and after digestion by a typical human microbiota in vitro [124]. The determination of risk assessments of cancer for environmental PAH mixtures the U.S. Environmental Protection Agency (EPA) utilized Benzo[a] pyrene as the reference compound. Recently, EPA changed the oral B[a]P cancer risk slope factor from 1 to 7.3 mg kg^{-1} day^{-1} on the basis of daily contact of 270–750 ng (adults in the USA, non-smoking), would correspond to a 3.9×10^{-6}–1.1×10^{-5} lifetime excess risk for developing cancer [125]. Naturally contaminated food products and charcoal-grilled, roasted or smoked food may be the most common route of B[a]P exposure in humans [126]. The literature studies and clinical trials showed that the toxic B[a]P compounds can potentially cause adenomas by targeting various organs of the human body, where the location of tumors was identified by route of its exposure. Oral administration and inhalation of B[a]P toxic compounds can cause cancers and tumors in lungs, liver, breast, and in gastrointestinal tract. However, B[a]P contaminated air reached the GIT via mucociliary clearance mechanism which subcequently enter and metabolized in intestinal enterocytes and liver hepatocytes by cytochrome P450-dependent monooxygenases and finally form diol-epoxide compounds which eventually bound to DNA and initiated carcinogenesis [4].

For other factors involved in B[a]P toxicity, uridine diphosphate (UDP), glutathione transferase (GNT), glucuronosyl transferase, methyl transferase, and epoxide hydrolase contribute to detoxification of PAHs which may be activated by prostaglandin synthase, lipoxygenase, or one-electron oxidation [127]. However, gut microbiota of human and rat could regenerate from its hepatic conjugate by detoxification process [5].

Recently various studies have been showing interest on the influence of B[a]P in gut microbiota. Administration of B[a]P in a dose-dependent manner showed alterations in the volatile matter framework and transcriptome of beneficial microbes of gut in an in vitro studies [4]. Another research showed that exposure to particulate matter, a potential mechanism to alter the gut microbiota through GIT and also explain the induced inflammation in GIT [128]. 7-hydroxybenzo[a]pyrene reported as a B[a]P derivative and also literature studies showed that human gut microbiota have ability to change B[a]P into estrogenic metabolites [4]. Gut remediation maybe a promising method using probiotics to alteration the composition and metabolism of the gut microbiota, upregulate B[a]P degradation, and downregulate B[a]P regeneration [4].

A study showed that PAHs removal depends on the pH of media, species and type of bacteria, and finally on the concentration of PAHs. Not only live but inactivated form of strains may also remove PAHs. Lactic acid bacteria (LABs) remove four polycyclic aromatic hydrocarbons (PAHs) namely, benzo(a)pyrene (B[a]P), benz(a) anthracene (B[a]A), chrysene (Chr) plus benzo(b), and *Lactobacillus acidophilus* LA-5 fluoranthene (BbF) from contaminated phosphate buffer saline (PBS) with a highest binding ability compare to *Bifidobacterium lactis* BB-12 which had the lowest rate [129].

5.1.3.2 The Effect of Nitro-PAHs on Gut Microbiota

Nitro-PAH comes from direct contact of humans with animals by oral, inhalation, ingestion and by dermal contact [130]. Oral ingestion is a significant route of human. Nitro-PAHs have been identified in foods such as fruits, vegetables, meat, tea leaves, plus in water (Table 5.3). It is reported in a study that the intake rate of 1-nitropyrene is about 9.7×10^{-7} mg kg^{-1} day^{-1}. Laboratory toxicity test showed that there is 100% development of hepatocellular carcinomas in the rats after the oral administration of 2-nitrofluorene in diet (at 2.37 mmol kg^{-1} diet) to rats [121].

The main nitro-PAHs in the environment are 2-nitrofluorene (NF) that can be used further as a model compound for nitro-PAHs. A study revealed the removal of NF when it is incubated with human feces in an in vitro experiment [141]. The reduction of NF to 2-aminofluorene has been reported in an in vivo study when NF is given to conventional rats. The reduction process is carried out by the intestinal bacteria, the formed compound was then further acetylated besides hydroxylated in the liver. The metabolic process is quantitatively the most eminent process which produce hydroxylated 2-acetylaminofluorene. Thc formation of hydroxylated nitrofluorenes by alternative metabolic pathways can directly cause mutagenicity [142].

Table 5.3 Adverse effect and aids of nitrate (NO_3) and nitrite (NO_2) in drinking water plus food on human health

Effects		
Adverse effect	List	References
Chronic effect (carcinogenic Effect of NO_3 and (NO_2	Gastrointestinal tract tumors	[131]
	Non-Hodgkin's lymphoma (NHL)	[132]
	Urinary tract tumors	[133]
	Brain tumors	[134]
	Pancreas tumors	[135]
Acute toxicity	Methaemoglobinaemia	[136]
	Mellitus diabetes	[137]
	Effect on the thyroid gland	[136]
Positive effect	Protective effects on the cardiovascular system	[138]
	Regulation of blood pressure plus maintain homeostasis of vessel	[139]
	NO_3 and NO_2 are existing in breast milk so offer nutritional with immunological aids to infants, however the source of NO_3 and NO_2 is unidentified in breast milk.	[140]
	NO_3 and NO_2 as preservatives are used to stabilize the color, fragrant, create, control food spoilage	[133]
	NO_3 as a factor that inhabits the growth of other micro organisms	[139]

5.1.4 Nitrite

Nitrite (NO_2) exist in various sources such as air, water, soil, and plants [143]. The conversion of nitrate to nitrite endogenously accounts for about 80–85% of composite systemic nitrite [144]. The overall consumption of nitrite by an individual is 1.2–3.0 mg per day [145] in which almost 93% is the converted form of nitrate [146]. The intake of high nitrogen containing nutritional sources and oxidation of NO inside the body are the other sources of nitrite in the body [147]. Nitrate is naturally existing in plants and prevalent in root and leafy vegetables (86%) like beet and lettuce, respectively. The prevalent sources of nitrite intake are vegetables (16%), cereals and baked products (34%), cured meats (39%) [138]. The high level of nitrate in drinking water is a major source. The main purpose of addition of nitrate and nitrite salts in cured meats like hot dogs, ham, and bacon, is to enhance the flavor, add color, and avert the avert spore-forming bacterium [133]. The nitrite taken from exogenous sources can undergo complete absorption in the duodenum and jejunum. Majority of the nitrite circulating systematically in the body transformed to NO and considered as a stable reservoir of NO [148].

In mammals the endogenous sources of NO_3 are mainly derived from the following: (1) oxidation of endogenous NO, (2) reduction of salivary nitrate reduction by commensals of the mouth and GI tract, (3) nutritional supplements like vegetables, meat, and potable water [131, 149]. The standard concentration of nitrate is highly regulated inside the body; however, the concentration can vary depending

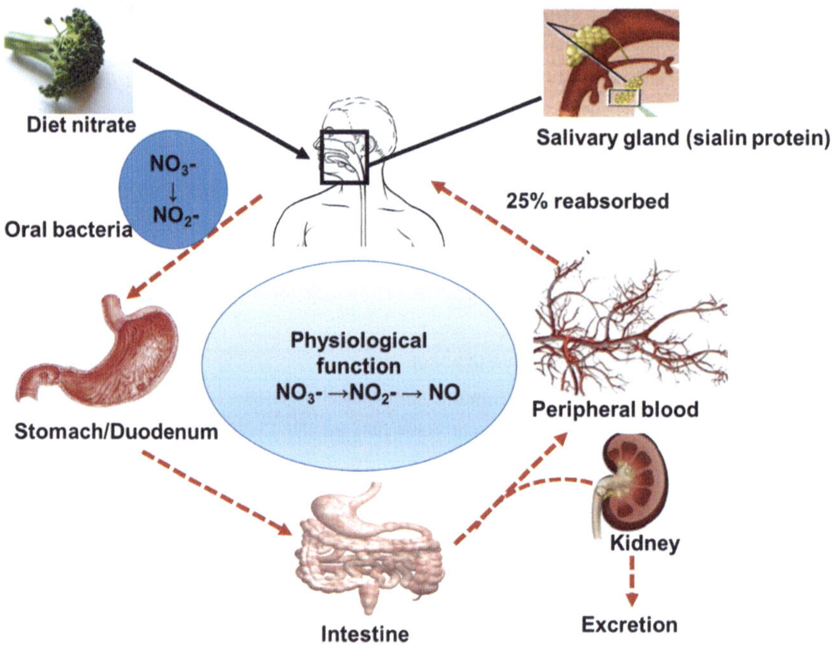

Fig. 5.3 Circulation of nitrate in the body

on the scavenging activity of NO by the various tissues (Fig. 5.3). In human plasma the accepted standard concentration of nitrite is 150–600 nM [150] and it can increase after consumption of nitrate rich products The enterosalivary circulation (almost 25%) of nitrate and its reduction to nitrate by the commensals of oral digestive system is the main route cause of increment of nitrate in the body [151].

Salivary glands are responsible for the reuse of dietary nitrate and sialin which play a vital role in the concentration and active transportation of nitrate. In salivary glands commensal oral bacteria converted nitrate into nitrite and later absorption takes place in stomach and intestine. Around 25% of nitrate reabsorbed by the salivary glands and rest of it excreted by the kidneys. Nitric oxide (NO^-), nitrite (NO_2^-), and nitrate (NO_3^-).

5.1.4.1 The Effect of Nitrite on Gut Microbiota

The bacteria of oral digestive system and the associated salivary gland plays an eminent role in process of nitrate conversion (NO_3–NO_2–NO) and its circulation. The absorption of nitrate in the stomach and intestine depends on its bio-availability. Almost 75% of it is excreted out in urine, whereas the rest nitrate content undergoes reabsorption by the salivary and biliary glands in the kidneys [152, 153]. The salivary glands account for the 25% of nitrate recycled under normal conditions which is almost 10 times more than the concentration of plasma [154]. The study conducted on salivary glands in 2012, discovered a nitrate transporter sialin in the membranes of mammalian cells. The discovery of sialin provides a framework to explore the metabolic effects of nitrate on human body [155, 156]. The facultative commensals of oral cavity situated in the posterior deep crypts of the tongue transform almost 5–7% of dietary nitrate to [157]. Afterward the formed nitrite in the stomach further transform to nitric oxide and systematically absorbed.

5.2 Conclusion

This chapter presents the various bidirectional interaction between environmental contaminants and gastrointestinal tract microbiota. The xenobiotic and environmental pollutants like pesticides, heavy metals, and PAHs significantly cause harmful effect on status of human and animal health. GI tract bacteria plays major role and have broad-spectrum enzymatic capabilities in remediation of these environmental pollutants. The metabolic activity of gut microbiota and contaminants-induced toxicity both affects the host organs by tissue damaging and other dysbiosis diseases. An unevenness of gut microbial community can lead to numerous diseases. Dietetic supplementation along with probiotics is an encouraging complement for efficaciously lowering the destruction made by environmental pollutants and by maintaining the gut microbiota of animals and humans. Further metagenomics studies should be accomplished for understanding the relationship between the composition of gut microbes and probiotics species under different nutrients and diet conditions. Moreover, the development of new probiotics and mixture of probiotics based on individual microbial composition would be an effective way for future studies for overall human health status.

References

1. Jin Y et al (2017) Effects of environmental pollutants on gut microbiota. Environ Pollut 222:1–9
2. Shaneeta Johnson KU, Moore C, Worthey A, Bendjemil S, Childs E, Hobson L, Danner O (2019) *The gut microbiota, obesity and the effect of dietary modulation and bariatric surgery on the microbiome: a review of the literature.* ECronicon
3. Shreiner AB, Kao JY, Young VB (2015) The gut microbiome in health and in disease. Curr Opin Gastroenterol 31(1):69–75
4. Defois C et al (2017) Environmental pollutant Benzo[a] Pyrene impacts the volatile Metabolome and Transcriptome of the human gut microbiota. Front Microbiol 8:1562
5. Claus SP, Guillou H, Ellero-Simatos S (2016) The gut microbiota: a major player in the toxicity of environmental pollutants? Npj Biofilms and Microbiomes 2(1):1–11
6. McCombe PA et al (2019) Gut microbiota in ALS: possible role in pathogenesis? Expert Rev Neurother 19(9):785–805
7. Anukam KC et al (2009) Probiotic lactobacillus rhamnosus GR-1 and lactobacillus reuteri RC-14 may help downregulate TNF-alpha, IL-6, IL-8, IL-10 and IL-12 (p70) in the neurogenic

bladder of spinal cord injured patient with urinary tract infections: a two-case study. Adv Urol 2009:680363
8. Amara AA, Shibl A (2015) Role of probiotics in health improvement, infection control and disease treatment and management. Saudi Pharm J 23(2):107–114
9. Abdin AA, Saeid EM (2008) An experimental study on ulcerative colitis as a potential target for probiotic therapy by lactobacillus acidophilus with or without "olsalazine". J Crohns Colitis 2(4):296–303
10. Devaraj NK et al (2019) The effects of probiotic supplementation on the incidence of Diarrhea in Cancer patients receiving radiation therapy: a systematic review with meta-analysis and trial sequential analysis of randomized controlled trials. Nutrients 11(12):2886
11. Tannock GW et al (2011) Testing probiotic strain Escherichia coli Nissle 1917 (Mutaflor) for its ability to reduce carriage of multidrug-resistant E. coli by elderly residents in long-term care facilities. J Med Microbiol 60(Pt 3):366–370
12. Buruiana DL et al (2015) Toxicity of heavy metals on the environment and human health. Ecol Econ Educ Legislation Ii:565–571
13. Singh R et al (2011) Heavy metals and living systems: an overview. Ind J Pharmacol 43 (3):246–253
14. Engwa GA et al (2019) Mechanism and health effects of heavy metal toxicity in humans, in poisoning in the modern world-new tricks for an old dog? IntechOpen, London
15. Hirner AV, Emons H (2004) Organic metal and metalloid species in the environment: analysis, distribution, processes and toxicological evaluation. Springer Science & Business Media, Berlin
16. Hughes MF (2002) Arsenic toxicity and potential mechanisms of action. Toxicol Lett 133 (2002):1–16
17. Satarug S et al (2010) Cadmium, environmental exposure, and health outcomes. Environ Health Perspect 118(2):182–190
18. Zhitkovich A (2011) Chromium in drinking water: sources, metabolism, and cancer risks. Chem Res Toxicol 24(10):1617–1629
19. Topcu A, Bulat T (2010) Removal of cadmium and lead from aqueous solution by Enterococcus faecium strains. J Food Sci 75(1):T13–T17
20. Monachese M, Burton JP, Reid G (2012) Bioremediation and tolerance of humans to heavy metals through microbial processes: a potential role for probiotics? Appl Environ Microbiol 78 (18):6397–6404
21. Jaishankar M et al (2014) Toxicity, mechanism and health effects of some heavy metals. Interdiscip Toxicol 7(2):60–72
22. Jacobs JA, Testa SM (2005) Overview of chromium (VI) in the environment: background and history. *Chromium (VI) Handbook:*1–21
23. Shekhawat K, Chatterjee S, Joshi BJIJOAR (2015) Chromium toxicity and its health hazards. Int J Adv Res 3(7):167–172
24. Matsumoto ST et al (2006) Genotoxicity and mutagenicity of water contaminated with tannery effluents, as evaluated by the micronucleus test and comet assay using the fish Oreochromis niloticus and chromosome aberrations in onion root-tips. Genet Mol Biol 29(1):148–158
25. Xia J et al (2018) Effects of short term lead exposure on gut microbiota and hepatic metabolism in adult zebrafish. Comp Biochem Physiol C 209:1–8
26. Ley RE et al (2006) Human gut microbes associated with obesity. Nature 444 (7122):1022–1023
27. Wu J et al (2016) Perinatal lead exposure alters gut microbiota composition and results in sex-specific bodyweight increases in adult mice. Toxicol Sci 151(2):324–333
28. Wu G et al (2017) Gut remediation: a potential approach to reducing chromium accumulation using lactobacillus plantarum TW1-1. Sci Rep 7(1):1–12
29. Eckburg PB et al (2005) Diversity of the human intestinal microbial flora. Science 308 (5728):1635–1638

30. Berg M et al (2016) Assembly of the Caenorhabditis elegans gut microbiota from diverse soil microbial environments. ISME J 10(8):1998–2009
31. Tang R et al (2019) Toxic responses of metabolites, organelles and gut microorganisms of Eisenia fetida in a soil with chromium contamination. Environ Pollut 251:910–920
32. Yao Q et al (2019) Effects of hexavalent chromium on intestinal histology and microbiota in Bufo gargarizans tadpoles. Chemosphere 216:313–323
33. Fengwei Tian YX, Li X, Zhai Q, Wang G, Zhang Q, Zhang H, Chen W (2015) Protective effects of lactobacillus plantarum CCFM8246 against copper toxicity in mice. PLoS One 10 (11):e0143318
34. Upreti RK et al (2005) A comparative study on rat intestinal epithelial cells and resident gut bacteria: (I) effect of hexavalent chromium. Toxicol Mech Methods 15(5):331–338
35. Shrivastava R, Upreti RK, Chaturvedi UC (2003) Various cells of the immune system and intestine differ in their capacity to reduce hexavalent chromium. FEMS Immunol Med Microbiol 38(1):65–70
36. Latha S, Vinothini G, Dhanasekaran D (2015) Chromium [Cr(VI)] biosorption property of the newly isolated actinobacterial probiont Streptomyces werraensis LD22. 3 Biotech 5 (4):423–432
37. Upreti RK et al (2011) In vitro development of resistance to arsenite and chromium-VI in lactobacilli strains as perspective attenuation of gastrointestinal disorder. J Environ Biol 32 (3):325
38. Sinha V et al (2011) Amplification of arsH gene in lactobacillus acidophilus resistant to arsenite. Biotechnology 10(1):101–107
39. Bhakta JN et al (2012) Characterization of lactic acid bacteria-based probiotics as potential heavy metal sorbents. J Appl Microbiol 112(6):1193–1206
40. Balakrishnan R et al (2013) Antioxidant activity of coated probiotic lactobacillus casei on chromium(VI) induced oxidative stress in rats. Proc Natl Acad Sci Ind Sect B 84(2):305–310
41. Bezerra RC (2014) *Efeito do probiótico após toxicidade hepática do dicromato de potássio em ratos*
42. Wu G et al (2017) Gut remediation: a potential approach to reducing chromium accumulation using lactobacillus plantarum TW1-1. Sci Rep 7(1):1–12
43. Karunarathne A et al (2019) How many premature deaths from pesticide suicide have occurred since the agricultural green revolution? Clin Toxicol 58(4):1–6
44. Organization, W.H (1990) Public health impact of pesticides used in agriculture. World Health Organization, Geneva
45. Zheng S et al (2016) Distribution and risk assessment of 82 pesticides in Jiulong River and estuary in South China. Chemosphere 144:1177–1192
46. Pirsaheb M et al (2015) Organochlorine pesticides residue in breast milk: a systematic review. Med J Islam Repub Iran 29:228
47. Thakur DS et al (2014) Glyphosate poisoning with acute pulmonary edema. Toxicol Int 21 (3):328
48. Mnif W et al (2011) Effect of endocrine disruptor pesticides: a review. Int J Environ Res Public Health 8(6):2265–2303
49. Kim K-H, Kabir E, Jahan SAJSotTE (2017) Exposure to pesticides and the associated human health effects. Sci Total Environ 575:525–535
50. El-Gendy KS et al (2010) The role of vitamin C as antioxidant in protection of oxidative stress induced by imidacloprid. Food Chem Toxicol 48(1):215–221
51. Chourasiya S et al (2015) Health risk assessment of organochlorine pesticide exposure through dietary intake of vegetables grown in the periurban sites of Delhi, India. Environ Sci Pollut Res 22(8):5793–5806
52. Jin Y et al (2015) The toxicity of chlorpyrifos on the early life stage of zebrafish: a survey on the endpoints at development, locomotor behavior, oxidative stress and immunotoxicity. Fish Shellfish Immunol 43(2):405–414

53. Andersson H (2014) *Pesticides and health: a review of evidence on health effects, valuation of risks, and benefit-cost analysis.* Advances in Health Economics and Health Services Research
54. Karami-Mohajeri S, Abdollahi MJH (2011) Toxic influence of organophosphate, carbamate, and organochlorine pesticides on cellular metabolism of lipids, proteins, and carbohydrates: a systematic review. Hum Exp Toxicol 30(9):1119–1140
55. Smulders CJ et al (2003) Selective effects of carbamate pesticides on rat neuronal nicotinic acetylcholine receptors and rat brain acetylcholinesterase. Toxicol Appl Pharmacol 193 (2):139–146
56. Gao B et al (2018) The Carbamate Aldicarb altered the gut microbiome, Metabolome, and Lipidome of C57BL/6J mice. Chem Res Toxicol 32(1):67–79
57. Soloneski S et al (2015) Carbamates: a study on genotoxic, cytotoxic, and apoptotic effects induced in Chinese hamster ovary (CHO-K1) cells. Toxicol in Vitro 29(5):834–844
58. Li Q et al (2015) Carbamate pesticide-induced apoptosis in human T lymphocytes. Int J Environ Res Public Health 12(4):3633–3645
59. Denison M et al (1998) Carbaryl, a carbamate insecticide, is a ligand for the hepatic ah (dioxin) receptor. Toxicol Appl Pharmacol 152(2):406–414
60. Lifshitz M et al (1997) Carbamate poisoning in early childhood and in adults. J Toxicol Clin Toxicol 35(1):25–27
61. Zheng T et al (2001) Agricultural exposure to carbamate pesticides and risk of non-Hodgkin lymphoma. J Occup Environ Med 43(7):641–649
62. Gammon DW et al (2019) Pyrethroid neurotoxicity studies with bifenthrin indicate a mixed Type I/II mode of action. Pest Manag Sci 75(4):1190–1197
63. Ensley SM (2018) Pyrethrins and pyrethroids. In: Veterinary toxicology. Elsevier, Oxford, pp 515–520
64. Chang J et al (2016) Bioaccumulation and enantioselectivity of type I and type II pyrethroid pesticides in earthworm. Chemosphere 144:1351–1357
65. Fai PBA, Kinfack JST, Towa YJTJE (2017) Acute effects of binary mixtures of type II pyrethroids and organophosphate insecticides on Oreochromis niloticus. Ecotoxicology 26 (7):889–901
66. Chrustek A et al (2018) Current research on the safety of Pyrethroids used as insecticides. Medicina-Lithuania 54(4):61
67. Bordoni L et al (2019) Early impairment of epigenetic pattern in neurodegeneration: additional mechanisms behind pyrethroid toxicity. Exp Gerontol 124:110629
68. Burns CJ, Pastoor TPJCrit (2018) Pyrethroid epidemiology: a quality-based review. Crit Rev Toxicol 48(4):297–311
69. Ullah S, et al (2019) Biomarkers of pyrethroid toxicity in fish. *Environ Chem Lett*: 1–29
70. Sullivan KM et al (2019) Bioabsorption and effectiveness of long-lasting permethrin-treated uniforms over three months among North Carolina outdoor workers. Parasit Vectors 12(1):52
71. Strungaru S-A et al (2019) Toxicity and chronic effects of deltamethrin exposure on zebrafish (Danio rerio) as a reference model for freshwater fish community. Ecotoxicol Environ Saf 171:854–862
72. Velmurugan G et al (2017) Gut microbial degradation of organophosphate insecticides-induces glucose intolerance via gluconeogenesis. Genome Biol 18(1):8
73. Zhao Y et al (2016) Effects of chlorpyrifos on the gut microbiome and urine metabolome in mouse (Mus musculus). Chemosphere 153:287–293
74. Gao B et al (2017) Sex-specific effects of organophosphate diazinon on the gut microbiome and its metabolic functions. Environ Health Perspect 125(2):198–206
75. Liu Z, Fu Z, Jin YJC (2017) Immunotoxic effects of atrazine and its main metabolites at environmental relevant concentrations on larval zebrafish (Danio rerio). Chemosphere 166:212–220
76. Nasuti C et al (2016) Changes on fecal microbiota in rats exposed to permethrin during postnatal development. Environ Sci Pollut Res 23(11):10930–10937

77. Daisley BA et al (2017) Neonicotinoid-induced pathogen susceptibility is mitigated by lactobacillus plantarum immune stimulation in a Drosophila melanogaster model. Sci Rep 7 (1):1–13
78. Kittle RP et al (2018) Effects of glyphosate herbicide on the gastrointestinal microflora of Hawaiian green turtles (Chelonia mydas) Linnaeus. Mar Pollut Bull 127:170–174
79. Dai P et al (2018) The herbicide glyphosate negatively affects midgut bacterial communities and survival of honey bee during larvae reared in vitro. J Agric Food Chem 66(29):7786–7793
80. Kan H et al (2015) Correlations of gut microbial community shift with hepatic damage and growth inhibition of Carassius auratus induced by pentachlorophenol exposure. Environ Sci Technol 49(19):11894–11902
81. Jin Y et al (2016) The fungicide imazalil induces developmental abnormalities and alters locomotor activity during early developmental stages in zebrafish. Chemosphere 153:455–461
82. Jin C et al (2018) Insights into a possible influence on gut microbiota and intestinal barrier function during chronic exposure of mice to imazalil. Toxicol Sci 162(1):113–123
83. Wu S et al (2018) Exposure to the fungicide propamocarb causes gut microbiota dysbiosis and metabolic disorder in mice. Toxicol Sci 237:775–783
84. Xu C et al (2014) Changes in gut microbiota may be early signs of liver toxicity induced by epoxiconazole in rats. Chemotherapy 60(2):135–142
85. Daisley BA et al (2018) Microbiota-mediated modulation of organophosphate insecticide toxicity by species-dependent interactions with lactobacilli in a Drosophila melanogaster insect model. Appl Environ Microbiol 84(9):e02820–e02817
86. Harishankar M, Sasikala C, Ramya MJB (2013) Efficiency of the intestinal bacteria in the degradation of the toxic pesticide, chlorpyrifos. 3 Biotech 3(2):137–142
87. Claus SP et al (2016) The gut microbiota: a major player in the toxicity of environmental pollutants? Npj Biofilms Microbiomes 2(1):1–11
88. Williams MK et al (2008) Changes in pest infestation levels, self-reported pesticide use, and permethrin exposure during pregnancy after the 2000–2001 US Environmental Protection Agency restriction of organophosphates. Environ Health Perspect 116(12):1681–1688
89. Morgan MK (2012) Children's exposures to pyrethroid insecticides at home: a review of data collected in published exposure measurement studies conducted in the United States. Int J Environ Res Public Health 9(8):2964–2985
90. Barr DB et al (2010) Urinary concentrations of metabolites of pyrethroid insecticides in the general US population: National Health and nutrition examination survey 1999–2002. Environ Health Perspect 118(6):742–748
91. Seth RK et al (2018) Increased butyrate priming in the gut stalls microbiome associated-gastrointestinal inflammation and hepatic metabolic reprogramming in a mouse model of gulf war illness. Toxicol Appl Pharmacol 350:64–77
92. Liu C et al (2018) Insertion of 275-bp SINE into first intron of PDIA4 gene is associated with litter size in Xiang pigs. Anim Reprod Sci 195:16–23
93. Wu P et al (2016) The fungicide propamocarb increases lignin by activating the phenylpropanoid pathway in Cucumis sativus L. Hortic Environ Biotechnol 57(5):511–518
94. Zhu W et al (2016) Gut microbial metabolite TMAO enhances platelet hyperreactivity and thrombosis risk. Cell 165(1):111–124
95. Bennett BJ et al (2013) Trimethylamine-N-oxide, a metabolite associated with atherosclerosis, exhibits complex genetic and dietary regulation. Cell Metab 17(1):49–60
96. Masson P (2011) Evolution of and perspectives on therapeutic approaches to nerve agent poisoning. Toxicol Lett 206(1):5–13
97. Bouhafs L et al (2015) Protective effects of probiotic lactobacillus plantarum BJ0021 on liver and kidney oxidative stress and apoptosis induced by endosulfan in pregnant rats. Ren Fail 37 (8):1370–1378
98. Zhang Y et al (2010) The antioxidative effects of probiotic lactobacillus casei Zhang on the hyperlipidemic rats. Eur Food Res Technol 231(1):151–158

99. Anderson RC et al (2010) Lactobacillus plantarum MB452 enhances the function of the intestinal barrier by increasing the expression levels of genes involved in tight junction formation. BMC Microbiol 10(1):316
100. Trinder M et al (2016) Probiotic lactobacillus rhamnosus reduces organophosphate pesticide absorption and toxicity to Drosophila melanogaster. Appl Environ Microbiol 82 (20):6204–6213
101. Chiocchetti GM et al (2019) Use of lactic acid bacteria and yeasts to reduce exposure to chemical food contaminants and toxicity. Crit Rev Food Sci Nutr 59(10):1534–1545
102. Zhu Y et al (2019) Health risk from dietary exposure to polycyclic aromatic hydrocarbons (PAHs) in a typical high cancer incidence area in Southwest China. Sci Total Environ 649:731–738
103. Roslund MI et al (2018) Half-lives of PAHs and temporal microbiota changes in commonly used urban landscaping materials. Peer J 6:e4508
104. Aly Salem DMS et al (2014) The monitoring and risk assessment of aliphatic and aromatic hydrocarbons in sediments of the Red Sea, Egypt. Egyp J Aquat Res 40(4):333–348
105. Albanese S et al (2015) Polycyclic aromatic hydrocarbons in the soils of a densely populated region and associated human health risks: the Campania plain (southern Italy) case study. Environ Geochem Health 37(1):1–20
106. Bandowe BA et al (2014) Polycyclic aromatic compounds (PAHs and oxygenated PAHs) and trace metals in fish species from Ghana (West Africa): bioaccumulation and health risk assessment. Environ Int 65:135–146
107. Lawal AT, Fantke P (2017) Polycyclic aromatic hydrocarbons. A review. Cogent Environ Sci 3(1):1339841
108. Li GL et al (2014) Carcinogenic and mutagenic potencies for different PAHs sources in coastal sediments of Shandong peninsula. Mar Pollut Bull 84(1–2):418–423
109. Kong S et al (2015) Variation of polycyclic aromatic hydrocarbons in atmospheric PM2.5 during winter haze period around 2014 Chinese spring festival at Nanjing: insights of source changes, air mass direction and firework particle injection. Sci Total Environ 520:59–72
110. Kuppusamy S et al (2016) Biodegradation of polycyclic aromatic hydrocarbons (PAHs) by novel bacterial consortia tolerant to diverse physical settings—assessments in liquid- and slurry-phase systems. Int Biodeterior Biodegradation 108:149–157
111. Leung AOW, Cheung KC, Wong MH (2013) Spatial distribution of polycyclic aromatic hydrocarbons in soil, sediment, and combusted residue at an e-waste processing site in Southeast China. Environ Sci Pollut Res 22(12):8786–8801
112. Yu W et al (2015) Environmental risk assessments and spatial variations of polycyclic aromatic hydrocarbons in surface sediments in Yangtze River estuary, China. Mar Pollut Bull 100(1):507–515
113. Kamal A et al (2014) Cancer risk evaluation of brick kiln workers exposed to dust bound PAHs in Punjab province (Pakistan). Sci Total Environ 493:562–570
114. Li Y et al (2017) Presence, distribution and risk assessment of polycyclic aromatic hydrocarbons in rice-wheat continuous cropping soils close to five industrial parks of Suzhou, China. Chemosphere 184:753–761
115. Devi NL et al (2016) Environmental carcinogenic polycyclic aromatic hydrocarbons in soil from Himalayas, India: implications for spatial distribution, sources apportionment and risk assessment. Chemosphere 144:493–502
116. Beriro DJ et al (2016) A review of the current state of the art of physiologically-based tests for measuring human dermal in vitro bioavailability of polycyclic aromatic hydrocarbons (PAH) in soil. J Hazard Mater 305:240–259
117. Appenzeller BM et al (2012) Simultaneous determination of nicotine and PAH metabolites in human hair specimen: a potential methodology to assess tobacco smoke contribution in PAH exposure. Toxicol Lett 210(2):211–219

118. Deziel NC et al (2013) A multi-day environmental study of polycyclic aromatic hydrocarbon exposure in a high-risk region for esophageal cancer in China. J Expo Sci Environ Epidemiol 23(1):52–59
119. Liu S et al (2008) Seasonal and spatial occurrence and distribution of atmospheric polycyclic aromatic hydrocarbons (PAHs) in rural and urban areas of the north Chinese plain. Environ Pollut 156(3):651–656
120. Adeniji AO, Okoh OO, Okoh AI (2019) Levels of polycyclic aromatic hydrocarbons in the water and sediment of Buffalo River estuary, South Africa and their health risk assessment. Arch Environ Contam Toxicol 76(4):657–669
121. Bandowe BAM, Meusel H (2017) Nitrated polycyclic aromatic hydrocarbons (nitro-PAHs) in the environment—a review. Sci Total Environ 581:237–257
122. Idowu O et al (2019) Beyond the obvious: environmental health implications of polar polycyclic aromatic hydrocarbons. Environ Int 123:543–557
123. Andersson H et al (2009) Low levels of the air pollutant 1-nitropyrene induce DNA damage, increased levels of reactive oxygen species and endoplasmic reticulum stress in human endothelial cells. Toxicology 262(1):57–64
124. Van de Wiele T et al (2005) Human colon microbiota transform polycyclic aromatic hydrocarbons to estrogenic metabolites. Environ Health Perspect 113(1):6–10
125. Madeen E et al (2019) Toxicokinetics of benzo [a] pyrene in humans: extensive metabolism as determined by UPLC-accelerator mass spectrometry following oral micro-dosing. Toxicol Appl Pharmacol 364:97–105
126. Zelinkova Z, Wenzl T (2015) The occurrence of 16 EPA PAHs in food—a review. Polycycl Aromat Compd 35(2–4):248–284
127. Shimada T (2006) Xenobiotic-metabolizing enzymes involved in activation and detoxification of carcinogenic polycyclic aromatic hydrocarbons. Drug Metab Pharmacokinet 21(4):257–276
128. Mutlu EA et al (2018) Inhalational exposure to particulate matter air pollution alters the composition of the gut microbiome. Environ Pollut 240:817–830
129. Yousefi M et al (2019) In vitro removal of polycyclic aromatic hydrocarbons by lactic acid bacteria. J Appl Microbiol 126(3):954–964
130. Ovrevik J et al (2013) Differential chemokine induction by 1-nitropyrene and 1-aminopyrene in bronchial epithelial cells: importance of the TACE/TGF-alpha/EGFR-pathway. Environ Toxicol Pharmacol 35(2):235–239
131. Bryan NS (2006) Nitrite in nitric oxide biology: cause or consequence? A systems-based review. Free Radic Biol Med 41(5):691–701
132. Cocco P et al (2003) Nitrate in community water supplies and incidence of non-Hodgkin's lymphoma in Sardinia, Italy. J Epidemiol Commun Health 57(7):510–511
133. Ma L et al (2018) Nitrate and nitrite in health and disease. Aging Dis 9(5):938–945
134. Paul RJR, Dich J, Hakulinen T (1999) Risk of colorectal and other gastro-intestinal cancers after exposure to nitrate, nitrite and n-nitroso compounds: a follow-up study. Int J Cancer 80:852–856, 2000
135. Gulis G, Czompolyova M, Cerhan JR (2002) An ecologic study of nitrate in municipal drinking water and cancer incidence in Trnava District, Slovakia. Environ Res 88(3):182–187
136. Volkmer BG et al (2005) Influence of nitrate levels in drinking water on urological malignancies: a community-based cohort study. BJU Int 95(7):972–976
137. Suthar S et al (2009) Nitrate contamination in groundwater of some rural areas of Rajasthan, India. J Hazard Mater 171(1–3):189–199
138. Kilfoy BA et al (2011) Dietary nitrate and nitrite and the risk of thyroid cancer in the NIH-AARP diet and health study. Int J Cancer 129(1):160–172
139. Parvizishad M et al (2017) A review of adverse effects and benefits of nitrate and nitrite in drinking water and food on human health. Health Scope 6(3):e14164
140. Sindelar JJ, Milkowski AL (2012) Human safety controversies surrounding nitrate and nitrite in the diet. Nitric Oxide 26(4):259–266

141. Hirayama K et al (2000) Effects of human intestinal flora on mutagenicity of and DNA adduct formation from food and environmental mutagens. Carcinogenesis 21(11):2105–2111
142. Möller L et al (1988) The role of the intestinal microflora in the formation of mutagenic metabolites from the carcinogenic air pollutant 2-nitrofluorene. Carcinogenesis 9(5):823–830
143. Gassara F et al (2016) Green alternatives to nitrates and nitrites in meat-based products–a review. Crit Rev Food Sci Nutr 56(13):2133–2148
144. Butler A (2015) Nitrites and nitrates in the human diet: carcinogens or beneficial hypotensive agents? J Ethnopharmacol 167:105–107
145. Sobsey M, Bartram S (2003) Water quality and health in the new millennium: the role of the World Health Organization guidelines for drinking-water quality. Forum Nutr 56:396–405
146. Sindelar JJ, Milkowski AL (2012) Human safety controversies surrounding nitrate and nitrite in the diet. Nitric Oxide 26(4):259–266
147. Archer DL (2002) Evidence that ingested nitrate and nitrite are beneficial to health. J Food Prot 65(5):872–875
148. Hunault CC et al (2009) Bioavailability of sodium nitrite from an aqueous solution in healthy adults. Toxicol Lett 190(1):48–53
149. Lauer T et al (2001) Plasma nitrite rather than nitrate reflects regional endothelial nitric oxide synthase activity but lacks intrinsic vasodilator action. Proc Natl Acad Sci U S A 98 (22):12814–12819
150. Janero DR et al (2004) Differential nitros(yl)ation of blood and tissue constituents during glyceryl trinitrate biotransformation in vivo. Proc Natl Acad Sci U S A 101(48):16958–16963
151. Kim-Shapiro DB et al (2005) The reaction between nitrite and hemoglobin: the role of nitrite in hemoglobin-mediated hypoxic vasodilation. J Inorg Biochem 99(1):237–246
152. Lundberg JO et al (2011) Roles of dietary inorganic nitrate in cardiovascular health and disease. Cardiovasc Res 89(3):525–532
153. Fritsch P et al (1985) Excretion of nitrates and nitrites in saliva and bile in the dog. Cardiovasc Res 23(7):655–659
154. Spiegelhalder B et al (1976) Influence of dietary nitrate on nitrite content of human saliva: possible relevance to in vivo formation of N-nitroso compounds. Food Cosmet Toxicol 14 (6):545–548
155. Lundberg JO (2012) Nitrate transport in salivary glands with implications for NO homeostasis. Proc Natl Acad Sci 109(33):13144–13145
156. Qu X et al (2016) From nitrate to nitric oxide: the role of salivary glands and oral bacteria. J Dent Res 95(13):1452–1456
157. Bryan NS, Ivy JL (2015) Inorganic nitrite and nitrate: evidence to support consideration as dietary nutrients. Nutr Res 35(8):643–654

Chapter 6
Environmental Pollutants that Can Be Metabolized by the Host, but Would Be Harmful to Humans (e.g., Causing Cancers, etc.)

Marwa M. El-Dalatony, Mostafa El-Sheekh, and Xiangkai Li

6.1 Introduction

Environmental pollutants are gradually increased and the term xenobiotics are commonly used in context of environmental pollution because they are synthetic compounds produced from industries and agriculture [1]. Human body has number of microorganisms commonly called as human microbiota [2, 3]. The diversity and functioning of this community depend upon body size, shape, and different environmental conditions (e.g., pH, oxygen, substrate availability, humidity, and temperature) at different sites [3]. Site-specific microbiome which associate with skin, respiratory tract, and gut are the first to encounter xenobiotics and mediate a pass to internal organ system [4]. Besides, most interaction between human microbiota and xenobiotics occurs in human gut [4, 5]. The anaerobic environment of the gut is well-suited for a hydrolytic and reductive metabolism. And this will generate low molecular weight non-polar products that can easily absorbed by host cells. In comparison, the absorbed non-polar xenobiotics are metabolized and transported in liver by a rich collection of conjugative enzymes and these hepatic metabolisms may generate high molecular weight polar metabolites. The latter reach to the gut, secreted via bile and in gut they can be re-metabolized by hydrolytic and reductive enzymes [5, 6]. Hence, xenobiotics are metabolized by gut microbiota and can exert an intense influence on the bioavailability and toxicity of xenobiotics entering in gut from different routes.

M. M. El-Dalatony (✉) · X. Li
School of Life Science, Lanzhou University, Lanzhou, Gansu, People's Republic of China
e-mail: marwa@hanyang.ac.kr

M. El-Sheekh
Faculty of Science, Botany Department, Tanta University, Tanta, Egypt

X. Li, P. Liu (eds.), *Gut Remediation of Environmental Pollutants*,
https://doi.org/10.1007/978-981-15-4759-1_6

6.1.1 Probiotics and Gut Microbiota

Food and Agricultural Organization (FAO) of the United Nations and the World Health Organization (WHO) states that probiotics are supplements of feed and have so many benefits for human and affect the host by improving the microbial balance with immune system. Nobel laureate Elie Metchnikoff in 1907 introduced the concept of probiotics to the world of science. In his studies he reported that the longevity and viability of *Bulgarians* and *lactobacilli* with consumption of fermented milk products, which can be used as probiotics [7]. This study suggested that some microorganisms are beneficial for human health. From that onwards, probiotics had been widely consumed and marketed as functional food, Mechanisms of proboscis include stimulation of epithelial cells, immunomodulation, include manipulation of intestinal microbial communities, fortification of intestinal barriers, and differentiation [8]. Mostly probiotics are developed these days made from *Bifidobacteria*, *Lactobacilli*, and lactic acid bacteria, like *streptococci* and Lactococci. Other probiotic strains include microbial strains like *Bacillus*, *Escherichia*, and Propionibacterium and some yeast genera, mainly *Saccharomyces* [9].

From birth to adulthood there are many factors that may influence the gut microbiota which include diet during infancy that is the presence of antibiotics in food, exposure of antibiotics, from environmental conditions and mode of delivery [10]. The gut microbiota plays an essential role in shaping the intestinal mucus layer [11], which helps us to digest fibers and synthesize amino-acids and vitamins [12]. Such benifits help in immune system modulation, energy metabolism and storage, neurodevelopment and even regulate growth & behavior [13]. There are many diseases associated with the alteration of gut microbiota [14]. Gut microbiota dysbiosis is the major cause of obesity [15]. Although, gut microbiota is very sensitive toward the diet, drugs and environmental pollutants.

6.1.2 Classification of Probiotics

Most of the microorganisms can be used as probiotics [16]. Genus name (for example, *Lactobacillus*) is the first name given to the bacterial strains based on physical characteristics, metabolic needs, similarity of qualities and metabolic end products. Species is the second name of bacteria like *acidophilus*, based on the common characteristics and that will distinguish them from other species. Strain is the much more specific classification of bacterium which divide members of same species into subgroups and it is based on the properties that these bacteria have in common and distinct it from other species (e.g., strain LA5) [16, 17] (Table 6.1).

Table 6.1 Commonly used probiotic bacteria [16, 17].

Lactobacillus spp.	*Bifidobacterium* spp.	Others
L. casei (rhamnosus)	*B. longum*	*Escherichia coli*
L. bulgaricus	*B. breve*	*Saccharomyces cerevisiae*
L. plantarum	*B. infantis*	*Enterococcus faecalis*
L. reuteri	*B. bifidum*	*Bacillus cereus*
L. acidophilus	*B. adolescentis*	*Streptococcus thermophilus*

6.1.2.1 *Lactobacillus*

It involves various Gram-positive facultative anoxic or microaerophilic bacteria. These are the essential part of the lactic acid bacteria group (including *Enterococcus*, *Pediococcus*, *Lactobacillus*, *Lactococcus*, *Gonococcus*, *Streptococcus,* and *Leuconostoc* species) that can convert hexose sugars to lactic acid and produce an acid in the environment which can inhibit the growth of harmful species [18]. In humans, *Lactobacilli* are present in the GIT and vagina with *Bifidobacterium* which is one of the first bacteria colonized the infant gut after delivery [19].

6.1.2.2 *Bifidobacterium*

Bifidobacterium includes Gram-positive non-motile anoxic bacteria. They are endosymbiotic inhabitants of the vagina and gastrointestinal tract of humans [20]. Strains of the genus *Bifidobacterium* are also used as probiotics because they have resistance mechanism to bile salt and many beneficial effects on other probiotic bacteria, which are generated in the presence of biological fluid [21].

6.1.2.3 *Saccharomyces*

Saccharomyces contains several yeasts including: *Saccharomyces cerevisiae* used for making bread plus beer, *Saccharomyces bayanus* which is used for making wine, and *Saccharomyces boulardii* used in medicine as a probiotic [22].

6.1.2.4 *Bacillus*

Bacillus sp. are Gram positive, aerobes or facultative aerobes capable of spore formation. Various species of *Bacillus* have been reported to have potential such as *B. subtilis*, *B. cereus*, and *B. coagulans* [23]. The use of *B. coagulans* as a therapeutic like other probiotics strains such as *lactobacillus* and *Bifidobacterium* sp. has been reported, whereas presence of *B. coagulans* in the composition of normal gut microbes has not been reported [24].

6.1.2.5 *Escherichia*

Escherichia sp. comprises of Gram-negative bacteria belonging to *Enterobacteriaceae* family, mostly reported with virulent serotypes (*E. coli* O157: H7). *Escherichia coli* is commonly found in lower intestine as a normal microbe of gut microflora with a known probiotic strain: *Escherichia coli* Nissle 1917 (EcN). A study revealed the effect of *Escherichia coli* Nissle 1917 amalgamated with other probiotics strains on the treatment of constipation [25]. The effects of this strain on gastrointestinal disorder, Crohn's disease [26], ulcerative colitis, IBD, and colon cancer have been studied [27].

6.1.2.6 *Streptococcus* and *Enterococcus*

Streptococcus and *Enterococcus* genera belong to the category of lactic acid producing bacteria and are reported to have various species that can cause heath implications such as *Streptococcus pneumoniae*, *Streptococcus pyogenes*, and vancomycin-resistant *Enterococcus faecium* [28]. Some species of *Enterococcus* like *Enterococcus faecium* PC4.1 show commensal relationship with skin, mouth, and intestine [29]. The potential probiotic strains are *Streptococcus thermophilus*, *Enterococcus durans*, and *Lactobacillus delbrueckii* subsp. *bulgaricus* [30, 31]. The use of *Enterococcus faecium* as probiotics has a long history, and proved its effectivness against antibiotic-associated diarrhea [32], the opportunistic strains of the genus serve as a reservoir of virulence and antibiotic resistance in animal study models (animal study). The use of opportunistic strains of these genera is not categorized under (GRAS) for humans consumption, but can be used as probiotics for animals [33, 34].

6.1.2.7 *Lactococcus*

Lactococcus genus consists of Gram-positive, lactic acid producing bacteria used to produce fermented products in the dairy industry. The acidification property of these bacteria is helpful in preventing the spoilage of milk by inhibiting the growth of spoilage microorganisms. The other properties of some species like *Lactococcus lactis* subsp. *lactis* as a probiotic of niacin production and adhesion to vaginal epithelial cells have been studied. A study on the use of *Lactococcus lactis* subsp. *lactis* CV56 in combination with other probiotics to treat antibiotic-associated diarrhea has been given [35–37].

6.2 Function Mechanism of Probiotics

6.2.1 Gut Barrier Function

The gut barrier defense system consists of the secretory IgA, antimicrobial peptides, mucous layer, and the epithelial junctional adhesion complex [38]. The location of epithelial cells in the center stage of the barrier effect has been reported, these cells receive molecular signals from the lumen of gut and exchange them with the underlying cells of immune system. These cells can communicate with the whole organism by the circulation of signaling molecules. Gut barrier defense plays an eminent function in the pathogenesis of various diseases associated with the GI tract like irritable bowel syndrome (IBS), inflammatory bowel disease (IBD), infectious enterocolitis plus coeliac disease [39].

Studies conducted on the use of *L. rhamnosus* GG (LGG) and probiotic mix VSL#3 on mice and Caco-2 intestinal cells have shown the influence of the strain on epithelial cells of intestine to maintain the coherence of the epithelial barrier. The persistence of LGG in the GI tract was connected with its in vivo expression of pili containing a mucus-binding domain [40]. An in vitro study on LGG and its soluble factors (p75 and p40) has revealed the prevention of apoptosis in epithelial cells by activating anti-apoptotic Akt and suppressing NF-kB. In addition, an increase in the secretion of mucin by epithelial cells was observed [41].

The effect of *L. plantarum*, *L. casei*, *L. rhamnosus,* and *L. acidophilus*, on the stimulation of distinct pathways of gene-regulatory networks in the human mucosa has been reported. These regulations involve upregulation of an activator of NF-kB signaling cascade known as IL-1b, involved in the transcription of genes responsible for the maturation of B-cell and lymphogenesis, thus supporting the barrier function [42].

The effect of *Lactobacillus*, *Bifidobacterium*, and *Streptococcus* as probiotics on post-infectious irritable bowel syndrome (PI-IBS) caused by *Trichinella spiralis* showed positive results in a mouse model. *Bifidobacterium* or *Lactobacillus* treatment on PI-IBS mice showed reduction in the abdominal contractile response and withdrawal reflex score, D-lactate level, and reduced plasma diamine oxidase (DAO) concentration. The suppression of proinflammatory cytokine IL-17 and IL-6 has been reported after probiotic administration and enhancement in the expression of occludin and claudin proteins of tight junction of cells [43].

6.2.2 Production of Inhibitory Compounds by Probiotics

The antibacterial property of probiotics against Gram-negative and Gram-positive bacterial pathogens involves the production of various antibacterial substances. These substances include production of organic acids, bacteriocins, diacetyl, ethanol, hydrogen peroxide, and carbon dioxide [44, 45]. The mechanisms of action of bacteriocins to inhibit the growth of pathogens include the pore formation in the cell walls of targeted cells and inhibition of synthesis of cell wall. Nisin an antimicrobial compound associated with the formation of a complexes with the precursors of cell wall and lipid II, to inhibit the synthesis of cell walls, and also prevent pore formation in the membranes by removing complex aggregates and incorporates peptides. Bacteriocin production potential offers various advantages to the strains in complex microbial environments as they have antimicrobial properties and can inhibit the pathogens of GI tract [46, 47].

Lactobacillus acidophilus can produce various antimicrobial compounds such as acidolin. acidophillin, and lactocidin and *Lactobacillus planatarum* can produce another antimicrobial compound "lactolin" [48]. The effect of bacteriocin producing *Lactobacillus salivarius* UCC118 strain on *Listeria monocytogenes* infected mice have shown protective results. The effect of bacteriocin Abp118 on stimulating antimicrobial response was confirmed by this study, where *Lb. salivarius* showed antagonistic relationship with the pathogen [49]. The inhibition of *Helicobacter pylori*, *E. coli*, *Listeria monocytogenes*, Rotavirus, and *Salmonella* by *Lactobacilli* and *bifidobacteria* have been reported [50].

Several strains of *Bifidobacterium* (*B. bifidum* NCFB 1454) have shown the production of a unique bacteriocin (bifidocin B), effective against Gram-positive bacteria. A high inhibition rate of *E. coli* C1845 and *Salmonella enterica ser. Typhimurium* SL1344 by two *Bifidobacterium* strains has been studied [50]. Inhibition of *Yersinia enterocolitica* an entero pathogen by twenty strains of *Lactobacillus* has been reported in addition with the inhibition of *Listeria monocytogenes* by *Lactobacillus plantarum* C4 and *Salmonella enterica serovar Typhimurium* by *Lactobacillus casei*. The main mechanism of inhibition involves the elevation of pH mainly from dextrose fermentation by *Lactobacillus* [51] (Table 6.2).

6.2.3 Adhesion Mechanism of Probiotics

Attachment to intestinal mucosa, an important characteristic for probiotics, is required for its colonization in intestine along with antagonism towards pathogens and variation of immune system. Various Lactobacillus proteins accompanied by saccharide moieties and lipoteichoic acids can improve the adhesion to mucous and bacterial surface adhesions that facilitate adhesion to the mucous layer [50, 65]. Bacterial adhesins, mucus-binding protein (MUB), from *Lactobacillus reuteri* are reported [66]. Probiotics, such as *L. plantarum*, can prevent the attachment of enteropathogenic *E. coli* by induction of MUC2 and MUC3 mucins. Therefore, protection against pathogens is provided by glycocalyx overlying and increased mucous layers. Moreover, due to the attachment of probiotic organisms gut epithelial surfaces, the adhesion sites are blocked for pathogen colonization [67]. Upon the ingestion of *lactobacilli*, it competes for the binding sites due to which few sites are available for pathogenic bacteria. Attachment is facilitated by Mannose specific adhesion proteins, that also attaches to cell surface and are important for pathogens

Table 6.2 Example of different inhibitory compounds produced by probiotic strains [51].

Compound	Example	Strain	Spectrum	References
Bacteriocins	Pediocin PA-1	*Ped. acidilactici*	Broad spectrum: Gram-positive bacteria	[52]
	Nisin	*Lc. lactis subsp. lactis*	Broad spectrum: Gram-positive bacteria without nisinase	[53]
	Enterocin AS48	*Ent. Faecalis*	Gram-positive bacteria, *Salmonella enterica, Bacillus subtilis, E. coli, B. cereus, B. circulans, Enterococcus faecalis, C. bovis, Micrococcus lysodeikticus, S. aureus, Ent. faecium, Enterobacter cloacae, Klebsiella pneumoniae, Salmonella typhimurium, Pseudomonas fluorescens, P. aeruginosa, Corynebacterium glutamicum, Nocardia corallina, Mycobacterium phlei, Micrococcus luteus, Proteus incontans, shigella sonnei.*	[54, 55]
	Enterolysin A	*Ent. Faecalis*	*Lb. sakei, Lb. brevis, Lb. curvatus, Lc. cremoris, Lb. lactis, Ped. pentosaceus, Ped. acidilactici, Ent. faecium, Ent. faecalis, L. innocua, L. ivanovii, Bacillus subtilis, B. cereus, S. carnosus, Propionibacterium jensenii*	[56]
Bacteriocin-like inhibitory substance (BLIS)		*Lc. lactis* subsp. *lactis* CECT-4434	*Staphylococcus aureus*	[57]
		Ped. acidilactici Kp10	*L. monocytogenes*	[58]
		Leuc. mesenteroides 406	*L. monocytogenes*	[59]
Antibiotic	Reuterin	*Lb. reuteri* DSM 20016	*Gram-positive (Clostridium and Staphylococcus) and Gram-negative (Escherichia, Salmonella, Shigella) bacteria, against the yeast, Saccharomyces cerevisiae, and against the*	[60]

(continued)

Table 6.2 (continued)

Compound	Example	Strain	Spectrum	References
			protozoan, Trypanosoma cruzi	
	Reutericyclin	*Lb. reuteri*	*Gram-positive bacteria (Lactobacillus, Bacillus, Enterococcus, Staphylococcus, and Listeria)*	[61, 62]
Organic acids	Lactic acid, Acetic acid	LAB	Broad spectrum: Bacteria affected by pH	[63]
Hydrogen peroxide		*Ped. acidilacti, Leuc. mesenteroides, Lb. brevis, Lb. plantarum, Lb. casei*	Broad spectrum: Catalase negative bacteria	[63]
Others	Ethanol	*Bifidobacterium longum*	Broad spectrum: Bacteria affected by membrane	[64]
		Ent. Faecalis, Lb. acidophilus, Lb. fermentum, Lb. plantarum, Weissella confuse	Dissociations	
	Diacetyl	*Lb. plantarum, Lb. helveticus, Lb. bulgaricus, Ent. Faecalis, Leuc. mesenteroides*	*E. coli, Listeria, Yersinia, Salmonella, Aeromanas*	[62]
	CO_2	*Heterofermentative* LAB	Broad spectrum: Aerobic bacteria	[62]

binding in gut, facilitates the attachment of *L. plantarum* Lp6 onto rat mucus preventing pathogen colonization [68]. Acid resistant strains from *Bifidobacterium longum* and *B. catenulatum* are reported to have effective attachment properties to human intestinal mucus in comparison to acid-sensitive [69]. In *Bifidobacteria*, acid resistance improves functionality through enhancing stability plus improving surface properties.

Combination of probiotics with VSL#3 improves the mucins synthesis and facilitate expression of mucin gene, therefore, enhancing the bacterial attachment to the epithelium of intestine [70]. Keratinocyte cell death, due to *Staphylococcus aureus*, in undifferentiated and differentiated keratinocytes is reduced by potential probiotics, *Lactobacillus reuteri* ATCC 55730 and *Lactobacillus rhamnosus* AC413. Probiotic efficiency was higher for Keratinocyte survival when they were applied before or simultaneously with *S. aureus* infection. *S. aureus* needs $\alpha5\beta1$ integrin for attachment to keratinocytes, protective effect like probiotic was observed by blocking of $\alpha5\beta1$ integrin. The competition for the binding site between pathogens and *L. reuteri* might be the protection mechanism for keratinocytes. Therefore, inhibition of *S. aureus* colonization and infection prevention can be achieved by application of topical probiotic prophylactically [71].

6.3 Probiotics and Nutrients Competition

One of the mechanisms for inhibiting pathogens form colonization in human gut might be the nutrient competition. There are two different ways for such competition; firstly, preventing the nutrient and energy source uptake by pathogen which is required for growth and proliferation in human gut. Secondly, production of metabolites like short chain fatty acids (SCFAs) and organic acids through fermentation and metabolism which lowers the gut pH making it unfavorable for most of the pathogens, e.g. *E. coli* and *Salmonella* [50]. *Bifidobacterium adolescentis* S2-1 prevents the growth of *Porphyromonas gingivalis* by outcompeting it for vitamin K and other growth factors [72]. After the exposure to probiotic (*Lactobacillus paracasei* or *Lactobacillus rhamnosus*), changes in pathways such as short chain fatty acids (SCFA), amino acid, and methylamines metabolism were observed in mice (germ free) colonized with microbiota of human baby [73].

Probiotics, for example, *L. delbrueckii* and *L. acidophilus*, prevent the availability of ferric hydroxide to pathogens by binding them to its cell surface [74]. Probiotic strains and exert inhibitory effects on Biofilm formation of pathogenic *Listeria monocytogenes* and *Salmonella typhimurium* are inhibited by *L. rhamnosus* and *L. paracasei* probiotic through different mechanisms including competition, displacement, and exclusion. A decrease of more than three log cycles biofilm cells was observed for *L. monocytogenes* [75].

6.4 Probiotics and Immune System

Immune system is affected by various reported pathways due to potential application of probiotics [76, 77]. Stimulating specific and nonspecific immunity is one of the possible mechanisms through which probiotics helps to prevent the intestinal disease in host. LAB products have immunomodulatory action through Toll like Receptors (TLRs) expression regulation, inflammatory responses inhibition, Dendritic cells (DCs) activation, and Natural Killer (NK) cells, among innate immunity; lymphocytes propagation, balancing the response of T-helper (Th1/Th2) cells, specific IgA secretion, in further ways [78]. *Bacillus subtilis* B10 and *Saccharomyces boulardii* targets specific TLRs and associated factors, hence, having a major role in controlling immunological functions of chicken bone marrow DCs. Probiotics get attached to surface of DCs. Upregulation in expression level of MHC-II, CD40, CD80, and CD86 genes was observed. Additionally, the expression of TLR1, TLR2, TLR4, and TLR15 (chicken specific) was enhanced and increased in levels of downstream related factors TRAF6, MyD88, NFκ- B mRNA, and TAB1was observed [79].

Accumulation and growth of healthy microorganisms in gut result in maturation of the several immune mechanisms, especially, for the IgA and IgM secreting cells circulation. After preparing, Memory B besides T cells move towards the effector sites, actively proliferate, then local stimulation of various cytokines and secretory IgA generation. Probiotic stimulates the IgA production upon entering the gut. Studies in mice (kept germ free) evidenced the IgA production in immune system [80]. Several studies suggested that improvement of innate and adaptive immunity along with alleviate allergies, prevention of gastric mucosal lesion development, and put up defense against intestinal pathogen infection was observed due to lactic acid bacteria (LAB) such as *Bifidobacterium* and *Lactobacillus* and also due to their fermented products [78].

Feeding to 1.4 years old rats resulted in enhanced immunosenescence associated Th1/Th2 imbalance, higher resistance to *E. coli* infection of aged mice, and increased antioxidant capacity were observed as a result of feeding *Lactobacillus rhamnosus* to mice (16 months old). Increase in levels of IFN-γ and decrease in levels of IL-4 and IL-10 production, increase in phagocytosis and neutrophil respiratory burst enzymes with no aggravation in plasma levels of MCP-1 and TNF-α was observed in the mice feed with probiotic. IgE levels and IgG1/IgG2a ratio decreased along with increase in activities of antioxidant enzymes were found in the probiotic fed mice, *E. coli* translocation to the organs of the mice were also reduced significantly [81].

6.4.1 Degradation of Toxins Receptors through Probiotics

Enzymatic modification of toxin receptor is done by probiotics; host is protected from intestinal disease of *Clostridium difficile* due to modification in toxin receptor in intestinal mucosa by *Saccharomyces boulardii*. Various other reported mechanisms are decreasing toxin production, lowering gut pH and decrease of virulence [50]. Probiotics could change receptors for toxins as well as prevent against pathology caused by toxins. *Saccharomyces boulardii* have the ability to degrade toxin receptors for *Clostridium difficile* in ileum of rabbit and by polyamines production, it can prevent cholera-prompted secretion in jejunum of rat. Impact of a multi-strain probiotic plus synbiotic formulation (*Lactobacillus paracasei F8*, L. *plantarum* F44, *Bifidobacterium lactis* 8:8, *B. breve* 46, resistant starch, isomaltooligosaccharides, and galacto-oligosaccharides) was studied in *Clostridium difficile NAP1/027 infected* C57BL/6 mice. Upon the formulation feeding, *lactobacilli* and *bifidobacteria* counts increased without detecting any caecal toxins. *C. difficile* DNA copies were found in significantly decreased after the qPCR of caecal [82].

6.4.2 Probiotics Roles in Anti-Proliferative

Due to the reduction in putrefactive bacteria including *Bacteroides, Clostridium*, and *coliforms* species and increase in *lactobacilli* and *bifidobacteria* that facilitate in reducing risk for colorectal cancer, probiotics are supposed to have anti-cancer activity. Probiotic, *Lactobacillus salivarius* ssp. *Salivarius*, reduced prevalence of adenocarcinoma in colon of IL-10 knockout rats [83]. Probiotic, *Streptococcus thermophilus* strain TH-4 have an anti-inflammatory activity along with the ability of high folate production which is important in epithelial cells for DNA repair [84, 85].

6.5 Gut Microbiota Modulation

Human gut microbes always have been immersed in the regulation of various biological functions, varying from cognitive processes and energy regulation to improving host immunity against harmful microorganisms and also neutralization of toxins. The potential application of probiotics and prebiotics always involves in the maintaining of host ideal gut health, treating/preventing host recurring inflammatory, and immune system linked diseases [86]. Probiotics have a wide range of application in prevention and treatment of several diseases which are induced or associated with the dysbiosis of gut microbiota such as acute infectious diarrhea and antibiotic-associated diarrhea, and also other GI tract diseases like colic's or irritable bowel syndrome. At the time of treatment the gut microbial community makeup stays more steady and that it positively relates with recovery of disease symptoms [87].

6.6 Probiotics and Health

Probiotics enhance the nutritive and microbial balance of host gastrointestinal tract. Probiotics work as a carrier that transport their beneficial functional components to different target locations in the gastrointestinal tract. Ingestion of live probiotic strains has more effective results which varies from strain to strain [88]. Whereas, it is not always essential to accomplish profits [89].

6.6.1 Probiotics Role in the Treatment of Gastrointestinal Disorders

6.6.1.1 Antibiotic-Associated Diarrhea (AAD)

A systemic review study on treating of antibiotic-associated diarrhea (AAD) by usage of probiotics in aged patients (more than 65 years) and in adults (18 to 64 years) evaluated 30 random managed tests that fit in the previously developed inclusion measures. The clinical studies proposed that probiotic act as an adjuvant for antibodies which lower down the chances of antibiotic-associated diarrhea (AAD) in adults, but not in aged persons [90]. PROSPERO study proved that a number of probiotic strains such as *S.boulardii* and *lactobacillus rhamnosus* GG have involved in the prevention of antibiotic-associated diarrhea but other strains such as Lactobacillus bulgaricus, L. delbrueckii, and S.salivarius are not capable of preventing ADD [91–93].

6.6.1.2 Irritable Bowel Syndrome (IBS)

Several physiological, epidemiological, and clinical studied data have indicated that gut microbiota involves in the pathogenesis of irritable bowel syndrome, however, IBS pathophysiology still undiscovered [94, 95].

A functional study showed that altering the host gut microbes in conjugation with probiotics can influence some host intestinal functions, like sensitivity and motility, which seems to be related to the irritable bowel syndrome pathogenesis I [96]. A clinical experiment showed that the group of patients (35,624) that have intake of *B. infantis* significantly improved their disease symptoms in comparison to placebo. Moreover, the serum IL-10/IL12 ratio normalized, indicating that probiotic can helps in remission of proinflammatory state associated with irritable bowel syndrome [97, 98]. In addition, *L. plantarum* is better than placebo in remission of few symptoms in IBS patients. Specifically, the DSM 9843 strain radically decreased flatulence, and the 299 V and LPO1 strains appreciably lowered the intestinal pain [99–101].

6.6.1.3 Ulcerative Colitis

A clinical experiment showed that the mesalamine treatment with strain *Lactobacillus GG* might be more efficient than standard treatment for preventing the relapsing time of disease [102]. *E. coli* strain Nissle 1917 showed similar effective results as of 5-aminosalicyclates in averting the relapsing of ulcerative colitis in adults [103].

6.6.1.4 Crohn's Disease

Clinical experiments performed with *E-coli* strain *Nissle* 1917 and with distinct strains of *Lactobacillus* had not shown any higher effect than placebo in averting the occurrence of Crohn's disease [104, 105]. A studied proved that daily intake of 3 g mesalamine alone was less effective than 2 g daily intake of mesalamine along with *S. boulardii* in lowering the relapsing of Crohn's disease in patients. But later on a clinical study did not verify these results [106, 107].

6.6.1.5 Pouchitis

Pouchitis is an inflammatory condition of the ileal reservoir in patients with acute and chronic refractory ulcerative colitis experienced restorative proctocolectomy with ileal pouchanal anastomosis (IPAA) [108]. Several clinical trials with probiotics have been conducted that have shown their safety and effectiveness in sustaining the reduction of pouch inflammation, also antibiotic treatment attained subsequent, like 5-aminosalicyclic acid also helps in relapsing of chronic pouchitis and prevention of acute pouchitis [109, 110]. A systematic review from the Cochrane Collaboration showed that VSL#3 was very efficient in sustaining the reduction of chronic pouchitis and also in averting the onset of pouchitis than placebo [111].

6.6.2 Probiotics for Depression and Anxiety

Depression and anxiety are two most common human mental health conditions, with lifetime prevalence rates worldwide. Gut and brain interact with each other through a particular pathway called gut-brain axis pathway that includes immune, endocrine, and neural systems. Administration of probiotic mixture containing *Bifidobacterium longum* BL04, *L. plantrum* LP, *Lactobacillus fermentum* LF16, and *L. rhamnosus* LR06 was given to examine the effect of probiotics on depression and anxiety was reported. The study did not provide any positive effect on sleep quality and depressive mood state [112]. Thus more significant clinical trials are needed to explore the effect of probiotics on depression and anxiety.

6.6.3 Human Gut Microbial Community

Human gut microbiota is the microorganisms that live in the human gut. It is complex community of microbes—estimated to contain 200 trillion cells and containing greater than 1000 diverse microbial species Fig. 6.1. Human gut

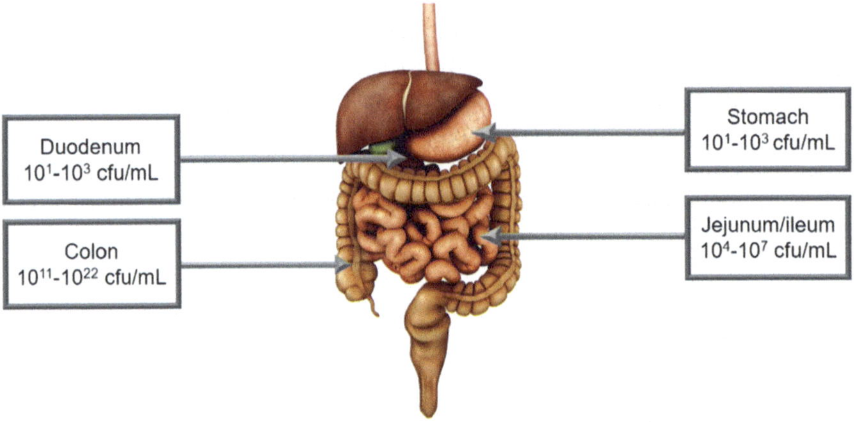

Fig. 6.1 Microbial density in the gut [114]

microbiota is composed of a wide range of bacteria, fungi, archaea, and viruses [113]. Gut microbiota—biome of microorganisms that live in the digestive tract of human beings whether on the intestinal mucosal surface or within the gut lumen.

Individual has their own stable fecal microbiota for lifetime and harbors different characteristic pattern of gut microbial flora. Around 90% of human gut microbiota are made up of *Bacteroidetes* and *Firmicutes*.

6.6.3.1 Function of Gut Microbiota

Intact microbiome is essential for the development of the GIT in many ways including—immune tolerance, the mucosa associated immune system, motility and vascularity, epithelial and barrier function. The microbiota which exhibiting commensalism in host provide homeostatic functions like immunomodulation, pathogen exclusion, upregulation of cytoprotective genes, regulation prevention of apoptosis, and maintenance of barrier function.

6.6.3.2 Metabolic Functions

N-digestible dietary residue fermentation e.g. cellulose, starch by aerobic bacteria, and short chain fatty acids (SCFAs), are the source for energy of both host and resident bacteria Gut Bacteroides involves in the breakdown of complex N-glycan with the help of enzymatic apparatus which is encoded by multiple co-regulated genetic loci [115]. Putrefaction of exogenous and endogenous protein (like sloughed epithelium and lysed bacteria) has been done by anacrobic bacteria, SCFAs as well as toxic substances like ammonia and amines [116].

6.6.3.3 Trophic Functions

Short chain fatty acids induce the differentiation and proliferation of epithelial cell. Moreover, butyrate promotes cells reversion from neoplastic to non-neoplastic phenotype (Fig. 6.2).

6.7 Development and Homeostasis of Immune System

Specialized epithelial cells (M cells), sample luminal antigens as well as the microflora transport them to the lymphoid follicles to develop tolerating anti-inflammatory response (Th2 response) through the production of IL 10 and TGFB. Due to the pertinacious interactions between the host and its bacteria the immunity of host constantly changed. Host microorganisms try to change the immune response by changing its surface antigenicity, so that organism can avoid detection by immunosurveillance and maintain predominance of ecological niche in intestinal tract. Bacteria commensalism have play an essential role in sustaining the intestinal epithelial homeostasis and these gut bacteria are recognized under normal steady-state conditions by TLRs. TLRs activation through commensal microflora is important for protection from gut injury and associated mortality [118].

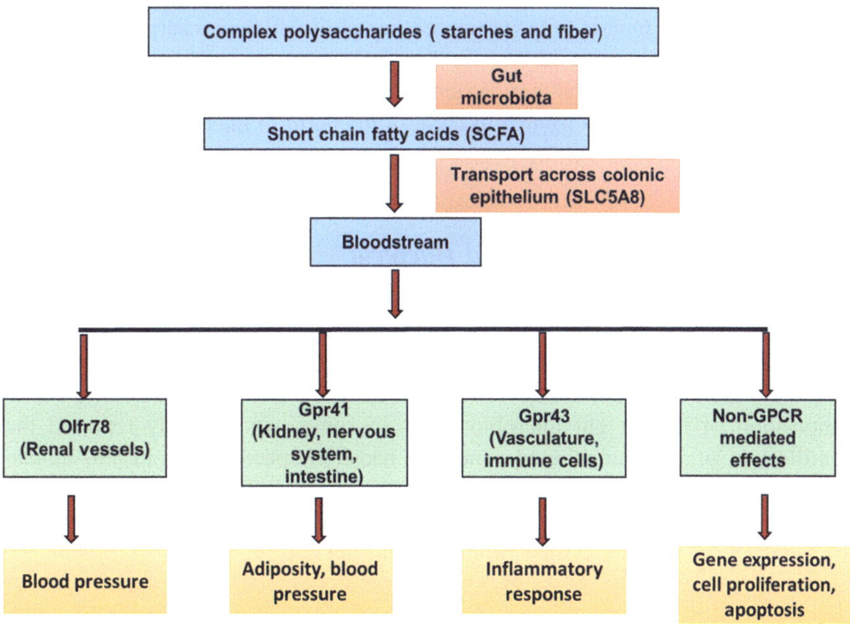

Fig. 6.2 Microbiota derived SCFAs and atherosclerosis [117]

Animal's colonization with major gut microbes, Bacteroides fragilis, physical and cellular maturation during immune system development is directed by a bacterial polysaccharide (PSA). During the colonization of B. fragilis, main activities of PSA are directing lymphoid organogenesis, correcting systemic T cell deficiencies and T (H)1/T(H)2 imbalances [119]. Communication between the host immune system and symbiotic microbiota facilitate by the bacterial metabolites and also affecting the balance between pro- and anti-inflammatory mechanisms [120]. Short chain fatty acids (SCFA), microbial metabolites regulate colonic Treg cell homeostasis [121].

6.7.1 Protective Function (Barrier Effect)

In barrier protective function microorganisms compete and attach to the brush border of host intestinal epithelial layer. Beneficial microorganisms compete for accessible nutrients and secrete antimicrobial (bacteriocins) [122].

6.7.2 Colonization Mechanism

Inflammation host responses change in microbiota composition and growth suppression induced by *Salmonella enterica* subspecies 1 *serovar Typhimurium* (S. Tm). Avirulent invGsseD mutant failed to trigger the colitis which was surpass by the gut microbiota in compare to wild type S. Tm. Inflammation can cause colonization resistance. Host immune defense system can alter the equilibrium between the pathogen and defensive microbiota in favor of the harmful microorganism [123].

6.7.3 Function of Uncultured Bacteria

The human gut microbial composition is associated with diseases and health of the host environment, but the awareness of different host microbial community is still needed for identifying the vast biological roles of the gut microbiota. The whole composition of human gut microbiota remains unknown. A study reported the identification of 1952 uncultured candidate bacterial species from 11,850 human gut microbiomes via reconstructing 92,143 metagenome-assembled genomes (Fig. 6.3). The identification of these species can help in understanding the interaction between probiotics and their beneficial effects [124].

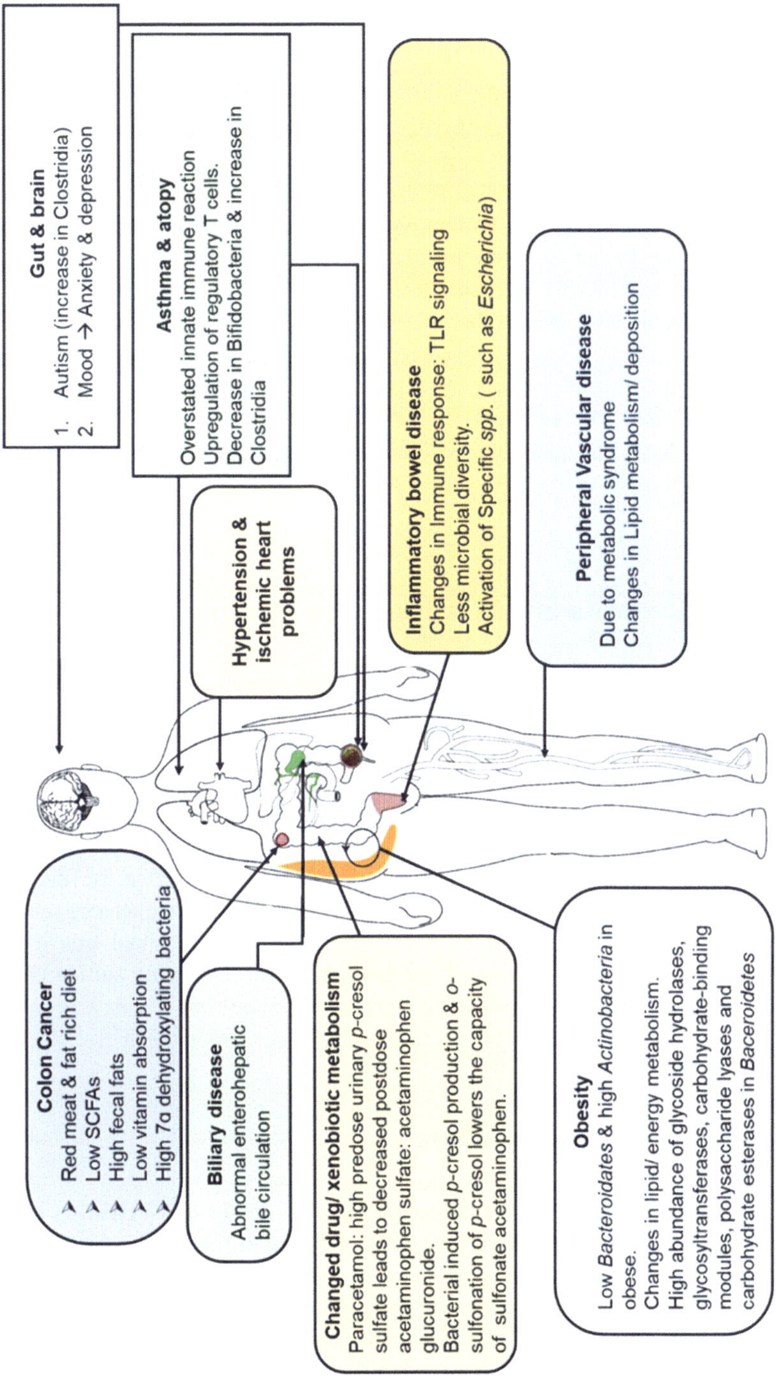

Fig. 6.3 Influence of gut microbial metabolism on human health [125]

6.8 The Gut Microbiota and Cancers

Colorectal cancer increases in human beings having age less than 50 years and it is related with human diet factors and daily eating habits which eventually affect the gut microbiota and CRC is the third most widespread cancer worldwide. In vitro experiments proliferation of CRC cells promoted by *F. nucleatum*. in mice, it is derived from the patient cells by CRC xenografts. Enterotoxigenic *Bacteroides fragilis* is the most long-studied human bacterial pathogen which causes diarrhea and inflammation in gastrointestinal tract of human beings. Enterotoxigenic *Bacteroides fragilis* (ETBF) increases colorectal cancer formation in mice. Currently, it was found in precancerous colonic lesions and biofilms coating human CRCs called adenomas (Fig. 6.4). Escherichia coli improve tumorigenesis in preclinical CRC experimental models by expressing the genomic island polyketide synthase (pks+) and are enriched in human colorectal cancer (CRC) tissues. Pks + *E. coli* secrete the genotoxin colibactin which caused alkylation in DNA, resulting in DNA adducts in colonic epithelial cells [126].

6.9 Gut Microbiota and Malabsorption Syndrome

Malabsorption syndrome is not exceptional, and it refers to the number of intestinal disorders which mimic the functional GI tract disorders. It is mainly due to the poor absorption of dietary carbohydrates, like fructose, lactose, etc. Occurrence and degree of malabsorption due to dietary lactose are widely diverse in the world with distinct population but most common in Asia than in America and Europe [127]. Number of host factors involves in the development of malabsorption such as degree of visceral hypersensitivity, host functional issues, cognitive dysfunction, colonic transit, host gut microbiota and also on the subtypes of microorganisms; bacteria such as *Methanobrevibacter smithii* effects on the intestinal transit due constipation and excess production of methane, however, hydrogen sulfide (H_2S) consider as a diarrhea biomarker [128].

6.10 Gut Microbiota and IBD

Irritable bowel disease related with the metabolic and compositional changes in the host intestinal microbiota. A study showed the effect on different microbial species of IBD suffering host, comprising decrease in *Dialister invisus*, *Bifidobacterium adolescentis, Faecalibacterium prausnitzii* and an increase in *Ruminococcus gnavus* and an unidentified member of Clostridium cluster XIVa [129]. A study revealed the wide range of data report about the host and microbial responses in 132 IBD patients,

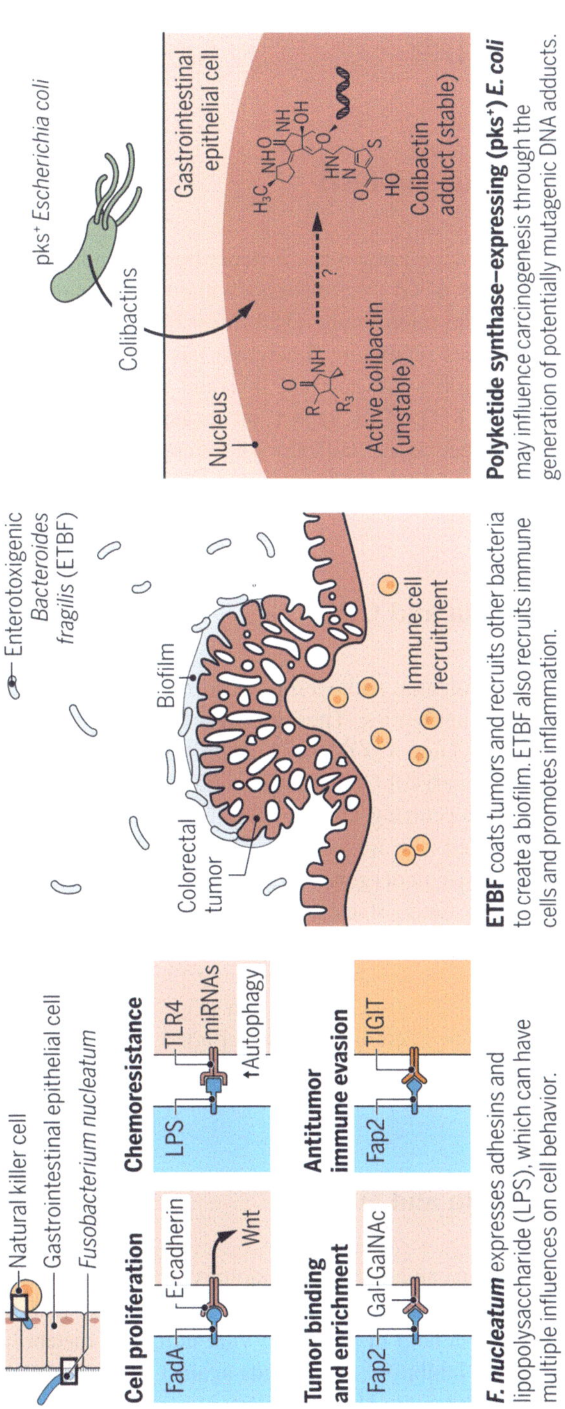

Fig. 6.4 Microorganisms may derive colorectal cancer [126]

showing the host immune factors, molecular functional profile, and gut microbiome in relation of metabolome [130].

6.11 Gut Microbiota and FBD

Functional bowel disorders are known as "irritable bowel syndrome" and they are very similar to the number of GI tract diseases without any clear pathogenesis. A profound sequencing of the microbiome (150-times fold as related to the human genome and bacterial genes regulating functions) has supported that the irritable bowel syndrome gut microbes are aberrant in count and has diverse number of bacterial families [113, 131]. This report presented that the *Firmicutes* and *Bacteroides* ratio might act as an indicator of microbial imbalance in irritable bowel syndrome [132].

6.12 Gut Microbiota and CDI

Clostridium difficile is a potential pathogen associated mostly with diarrhea caused by the frequent intake of antibiotics. The infections caused by *C. difficile* possess major health issues and are known as *Clostridium difficile* infections (CDI). The role of gut microbes in pathogenesis of CDI grabs the attention of researchers [133]. The patients suffering from reoccurring CDI have shown alterations in gut microbial composition, also associated with frequent intake of antibiotics. A study conducted on CDI patients who have undergone fecal microbiota transplantation (FMT), reduction in Firmicutes and Bacteroidetes population, and increment in Proteobacteria was observed in pre-FMT fecal samples [134]. Another study on CDI patients showed decrease in lactate producing phylotypes and opportunistic pathogens associated with endotoxin production (Fig. 6.5). An increment in the butyrate-producing anaerobic bacteria was also reported when compared to healthy control groups [135].

6.13 Gut Microbiota and Health

The microbes of human gut can affect the physiology of host in various dimensions and their interaction built a beneficial relationship for both host and gut microbes. Mutually beneficial bacteria help in providing vital nutrients, metabolize the complex compounds, produce inhibitory compounds against pathogens, and help in the formation of intestinal architecture [137] (Table 6.3).

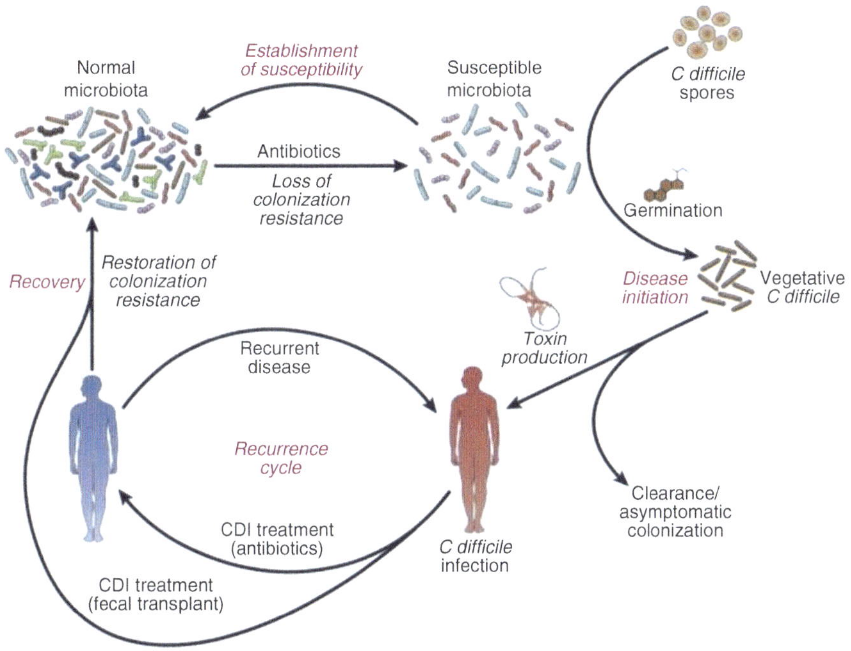

Fig. 6.5 Human gut microbiota and diseases [136]

6.13.1 Immune Regulation

Gut microbes can stimulate the normal development of host humoral and cellular mucosal immunity. Hematopoietic and non-hematopoietic cells of innate immunity can recognized the metabolites and signals of microbes and converted into physiological functions [151]. Clinical studies reported that the GF mice have showed defects in the formation of antibodies and gut-associated lymphoid tissues as comparison to normal mice [152]. A study has showed that the tolerogenic responses produced by gut microbes affect the gut dendritic cells and ceased the anti-inflammatory pathway of Th17 helper cells [153].

6.13.2 Drug Metabolism by Gut Microbiota

Microbiome-encoded enzymes elucidate the drug-metabolizing activities of host gut microbes and different communities on the basis of their genomic structural content and significantly affect the intestinal and systemic drug metabolism of mice [154].

Table 6.3 Gut microbiota, their metabolites and function [137].

Bacteria	Metabolites	Functions	References
Lactobacilli, Bifidobacterium	Vitamins: vit. B, K, biotin, riboflavin, folate, thiamine	Cofactor: Enzymatic reactions, regulate cell proliferation, enhance immune function.	[138, 139]
Clostridium, Bifidobacterium, Lactobacillus, Enterobacter, Roseburia	Acylglycerols, conjugated fatty acids, cholesterol, sphingomyelin, phosphatidylcholine, triglycerides	Improve intestinal permeability, decrease host fat mass and body weight, bile acid and production.	[140]
Clostridium, Bifidobacterium, Lactobacillus, Enterobacter, Bacteroids	Bile acids: glycocholate, cholate, etc.	Maintenance of intestinal barrier functions enhance lipid absorption, bile acid accumulation by some *Bifidobacteria.*	[141–143]
Clostridium, Bifidobacterium sp, coprococcus, roseburia	SCFAs: acetate, hexanoate, butyrate, propionate, isobutyrate	Lower the colonic pH, lower the level of cholesterol, pathogen inhibition, stimulate Na and H_2O absorption	[143, 144]
Lactobacillus, Bifidobacterium, Clostridium difficile, F. prausnitzii	Phenyl derivatives, benzoyl, phenol	Chronic diabetes and hepatities, asthma indication (urinary 3-Nitrotyrosine and 3-Nitro-4-hydroxyphenylacetic acid), obesity and hypertension biomarkers in humans.	[145]
Firmicutes, Actinobacteria, Proteobacteria, Bifidobacterium, Faecalibacterium prausnitzii	Choline metabolites: betaine, dimethylglycine, methylene, dimethyline, trimethyline	Neurotransmission, methyl transfer, cell membrane functioning	[146]
Clostridium sporogenes, E-coli	Indole derivatives	Protection against stress-induced GI epithelial damage	[147]
Lactobacillus acidophilus, Bacteroids fragilis	Polysaccharide A and B, Exopolysaccharides	Ceases cytokines levels, decreased neutrophil infiltration, host immune modulation.	[148]
Clostridium saccharolyticum, Campylobacter jejuni	Polyamines: cadaverine, spermine, spermidine, putrescine	Cell growth, apoptosis, increased calcium ion accumulation in mitochondria	[149]
Lactobacillus paracasei, Lactobacillus brevis	Gamma aminobutyric acid (GABA)	Inhibits CNS functions, decreases weight loss, promotes diuresis and hypotension	[150]

6.13.3 Bacterial Metabolite Enhances Athletic Performance

Veillonella strain enhance the mice treadmill run time and also increases the specific run time of marathon athletes. *V. atypica* improves the athlete's performance during physical activities (running) by metabolic conversion of lactate into propionate, hence consider as a natural microbiome-encoded enzymatic process [155].

6.13.4 Alleviation of Food Allergy (FA)

In food allergic infants dysbiotic fecal microbiota developed with in time but unsuccessful in mice. Therapy with *Clostridiales* strains, either as a monotherapy with Subdoligranulum variable or consortium, suppressed food allergy in mice. However, immunomodulatory *bacteroidales* consortium bacteriotherapy induced expression by regulator T (Treg) cells of the transcription factor ROR γt in a My D88-dependent manner, which was less in food allergic mice plus infants and futilely persuaded by their microbiota [156].

6.14 Conclusions

Industrial, agricultural, and domestic use of synthetic compounds produce large amount of environmental pollutants. From past several decades' environmental pollutants cause various health hazards and these pollutants can alter the functioning of gut microbiota. Use of probiotics will protect against the toxicity caused by these pollutants. There are number of bacterial, yeast, and fungal species which are used as probiotics. Various types of inhibitory compounds produced by probiotics shows antagonistic effect against pathogenic strains. It has been stated that probiotics produce extensive range of different *bacteriocins* such as *nicin* which constitute the major mechanism of antimicrobial act. *Lactobacilli* and *bifidobacteria* genera have been informed to produce *bacteriosins*, *lactolin*, *acidophillin acidolin*, and *lactocidin*, protection against infection with the foodborne pathogens. The identification of these species may help in understanding the interaction between probiotics and benefits with probiotics. Probiotics may increase the microbiological and nutritional balance of the gastrointestinal tract and used for the treatment of various gastrointestinal disorders like irritable bowel syndrome, Crohn's disease, pouchitis, antibiotic-associated diarrhea. Probiotics also used for enhancing the immune system by improving gut microbiota. It is concluded that the probiotics are essential for immune regulation, improve gut microbiota and for the treatment of gastrointestinal disorders.

References

1. Atashgahi S et al (2018) Prospects for harnessing biocide resistance for bioremediation and detoxification. Science 360(6390):743–746
2. Huttenhower C et al (2012) Structure, function and diversity of the healthy human microbiome. Nature 486(7402):207–214
3. Ding T, Schloss PD (2014) Dynamics and associations of microbial community types across the human body. Nature 509(7500):357–360
4. Dietert RR, Silbergeld EK (2015) Biomarkers for the 21st century: listening to the microbiome. Toxicol Sci 144(2):208–216
5. Sousa T et al (2008) The gastrointestinal microbiota as a site for the biotransformation of drugs. Int J Pharm 363(1–2):1–25
6. Koppel N, Rekdal VM, Balskus EP (2017) Chemical transformation of xenobiotics by the human gut microbiota. Science 356(6344):1246–1257
7. Metchnikoff E (1907) The prolongation of life: optimistic studies, trans. P. Chalmers Mitchell. New York: GP Putnam's Sons. Harvard
8. Thomas CM, Versalovic JJGM (2010) Probiotics-host communication: modulation of signaling pathways in the intestine. 1(3):148–163
9. He MQ, Shi BY (2017) Gut microbiota as a potential target of metabolic syndrome: the role of probiotics and prebiotics. Cell Biosci 7:54
10. Jandhyala SM et al (2015) Role of the normal gut microbiota. World J Gastroenterol 21 (29):8787–8803
11. Jakobsson HE et al (2015) The composition of the gut microbiota shapes the colon mucus barrier. EMBO Rep 16(2):164–177
12. Spanogiannopoulos P et al (2016) The microbial pharmacists within us: a metagenomic view of xenobiotic metabolism. Nat Rev Microbiol 14(5):273–287
13. Charbonneau MR et al (2016) Sialylated Milk oligosaccharides promote microbiota-dependent growth in models of infant undernutrition. Cell 164(5):859–871
14. Lange K et al (2016) Effects of antibiotics on gut microbiota. 34(3):260–268
15. Fei N, Zhao LP (2013) An opportunistic pathogen isolated from the gut of an obese human causes obesity in germfree mice. ISME J 7(4):880–884
16. Pizzorno JE, Murray MT, Joiner-Bey H (2016) The clinician's handbook of natural medicine e-book. Elsevier Health Sciences
17. Rao V, Rao L (2016) Probiotics and prebiotics in human nutrition and health. BoD–Books on Demand
18. Makarova K et al (2006) Comparative genomics of the lactic acid bacteria. Proc Natl Acad Sci U S A 103(42):15611–15616
19. Walker WA (2013) Initial intestinal colonization in the human infant and immune homeostasis. Ann Nutr Metab 63(Suppl. 2):8–15
20. Chen JJ, Cai W, Feng Y (2007) Development of intestinal bifidobacteria and lactobacilli in breast-fed neonates. Clin Nutr 26(5):559–566
21. Ruiz L, Margolles A, Sanchez B (2013) Bile resistance mechanisms in Lactobacillus and Bifidobacterium. Front Microbiol 4:396
22. Fijan S (2014) Microorganisms with claimed probiotic properties: an overview of recent literature. Int J Environ Res Public Health 11(5):4745–4767
23. Elshaghabee FMF et al (2017) Bacillus as potential probiotics: status, concerns, and future perspectives. Front Microbiol 8:1490
24. Konuray G, Erginkaya ZJF (2018) Potential use of Bacillus coagulans in the food industry. Foods 7(6):92
25. Chmielewska A, Szajewska H (2010) Systematic review of randomised controlled trials: probiotics for functional constipation. World J Gastroenterol 16(1):69–75
26. Xia P, Zhu J, Zhu G (2013) Escherichia coli Nissle 1917 as safe vehicles for intestinal immune targeted therapy—a review. 53(6):538–544

27. Behnsen J et al (2013) Probiotics: properties, examples, and specific applications. Cold Spring Harb Perspect Med 3(3):a010074
28. Hutkins RW, Goh Y (2014) *Streptococcus: Streptococcus thermophilus*. In: Encyclopedia of food microbiology: second edition. Elsevier Inc., pp 554–559
29. Hadji-Sfaxi I et al (2011) Antimicrobial activity and safety of use of *Enterococcus faecium* PC4.1 isolated from Mongol yogurt. Food Control 22(12):2020–2027
30. Garcia EF et al (2016) Identification of lactic acid bacteria in fruit pulp processing byproducts and potential probiotic properties of selected lactobacillus strains. Front Microbiol 7:1371
31. Pieniz S et al (2013) Production of selenium-enriched biomass by enterococcus durans. Biol Trace Elem Res 155(3):447–454
32. Hempel S et al (2012) Probiotics for the prevention and treatment of antibiotic-associated diarrhea a systematic review and meta-analysis. JAMA 307(18):1959–1969
33. DiRienzo DB (2014) Effect of probiotics on biomarkers of cardiovascular disease: implications for heart-healthy diets. Nutr Rev 72(1):18–29
34. Bednorz C et al (2013) Feeding the probiotic *Enterococcus faecium* strain NCIMB 10415 to piglets specifically reduces the number of Escherichia coli Pathotypes that adhere to the gut mucosa. Appl Environ Microbiol 79(24):7896–7904
35. Yang X, Wang Y, Huo GJGA (2013) Complete genome sequence of *Lactococcus lactis* subsp. *lactis* KLDS4. 0325. Genome Announc 1(6):e00962-13
36. Gao Y et al (2011) Complete genome sequence of Lactococcus lactis subsp. lactis CV56, a probiotic strain isolated from the vaginas of healthy women. J Bacteriol 193(11):2886–2887
37. Johnston BC et al (2011) Probiotics for the prevention of pediatric antibiotic-associated diarrhea. Cochrane Database Syst Rev (11):CD004827
38. McGuckin MA et al (2009) Intestinal barrier dysfunction in inflammatory bowel diseases. 15 (1):100–113
39. Blaut M, Klaus S (2012) Intestinal microbiota and obesity. In: Appetite control. Springer, pp 251–273
40. Lebeer S et al (2012) Functional analysis of Lactobacillus rhamnosus GG pili in relation to adhesion and immunomodulatory interactions with intestinal epithelial cells. Appl Environ Microbiol 78(1):185–193
41. Yan F, Polk DB (2002) Probiotic bacterium prevents cytokine-induced apoptosis in intestinal epithelial cells. J Biol Chem 277(52):50959–50965
42. van Baarlen P et al (2011) Human mucosal in vivo transcriptome responses to three lactobacilli indicate how probiotics may modulate human cellular pathways. Proc Natl Acad Sci U S A 108:4562–4569
43. Wang H et al (2014) Are there any different effects of Bifidobacterium, Lactobacillus and Streptococcus on intestinal sensation, barrier function and intestinal immunity in PI-IBS mouse model? PLoS One 9(3):e90153
44. Liao SFF, Nyachoti M (2017) Using probiotics to improve swine gut health and nutrient utilization. Anim Nutr 3(4):331–343
45. Razdan K, Parihar J, Bajaj B (2012) Isolation and characterization of a lipolytic and phytase producing probiotic for potential application in poultry feed. 2:369–377
46. Nielsen DS et al (2010) The effect of bacteriocin-producing Lactobacillus plantarum strains on the intracellular pH of sessile and planktonic Listeria monocytogenes single cells. 141:S53–S59
47. Hassan M et al (2012) Natural antimicrobial peptides from bacteria: characteristics and potential applications to fight against antibiotic resistance. J Appl Microbiol 113(4):723–736
48. Vilà i Miquel B, Esteve-Garcia E, Brufau de Barberà J (2010) Probiotic microorganisms: 100 years of innovation and efficacy. Modes of action
49. Corr SC et al (2007) Bacteriocin production as a mechanism for the antiinfective activity of Lactobacillus salivarius UCC118. 104(18):7617–7621
50. Bermudez-Brito M et al (2012) Probiotic mechanisms of action. 61(2):160–174

51. Bujalance C et al (2014) Lack of correlation between in vitro antibiosis and in vivo protection against enteropathogenic bacteria by probiotic lactobacilli. Res Microbiol 165(1):14–20
52. Reis JA et al (2012) Lactic acid bacteria antimicrobial compounds: characteristics and applications. Food Eng Rev 4(2):124–140
53. Shin JM et al (2016) Biomedical applications of nisin. J Appl Microbiol 120(6):1449–1465
54. Karpiński TM, Szkaradkiewicz AKJPJM (2013) Characteristic of bacteriocines and their application. Pol J Microbiol 62(3):223–235
55. Burgos MJG et al (2014) The cyclic antibacterial peptide Enterocin AS-48: isolation, mode of action, and possible food applications. Int J Mol Sci 15(12):22706–22727
56. Karpinski TM, Szkaradkiewicz AK (2013) Characteristic of bacteriocines and their application. Pol J Microbiol 62(3):223–235
57. Vera ECS et al (2018) Optimization of biosurfactant and bacteriocin-like inhibitory substance (BLIS) production by Lactococcus lactis CECT-4434 from agroindustrial waste. Biochem Eng J 133:168–178
58. Wong FWF et al (2017) Recovery of a bacteriocin-like inhibitory substance from Pediococcus acidilactici Kp10 using surfactant precipitation. Food Chem 232:245–252
59. Arakawa K et al (2016) Production of a bacteriocin-like inhibitory substance by Leuconostoc mesenteroides subsp. dextranicum 213M0 isolated from Mongolian fermented mare milk, airag. 87(3):449–456
60. Stevens M et al (2011) The potential of reuterin produced by Lactobacillus reuteri as a broad spectrum preservative in food. In: Protective cultures, antimicrobial metabolites and bacteriophages for food and beverage biopreservation. Elsevier, pp 129–160
61. Rattanachaikunsopon P, Phumkhachorn P (2010) Lactic acid bacteria: their antimicrobial compounds and their uses in food production 1(4):218–228
62. Singh VP (2018) Recent approaches in food bio-preservation-a review. 8(1):104–111
63. Whittenbury RJM (1964) Hydrogen peroxide formation and catalase activity in the lactic acid bacteria. 35(1):13–26
64. Elshaghabee FMF et al (2016) Ethanol production by selected intestinal microorganisms and lactic acid bacteria growing under different nutritional conditions. Front Microbiol 7:47
65. Van Tassell ML, Miller MJ (2011) Lactobacillus adhesion to mucus. 3(5):613–636
66. Buck BL et al (2005) Functional analysis of putative adhesion factors in Lactobacillus acidophilus NCFM. Appl Environ Microbiol 71(12):8344–8351
67. Ohland CL, MacNaughton WK (2010) Probiotic bacteria and intestinal epithelial barrier function. Am J Physiol Gastrointest Liver Physiol 298(6):G807–G819
68. Sun J et al (2007) Factors involved in binding of Lactobacillus plantarum Lp6 to rat small intestinal mucus. Lett Appl Microbiol 44(1):79–85
69. Collado MC et al (2006) Adhesion properties and competitive pathogen exclusion ability of Bifidobacteria with acquired acid resistance. J Food Prot 69(7):1675–1679
70. Caballero-Franco C et al (2007) The VSL# 3 probiotic formula induces mucin gene expression and secretion in colonic epithelial cells. 292(1):G315–G322
71. Prince T, McBain AJ, O'Neill CA (2012) Lactobacillus reuteri protects epidermal keratinocytes from Staphylococcus aureus-induced cell death by competitive exclusion. Appl Environ Microbiol 78(15):5119–5126
72. Hojo K et al (2007) Reduction of vitamin K concentration by salivary Bifidobacterium strains and their possible nutritional competition with Porphyromonas gingivalis. J Appl Microbiol 103(5):1969–1974
73. Martin FPJ et al (2008) Probiotic modulation of symbiotic gut microbial–host metabolic interactions in a humanized microbiome mouse model. 4(1):157
74. Elli M et al (2000) Iron requirement of Lactobacillus spp. in completely chemically defined growth media. J Appl Microbiol 88(4):695–703
75. Woo J, Ahn J (2013) Probiotic-mediated competition, exclusion and displacement in biofilm formation by food-borne pathogens. 56(4):307–313

76. van Hemert S, Verwer J, Schütz B (2013) Clinical studies evaluating effects of probiotics on parameters of intestinal barrier function. 3:212–221
77. Hyland NP, Quigley EM, Brint E (2014) Microbiota-host interactions in irritable bowel syndrome: epithelial barrier, immune regulation and brain-gut interactions. World J Gastroenterol 20(27):8859–8866
78. Tsai YT, Cheng PC, Pan TM (2012) The immunomodulatory effects of lactic acid bacteria for improving immune functions and benefits. Appl Microbiol Biotechnol 96(4):853–862
79. Rajput IR et al (2014) Saccharomyces boulardii and Bacillus subtilis B10 modulate TLRs mediated signaling to induce immunity by chicken BMDCs. J Cell Biochem 115(1):189–198
80. Ng SC et al (2009) Mechanisms of action of probiotics: recent advances. Inflamm Bowel Dis 15(2):300–310
81. Sharma R et al (2014) Improvement in Th1/Th2 immune homeostasis, antioxidative status and resistance to pathogenic E. coli on consumption of probiotic Lactobacillus rhamnosus fermented milk in aging mice. Age 36(4):9686
82. Kondepudi KK et al (2014) A novel multi-strain probiotic and synbiotic supplement for prevention of Clostridium difficile infection in a murine model. 58(10):552–558
83. O'Shea EF et al (2012) Production of bioactive substances by intestinal bacteria as a basis for explaining probiotic mechanisms: Bacteriocins and conjugated linoleic acid. Int J Food Microbiol 152(3):189–205
84. Van Guelpen B et al (2006) Low folate levels may protect against colorectal cancer. Gut 55 (10):1461–1466
85. Tooley KL et al (2006) Oral ingestion of Streptococcus thermophilus diminishes severity of small intestinal mucositis in methotrexate treated rats. Cancer Biol Ther 5(6):593–600
86. Lin C-S et al (2014) Impact of the gut microbiota, prebiotics, and probiotics on human health and disease. 37(5):259–268
87. Ceapa C et al (2013) Influence of fermented milk products, prebiotics and probiotics on microbiota composition and health. Best Pract Res Clin Gastroenterol 27(1):139–155
88. Islam SUJM (2016) Clinical uses of probiotics. 95(5):e2658
89. Sullivan Å, Nord C (2005) Probiotics and gastrointestinal diseases. 257(1):78–92
90. Jafarnejad S et al (2016) Probiotics reduce the risk of antibiotic-associated diarrhea in adults (18–64 years) but not the elderly (> 65 years) a meta-analysis. 31(4):502–513
91. Szajewska H, Kołodziej M (2015) Systematic review with meta-analysis: Lactobacillus rhamnosus GG in the prevention of antibiotic-associated diarrhoea in children and adults. 42 (10):1149–1157
92. Szajewska H, Kołodziej M (2015) Systematic review with meta-analysis: Saccharomyces boulardii in the prevention of antibiotic-associated diarrhoea. 42(7):793–801
93. Patro-Golab B, Shamir R, Szajewska H (2015) Yogurt for treating antibiotic-associated diarrhea: systematic review and meta-analysis. Nutrition 31(6):796–800
94. Ringel Y, Carroll IM (2009) Alterations in the intestinal microbiota and functional bowel symptoms. 19(1):141–150
95. Salonen A, de Vos WM, Palva A (2010) Gastrointestinal microbiota in irritable bowel syndrome: present state and perspectives. Microbiology 156:3205–3215
96. Moayyedi P et al (2010) The efficacy of probiotics in the treatment of irritable bowel syndrome: a systematic review. Gut 59(3):325–332
97. Whorwell PJ et al (2006) Efficacy of an encapsulated probiotic Bifidobacterium infantis 35624 in women with irritable bowel syndrome. Am J Gastroenterol 101(7):1581–1590
98. O'Mahony L et al (2005) Lactobacillus and Bifidobacterium in irritable bowel syndrome: symptom responses and relationship to cytokine profiles. Gastroenterology 128(3):541–551
99. Nobaek S et al (2000) Alteration of intestinal microflora is associated with reduction in abdominal bloating and pain in patients with irritable bowel syndrome. Am J Gastroenterol 95(5):1231–1238
100. McFarland LV, Dublin S (2008) Meta-analysis of probiotics for the treatment of irritable bowel syndrome. World J Gastroenterol 14(17):2650–2661

101. Carroll IM et al (2011) Molecular analysis of the luminal- and mucosal-associated intestinal microbiota in diarrhea-predominant irritable bowel syndrome. Am J Physiol Gastrointest Liver Physiol 301(5):G799–G807
102. Zocco MA et al (2006) Efficacy of lactobacillus GG in maintaining remission of ulcerative colitis. Aliment Pharmacol Ther 23(11):1567–1574
103. Kruis W et al (2004) Maintaining remission of ulcerative colitis with the probiotic Escherichia coli Nissle 1917 is as effective as with standard mesalazine. Gut 53(11):1617–1623
104. Prantera C et al (2002) Ineffectiveness of probiotics in preventing recurrence after curative resection for Crohn's disease: a randomised controlled trial with lactobacillus GG. Gut 51 (3):405–409
105. Guslandi M (2015) Role of probiotics in Crohn's disease and in Pouchitis. J Clin Gastroenterol 49:S46–S49
106. Guslandi M et al (2000) Saccharomyces boulardii in maintenance treatment of Crohn's disease. Dig Dis Sci 45(7):1462–1464
107. Kollman KA, Goulet O, Vanderhoof JA (2001) Saccharomyces boulardii does not stimulate mucosal hyperplasia after intestinal resection in the rat. J Pediatr Gastroenterol Nutr 32 (4):454–457
108. McLaughlin S et al (2008) Restorative proctocolectomy, indications, management of complications and follow-up–a guide for gastroenterologists. 27(10):895–909
109. Shen J, Zuo ZX, Mao AP (2014) Effect of probiotics on inducing remission and maintaining therapy in ulcerative colitis, Crohn's disease, and Pouchitis: meta-analysis of randomized controlled trials. Inflamm Bowel Dis 20(1):21–35
110. Persborn M et al (2013) The effects of probiotics on barrier function and mucosal pouch microbiota during maintenance treatment for severe pouchitis in patients with ulcerative colitis. Aliment Pharmacol Ther 38(7):772–783
111. Holubar SD et al (2010) Treatment and prevention of pouchitis after ileal pouch-anal anastomosis for chronic ulcerative colitis. (6):CD001176
112. Marotta A et al (2019) Effects of probiotics on cognitive reactivity, mood, and sleep quality. Front Psych 10:164
113. Qin JJ et al (2010) A human gut microbial gene catalogue established by metagenomic sequencing. Nature 464(7285):59–65
114. O'Hara AM, Shanahan F (2006) The gut flora as a forgotten organ. EMBO Rep 7(7):688–693
115. Briliute J et al (2019) Complex N-glycan breakdown by gut Bacteroides involves an extensive enzymatic apparatus encoded by multiple co-regulated genetic loci. Nat Microbiol 4 (9):1571–1581
116. Rowland I et al (2018) Gut microbiota functions: metabolism of nutrients and other food components. Eur J Nutr 57(1):1–24
117. Org E, Mehrabian M, Lusis AJ (2015) Unraveling the environmental and genetic interactions in atherosclerosis: central role of the gut microbiota. 241(2):387–399
118. Mazmanian SK et al (2005) An immunomodulatory molecule of symbiotic bacteria directs maturation of the host immune system. Cell 122(1):107–118
119. Rakoff-Nahoum S et al (2004) Recognition of commensal microflora by toll-like receptors is required for intestinal homeostasis. Cell 118(2):229–241
120. Arpaia N, Barton GM (2013) The impact of toll-like receptors on bacterial virulence strategies. Curr Opin Microbiol 16(1):17–22
121. Smith PM et al (2013) The microbial metabolites, short-chain fatty acids, regulate colonic Treg cell homeostasis. 341(6145):569–573
122. Mathipa MG, Thantsha MS (2017) Probiotic engineering: towards development of robust probiotic strains with enhanced functional properties and for targeted control of enteric pathogens. Gut Pathog 9:28
123. Stecher B et al (2007) Salmonella enterica serovar typhimurium exploits inflammation to compete with the intestinal microbiota. PLoS Biol 5(10):2177–2189

124. Almeida A et al (2019) A new genomic blueprint of the human gut microbiota. Nature 568 (7753):499–504
125. Kinross JM, Darzi AW, Nicholson JK (2011) Gut microbiome-host interactions in health and disease. Genome Med 3:14
126. Garrett WS (2019) The gut microbiota and colon cancer. Science 364(6446):1133–1135
127. Ghoshal UC, Ghoshal U (2020) Investigations for dietary carbohydrate malabsorption and gut microbiota, pp 359–370
128. Sharma A et al (2014) Fructose malabsorption is not uncommon among patients with irritable bowel syndrome in India: a case–control study. Indian J Gastroenterol 33(5):466–470
129. Joossens M et al (2011) Dysbiosis of the faecal microbiota in patients with Crohn's disease and their unaffected relatives. Gut 60(5):631–637
130. Lloyd-Price J et al (2019) Multi-omics of the gut microbial ecosystem in inflammatory bowel diseases. Nature 569(7758):655–662
131. Rajilic-Stojanovic M et al (2011) Global and deep molecular analysis of microbiota signatures in fecal samples from patients with irritable bowel syndrome. Gastroenterology 141 (5):1792–1801
132. Jeffery IB et al (2012) An irritable bowel syndrome subtype defined by species-specific alterations in faecal microbiota. Gut 61(7):997–1006
133. Lessa FC et al (2015) Burden of *Clostridium difficile* infection in the United States. N Engl J Med 372(24):2369–2370
134. Weingarden AR et al (2014) Microbiota transplantation restores normal fecal bile acid composition in recurrent Clostridium difficile infection. Am J Physiol Gastrointest Liver Physiol 306(4):G310–G319
135. Antharam VC et al (2013) Intestinal Dysbiosis and depletion of Butyrogenic Bacteria in Clostridium difficile infection and nosocomial diarrhea. J Clin Microbiol 51(9):2884–2892
136. Britton RA, Young VB (2012) Interaction between the intestinal microbiota and host in Clostridium difficile colonization resistance. Trends Microbiol 20(7):313–319
137. Round JL, Mazmanian SK (2009) The gut microbiota shapes intestinal immune responses during health and disease (vol 9, pg 313, 2009). Nat Rev Immunol 9(8):600
138. Nicholson JK et al (2012) Host-gut microbiota metabolic interactions. Science 336 (6086):1262–1267
139. LeBlanc JG et al (2013) Bacteria as vitamin suppliers to their host: a gut microbiota perspective. Curr Opin Biotechnol 24(2):160–168
140. Martins SV et al (2015) Adipocyte membrane glycerol permeability is involved in the anti-adipogenic effect of conjugated linoleic acid. Biochem Biophys Res Commun 458 (2):356–361
141. Daliri EB-M, Lee BH (2015) New perspectives on probiotics in health and disease. 4(2):56–65
142. de Diego-Cabero N et al (2015) Bile acid mediated effects on gut integrity and performance of early-weaned piglets. BMC Vet Res 11:111
143. Kelly CJ et al (2015) Crosstalk between microbiota-derived short-chain fatty acids and intestinal epithelial HIF augments tissue barrier function. Cell Host Microbe 17(5):662–671
144. den Besten G et al (2015) Short-chain fatty acids protect against high-fat diet–induced obesity via a PPARγ-dependent switch from lipogenesis to fat oxidation. 64(7):2398–2408
145. Chao M-R et al (2015) Simultaneous detection of 3-nitrotyrosine and 3-nitro-4-hydroxyphenylacetic acid in human urine by online SPE LC-MS/MS and their association with oxidative and methylated DNA lesions. 28(5):997–1006
146. Craciun S, Balskus EP (2012) Microbial conversion of choline to trimethylamine requires a glycyl radical enzyme. Proc Natl Acad Sci U S A 109(52):21307–21312
147. Shimada Y et al (2013) Commensal bacteria-dependent indole production enhances epithelial barrier function in the colon. PLoS One 8(11):e80604
148. Li L et al (2014) Immunoregulatory effects on Caco-2 cells and mice of exopolysaccharides isolated from Lactobacillus acidophilus NCFM. Food Funct 5(12):3261–3268

149. Graham SF et al (2015) Untargeted metabolomic analysis of human plasma indicates differentially affected polyamine and L-Arginine metabolism in mild cognitive impairment subjects converting to Alzheimer's disease. PLoS One 10(3):e0119452
150. Pandeya DR et al (2012) Host-microbial interaction in the mammalian intestine and their metabolic role inside. Biomed Res 23(1):9–21
151. Thaiss CA et al (2016) The microbiome and innate immunity. Nature 535(7610):65–74
152. Madsen KL et al (1999) Lactobacillus species prevents colitis in interleukin 10 gene–deficient mice. 116(5):1107–1114
153. Magrone T, Jirillo E (2013) The interplay between the gut immune system and microbiota in health and disease: Nutraceutical intervention for restoring intestinal homeostasis. Curr Pharm Des 19(7):1329–1342
154. Zimmermann M et al (2019) Mapping human microbiome drug metabolism by gut bacteria and their genes. Nature 570(7762):462–467
155. Scheiman J et al (2019) Meta-omics analysis of elite athletes identifies a performance-enhancing microbe that functions via lactate metabolism. Nat Med 25(7):1104–1109
156. Abdel-Gadir A et al (2019) Microbiota therapy acts via a regulatory T cell MyD88/ROR gamma t pathway to suppress food allergy. Nat Med 25(7):1164–1174

Chapter 7
Gut Remediation: Back to the Future

Zhenmin Ling, Yiming Jiang, and Xiangkai Li

7.1 Healthy Risk to Humans from Environmental Pollution

Nowadays, global environmental contamination of pollutants is a serious problem [1]. The reasons are that the increasing types and concentrations of pollutants typically derive from booming anthropogenic activities, industrialization, and urbanization [2]. The primary sorts of environmental pollutants vary, such as pesticide residual, agricultural chemicals, antibiotics residual, refractory organic pollutants, heavy metals. Based on the niche, environmental pollutants can be divided into air pollutants, water pollutants, and soil pollutants. The sources of one pollutant in the natural environment are complexed. More seriously, the pollutants do not only appear alone but also transfer and influence among them. For example, heavy metals in the cropland derive from mining activities, irrigation, solid-waste disposal, pesticides, fertilizers, and atmospheric deposition [3].

Additionally, heavy metals in the natural environment, especially in soil and water, are difficult to be removed with current remediation methods. Even worse, the crops grew or exposed to the contaminated soils or water will be influenced. Consequently, they will be consumed directly or indirectly by humans that may cause health problems. The various environmental pollutants are threatening human beings' survival and health, food safety, and ecological balance dramatically. The

Zhenmin Ling and Yiming Jiang are contributed equally to this work.

Z. Ling · X. Li (✉)
Ministry of Education Key Laboratory of Cell Activities and Stress Adaptations, School of Life Science, Lanzhou University, Lanzhou, Gansu, PR China
e-mail: xkli@lzu.edu.cn

Y. Jiang
Institute of Virology (VIRO), Helmholtz Zentrum München, German Research Center for Environmental Health, Neuherberg, Germany

X. Li, P. Liu (eds.), *Gut Remediation of Environmental Pollutants*,
https://doi.org/10.1007/978-981-15-4759-1_7

hazards of environmental contaminants to human health are mainly manifested in three aspects:

1. Specific damage. It manifests mostly in the following types: (i) Acute and subacute poisoning; (ii) Chronic poisoning, which was mainly caused by long-term effects of environmental pollutants after entering the environment for several years; (iii) Carcinogenicity: 90% of them are related to chemical factors, 5% to physical factors, and 5% to biological factors (fungi, viruses, parasites); (iv) Teratogenicity; (v) Mutagenicity; and (vi) Sensitization.
2. Non-specific damage. The main manifestations are the increasing incidence of common diseases, decreased body resistance, and reduced labor capacity.
3. Complex diseases caused by environmental pollution. For example, (i) Infectious diseases such as typhoid, cholera, dysentery; (ii) public hazards such as "Minamata disease" and "Pain disease"; (iii) Occupational diseases such as silicosis and lead poisoning; and (iv) Food-borne diseases such as bacterial, chemical food poisoning, puffer and mushroom poisoning, and infectious and poisoning caused by various pathogenic factors of food pollution.

A classic example of the health effects was the itai-itai disease that happened in Japan in the 20th century, caused by the consumption of crops, e.g., rice and soybean grown in areas polluted with heavy metals cadmium (Cd) [4].

7.2 Intestinal Microbiota and Microbiota Targeted Therapies

Only 10% of us are human [5], while the rest 90% are microbial cells, including bacteria, fungi, archaea, and single-celled eukaryotes, or micro-living entities such as viruses [6]. The microorganisms parasitize on the human body, such as skin, genitourinary tract, gastrointestinal tract, and respiratory tract. After a long period of natural selection, the microorganisms parasitize on the human body have formed a symbiotic relationship with the host, interrelated, and interacted with each other. Meanwhile, these microorganisms also interact with each other, building a community and ecosystem, occupying different niches, creating a very stable microbial environment together [7].

Medical clinical data showed that around 60% of fecal solids of the fecal mass consist of microorganisms from the human large intestine. With the deepening of research, more and more researchers have recognized that the three main functions of gut microorganisms, the metabolic function, nutritional function, and protective function [8, 9]. It is debatable how many microbial species in intestinal. Typically, there are 500-1000 species of intestinal bacteria considered in animals, while among them, 300-500 different specics for humans [10]. However, some published papers use a multidisciplinary approach to conclude that the gut contains more than 35000 bacterial species [11]. Regardless of the difference of numbers, it is consensus that the quantity of microorganisms in the stomach and small intestine is low. However, to be compared, there are a tremendous amount of microorganisms existed in the colorectal tract, and in where they constitute a complex and dynamic microbial ecosystem that mainly carries out the performance of decomposing nutrients and help our absorption [8].

Many species of bacteria have evolved and adapted to live and grow in the human intestine, such as strict anaerobes, facultative anaerobes, and aerobes. Among them, the strictly anaerobic bacteria are 2-3 orders of magnitude more than the facultatively anaerobic bacteria and aerobic bacteria[12]. Up to now, more than 50 phyla have been found in viscera, among which *Bacteroides* and *Firmicutes* are dominant, while *Proteobacteria*, *Verrucomibia*, *Actinobacteria*, *Fusobacteria*, and *cyanobacteria* are next.

The health of the gut is determined by the balance of all microbes that inhabit there. That is, when the intestinal microbial community is in disorder, exogenous pathogens can "take the opportunity to enter" and colonize and multiply in the intestinal tract of the body, causing inflammation [13]. Many diseases are accompanied by similar inflammation, and patients are usually given antibiotics in clinical practice. Later, studies found that antibiotics would cause some damage to the patient's body, and some patients showed resistance to antibiotics. Therefore, the method of "microbiota targeted therapies" has been proposed. At present, there are two methods: one is fecal bacteria transplantation, while another is probiotic treatment.

7.2.1 Fecal Bacteria Transplantation

The way of fecal bacteria transplantation traced to the 4th century, when Ge Hong, a famous Chinese pharmacist, used feces as medicine to treat patients with food poisoning or severe diarrhea. In the 16th century, Li Shizhen, a renowned medical scientist in Ming Dynasty, described in detail the manufacturing method of the fecal medicaments and named it 'Yellow Dragon Decoction,' which was used for the treatment of diarrhea, abdominal pain, vomiting, and constipation [10]. It was not until modern times this therapy began to attract people's attention when people started to call it "fecal bacteria transplantation," "fecal bacteria treatment," or "intestinal microbial transplantation." It is defined that infusing a healthy individual's intestinal microorganisms into a patient's intestine to treat a specific disease. In the treatment, a healthy person is a donor while a patient is a subject. More and more clinical trials reported that the stool microorganisms from the healthy human could become assistants in the treatment and recovery of some diseases, such as immunoregulation, microbial regulation, and metabolic regulation [14]. Mountains of studies reported that fecal transplantation could repair the intestinal microbial disorder and pseudomembranous colitis caused by the infection of *Clostridium difficile* [9]. Therefore, fecal bacteria transplantation is an efficient treatment for certain diseases related to gut disorder with a low treatment cost.

7.2.2 Gut Probiotics Therapy

Along with the long-term biological evolution, the microbial variation and taxonomic diversity among mammalian gut microbiomes showed convergence. However, the same rates of microbiome divergence without dietary transitions [15]. That means the gut microorganisms showed their stability and dependence on the specific mammal host, especially for the human being. Moreover, more in-depth studies have demonstrated that gut microorganisms have some amortization function for the hazards expose and risk. Notably, the gut probiotics not only can reduce risk but also can regulate the bacterial community, which will assist the human gut health for the alimentation [14, 16]. Probiotics are known to have a role in the prevention or treatment of some diseases [17]. Thus, it is easy to understand that probiotic therapy is to take probiotics to regulate the intestinal flora, inhibit the growth and reproduction of harmful bacteria, and then achieve the purpose of treatment.

7.3 Gut Remediation and Its Application

7.3.1 Enlightenment Age of Gut Remediation

It is not only important to monitor the pollutants, but also essential to degrade them or remediate them for their potential ecological risk. Microbial remediation, which initially refers to the process of removing contaminants from soil and water or making them harmless through the action of microorganisms, has been widely used in the prevention of pollutants in a natural environment. Generally, it includes pollutant degradation or detoxification under natural or human-made control conditions. As the microbial biotechnology developed to solve the environmental contamination, such as the refractory pollutants, e.g., Polycyclic aromatic hydrocarbons; antibiotic residue; and heavy metals, e.g., cadmium (Cd), chromium (Cr), arsenic (As), lead (Pb), nickel (Ni), mercury (Hg), and their derivatives, e.g., methylmercury (MeHg) [18–27]. Traditional physical and chemical remediation methods can effectively alleviate environmental pollution, however, they also release secondary contaminants to the environment and increase the cost of use [28]. In contrast, bioremediation is more natural to use because of its low economic cost, high efficiency, and less environmental pollution [29]. Thus, microbial remediation is becoming increasingly important in bioremediation and ecological safety.

With the development of the omics technology and DNA sequencing technology, the more precious evolution principle of mammal gut microorganisms and more accurate succession rules of the gut community have been determined. Meanwhile, more and more functional gut microorganisms and novel genes are detected, isolated, and applied in the prevention and treatment of environmental contamination. Jayanta Kumar Biswas et al. implied that the bacteria *Bacillus licheniformis*, which was isolated from the gut of earthworm (*Metaphire posthuma*), synthesized extracellular polymeric substance for remediation of Cu(II) and Zn(II). Besides, the strain showed a maximum tolerance of 8 and 6 mM for Cu(II) and Zn(II), respectively. It removed 34.5 % of Cu(II) and 54.4 % of Zn(II) at 25 mg L-1 after 72 and 96 h incubation, respectively [30]. Cai et al. have reported that fly larvae gut microorganisms play a vital role in the degradation of organic contaminants, and some bacteria can degrade the tetracycline in vitro [31, 32]. In consequence, the successful application of functional microbial strains isolated from the intestinal tract in natural pollution remediation provides new ideas for microbial remediation in vivo.

7.3.2 Gut Remediation

Long-term exposure to the hazards would finally affect the final receptor—the human being. To this purpose, it is vital to building a robust and solidus line of defense for environmental contamination. People have been trying to reduce the level of toxins in the diet. However, it is often not very successful. Therefore, it is necessary to study a new method for host reducing pollutants.

The gut remediation is precisely the ecological safety novel method for the recovery of environmental pollutants in vivo. It derived from the microbial remediation, which refers to the process of removing contaminants or making them harmless through the action of microorganisms in vivo.

Many studies have shown that there is a close interaction between microbial degradation of pollutants, host, and intestinal microbiota [33]. The intestinal microbiota is a definite health asset that crucially influences the healthy structural and functional development of the mucosal immune system. Only when the intestinal microbial ecosystem is relatively stable can it perform normal physiological functions. Generally, the human body is stimulated by various internal and external stimuli every day. The intestinal microbial ecosystem has a particular capacity of reducing stimulation and regulation, which keeps the dynamic balance of the system, which is also the need for the body to maintain routine work. Here taking the environmental risk pollutant PAHs as an example, there are millions of microbial decomposers in earthworm gut have the function of PAHs degradation [34]. Some of these microorganisms, such as *Pseudomonas*, *Alcaligenes*, and *Acidobacterium*, participate in degrading hydrocarbons. Moreover, some fungi, such as *Penicillium*, *Aspergillus*, and *Mucor*, are found in the gut of earthworm, and they are known to degrade hydrocarbons [35].

7.3.3 Probiotics in Gut Remediation

The gut remediation is one method to establish a protective barrier towards the risk of environmental contaminants in the human body. This technology depends on the gut probiotics functions, which include the adsorption of the pollutants, protection, and remission from the contamination, and the regulation of the gut microbial community. The adsorption contains the cell wall adsorption, extracellular polymeric substances adsorption, S-layer adsorption. The peptidoglycans and phosphoteichoic acid polymers on the cell walls of *Lactobacillus rhamnosus* and some *Bifidobacterium longum* have a strong ability to adsorb metal cations [36, 37]. Many S-layer proteins (S-layer) on the surface of *lactobacilli* can bind heavy metals with a large number of negatively charged functional groups such as COO- [38]. The ability of the S-layer protein of *Lactobacillus* can adsorb cadmium on the cell surface [39]. Additionally, it is a considerable improvement to search for the food-grade microorganisms that can be delivered to the gastrointestinal tract, and that can sequester toxins [40]. Thus, the probiotics are the best choice for gut remediation and will be the core of great potential.

Furthermore, the mechanisms of probiotics in gut remediation were demonstrated, including the following aspects:

1. *Protection and remission*

The protection and remission mainly refer to that the probiotics protect cell function. Some probiotics, such as *Lactobacillus reuteri*, are involved in the regulation of intestinal flora metabolism, and increase the production of short-chain fatty acids and other organic acids [41]. These organic acids help to increase the solubility of divalent mineral elements such as Ca (II), Mg (II), and Fe (II) and form a competitive relationship between metal ions, thus reducing the absorption of divalent heavy metal ions in the small intestine [42]. Moreover, some *Lactobacillus plantarum* can maintain the intestinal barrier function and lessen the accumulation of heavy metals in the intestinal tract in various ways. For example, they can reduce the apoptosis induced by heavy metals, alleviate the toxicity induced by heavy metals, alleviate oxidative stress and inflammatory reaction, reverse the damage of tight junction, and reduce the permeability of intestinal epithelial cells to [43, 44].

2. *Regulation*

The regulation means the probiotics will regulate the gut microbial function. As the first line of defense to control environmental pollutants entering the body, the integrity of the intestinal barrier depends on the interaction between intestinal microorganisms and host. Wu et al. have found that the chromium caused specific changes in the overall structure of intestinal microflora in mice. In meanwhile, *Lactobacillus plantarum* tw1-1 restored 49 of 79 OTUs with relative abundance changes. It enhanced the reduction ability of intestinal microflora to chromium (VI), and this research conclusion also suggested that the regulatory mechanism of probiotics on intestinal microbiota [45].

The biotechnology (i.e., meta-omics, including metagenomic, metatranscriptomic, and metabonomic) developed and bioremediation demanded. The new technology will be further used to build our understanding of host–gut probiotics interaction, as well as to provide meaningful insights into the mechanism of gut probiotics and clarify the causal relationship between gut probiotics and the related symptoms more deeply [46].

7.4 Gut remediation for various pollutant treatment

7.4.1 Gut remediation for heavy metals

In recent years, more and more attention has been paid to the effects of heavy metals exposure in the gut. [43, 47]. The intestinal microorganisms can inhibit heavy metals absorption by other ways besides heavy metals binding, with a focus on the protection of the gut barrier. Therefore, it is important to protect the gut barrier against heavy metals toxicity and to inhibit intestinal heavy metals absorption. The adsorption of heavy metals through the intestinal microorganism was cost-effective, and the adsorption efficiency was very high. Also, gut remediation has provided a new design method for the decrease of heavy metals. Therefore, using intestinal microbes to reduce the accumulation of heavy metals in the body is a useful technique.

7.4.1.1 Gut remediation for cadmium (Cd)

Cd is a kind of environmental pollutant, which is harmful to human and animal health. One study demonstrates that oral administration probiotics can avoid Cd absorption by intestinal Cd sequestration and protecting the gut barrier [43]. Therefore, a probiotic strain with Cd-binding ability is possibly used as an additive for the prevention of Cd absorption in the gut. *Lactobacillus plantarum* is one of the probiotics in the animal intestine. In the in vitro assay, the *L. plantarum* could alleviate the cytotoxicity of Cd in the human intestinal cell line HT-29. *L. plantarum* CCFM8610 that can bind Cd according to a previous study, which also prevented Cd absorption in mice [43]. In a mouse model, *L. plantarum* CCFM8610 increased Cd levels in feces and reduced Cd accumulation in the organs and tissues of Cd-exposed mice. Cd absorption is inhibited by *L. plantarum* in the intestines of mice, and the main reason is the strain possessing the Cd-binding ability [43, 48]. Zhai et al. demonstrated that probiotics could significantly inhibit and control Cd absorption in the gut by protecting the intestinal barrier. Also, the protection associated with the alleviation of oxidative stress caused by Cd [43]. Therefore, gut remediation is a promising biotechnology for reducing Cd.

7.4.1.2 Gut remediation for mercury (Hg)

Hg contamination attracts worldwide attention, which poses a threat to humans and aquatic organisms. [49]. Environmental Hg contamination is an urgent global

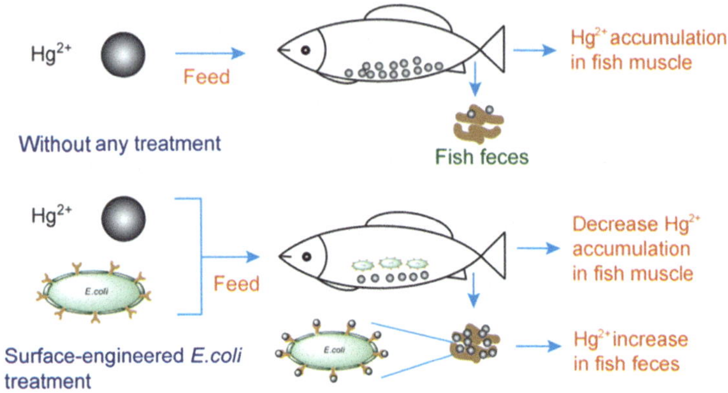

Fig 7.1 Surface-engineered *E. coli* reduced the Hg(II) accumulation in fish. Reference: [52]

problem. Hg in the food chain is mainly present in inorganic methylmercury (MeHg) or mercury ions (Hg(II)) form [50]. The main sources of Hg exposure in nature are fishes due to the rapid biomagnification of Hg in the aquatic environment; therefore, the gut of fish is the first organ to be contaminated with Hg. *Escherichia coli* is one of the animal intestinal microorganisms, and it has a significant proportion [51]. A study shows that Hg-binding peptide was displayed on the cell surfaces of *E. coli*, forming surface-engineered *E. coli*, which promoted adsorption of Hg(II) [52]. The surface-engineered *E. coli* were fed to fish *Carassius auratus*, and the *E. coli* colonized in the fish intestine (Fig. 7.1). The engineered bacteria-fed fish decreased about 51.1% accumulation of Hg(II) compared to the fish without the intake of the surface-engineered *E. coli* [52]. It indicates that the engineered bacteria bound to Hg (II) in the fish intestines were excreted in the fish feces. That is, the engineered *E. coli* significantly alleviated the toxicity of Hg(II) to fish by adsorbing Hg (II) [52]. Furthermore, the changes of microbial diversity in the intestine caused by Hg(II) exposure were mitigated by Hg-binding bacteria, thereby protecting the microbial community structure of the intestine. The engineered strain in fish intestines accelerated the Hg(II) excretion and reduced the Hg(II) level in the muscle tissue. Thus, Hg(II) contamination in fish was controlled by the engineered bacteria, which prevented the Hg(II) toxicity in fish. The strategy using engineered bacteria is an interesting approach for limiting Hg(II) pollution in fish.

Previous work explored that the lactic acid bacteria reduced Hg(II) after emulated gastrointestinal digestion [40]. The lactic acid bacteria with good Hg(II)-binding ability and Hg(II) were mixed to gastrointestinal digestion (Jadán-Piedra, Alcántara et al. 2017). The lactic acid bacteria decreased the Hg(II) more than 72% under emulated gastrointestinal digestion conditions [40]. Besides, one study shows that people who regularly consume fish after intake of yogurt with *L. rhamnosus* strain were investigated, and their blood levels of Hg were determined [47]. It was found that a low significant decrease in Hg levels obtained in these people (Bisanz et al., 2014). The work shows that lactic acid bacteria are capable of binding Hg(II) under gastrointestinal digestion. Therefore, the bacteria could be sufficient to inhibit the absorption of Hg(II) when it is ingested [40]. These successful research cases suggest that the intestinal microorganisms are a promising approach for reducing Hg(II).

7.4.1.3 Gut remediation for MeHg

Some anaerobic bacteria produce MeHg, and it is one of the most toxic forms of Hg and it can damage the nervous system of animals [53, 54]. The fish exposure to MeHg has been widely concerned due to its rapid biomagnification [47]. One study showed that human exposure to the MeHg main is due to fish and seafood consumption [55]. MeHg can continue to accumulate in fish in surrounding waters [56, 57]. People who consume fish polluted with MeHg are prone to Minamata disease [58]. The human exposure to MeHg is due to the ingestion of MeHg polluted fish [59]. A novel MeHg-binding peptide was displayed on the cell surfaces of *E. coli* W-1, which was isolated from healthy fish feces [60]. The cell-surfaced displayed cells efficiently removed MeHg, and the adsorption of MeHg in the engineered *E. coli* strain was fourfold higher than that in the unmodified *E. coli* strain [60]. Fish *C. auratus* as a model was fed with surface-engineered *E. coli*. The result showed that MeHg concentration was decreased by about 36.3 % in muscle tissue of fish, whereas MeHg concentration was increased in the fish feces compared with the control group [60]. The surface-engineered strain in the intestine adsorbed MeHg and avoided its absorption by muscles, and these bacteria binding MeHg were excreted in the fish feces [60]. The engineered *E. coli* with MeHg-binding ability prevented fish against MeHg toxicity, which can reduce MeHg accumulation in fish.

The intestinal ecology of animals is significantly affected by chronic oral exposure to heavy metals [43]. About 98% of the oral administration MeHg is absorbed into the intestine of an animal when MeHg enters the food chain [61]. MeHg always caused *Shewanella* to overgrow, which is harmful to the experimental fish and changes the composition of microbial communities [60]. Engineered *E. coli* inhibits fish intestinal absorption of MeHg and alleviates MeHg toxicity by reducing MeHg [62]. Using gut remediation for detoxication is a new method to remove MeHg in fish.

Parachlorella beyerinckii CK-5 that is a kind of unicellular green algae, can reduce MeHg in mice gut [61]. In this work, mice were orally administered MeHg chloride with or without *P. beyerinckii* powder [61]. After oral administration of *P. beyerinckii*, feces and urine of mice were collected, and total Hg concentration in these samples was tested. It was found that some Hg was excreted in feces and urine of the group with *P. beyerinckii*. The excretion of Hg was increased by about 1.9 and 2.2 times in feces and urine compared to that of the MeHg group (without *P. beyerinckii*) [61]. These results indicated that oral administration of *P. beyerinckii* might accelerate the excretion of MeHg in feces and urine [40, 61]. Therefore, it is feasible to reduce the toxicity of methylmercury through intestinal repair in the future.

7.4.1.4 Gut remediation for lead (Pb)

The heavy metal Pb in the environment is a threat to human health and causes some dysfunctions in animals. Pb intake by animals first passes through the intestine [63]. Intestinal bacteria play an essential role in Pb absorption, bioaccumulation, and excretion. In a previous study, the relationship between intestinal microbiota and Pb toxicity was assessed in a mouse model [63]. In the first place, one group of mice was fed with a broad-spectrum antibiotic cocktail to deplete their gut microbes and then was orally exposed to Pb for three days [63]. Compared to the control mice, Pb concentrations in primary organs and the blood were increased, and Pb fecal concentrations were decreased in antibiotic-treated mice; this conclusion shows that intestinal microbiota protected the Pb absorption from acute oral Pb exposure [63]. In the next place, three Pb-intolerant intestinal microbes, *Oscillibacter ruminantium*, *Faecalibacterium prausnitzii*, and *Akkermansia muciniphila*, were fed to mice, and the effects for Pb toxicity were assessed. Mice fed with *O. ruminantium* significantly reduced Pb levels in kidney and blood [63]. *F. prausnitzii* treatment effectively accelerated the Pb excretion in feces and decreased Pb levels in primary organs and the blood [63]. The above researches indicate the potential for reducing Pb toxicity by the regulation of intestinal microbiota.

7.4.2 Gut remediation for organic pollutants

7.4.2.1 Gut remediation for erythromycin

Antibiotics contamination is a threat to humans and the environment [64, 65]. With the rapid development of agriculture, antibiotics spread into the environment. The recalcitrant primary antibiotics contaminants are from livestock manure [66, 67]. Erythromycin is one of the common-used antibiotics, which is always absorbed in the small intestine, and the residue is excreted in feces [68]. Besides, it is reported that about half of the orally administered erythromycin is discharged into the feces of animal [69].

E. coli in the large intestine of some animals is one of the gut microbiota, and it plays a significant role in the gut [70]. There are study reports that using engineered *E. coli* reduces erythromycin [71]. *Erythromycin esterase* with the ability of erythromycin degradation displayed on the cell surface of *E. coli* formed a surface-displayed strain, which exhibited a high erythromycin esterase activity and stability in degrading erythromycin [71]. The surface-engineered bacteria were fed to female mice and were found to colonize in the large intestine and decrease 83.43% of erythromycin in the feces compared with that in the control group (Fig. 7.2) [71]. The engineered *E. coli* not only eliminated antibiotics from the pollution source but also helped to accelerate subsequent treatment due to the elimination of antibiotics. In the large intestine of mice, the engineered *E. coli* could eliminate a large

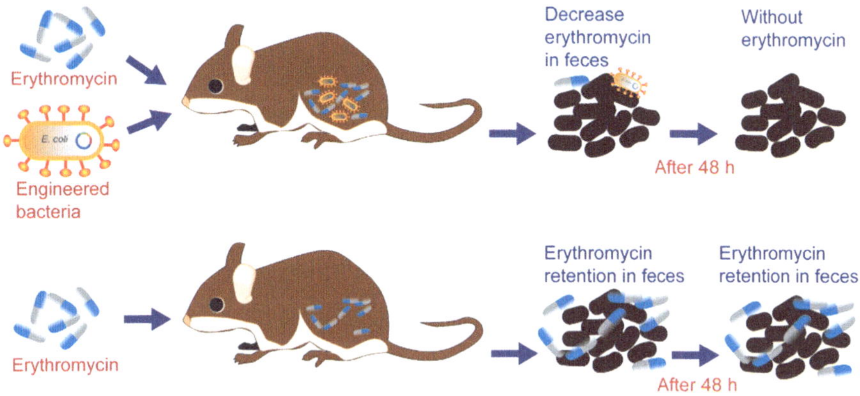

Fig. 7.2 Engineered *E. coli* reduce erythromycin in animal feces. Reference: [71].

number of residual erythromycin, thus decreasing erythromycin in mice feces. The work provides a novel strategy to reduce antibiotics by gut remediation from source and prevent antibiotics release to the environment. Therefore, using intestinal microbes to decrease antibiotics spread into the environment is an exciting method. Surface-engineered bacteria may be very promising for remediation antibiotics (Singh et al., 2011).

7.4.2.2 Gut remediation for tetracycline

It has been reported that gut microorganisms of fly larvae play a crucial role in the degradation of organic contaminants [31]. In other words, some functional bacteria in the gut can remediate pollutants [31]. One study shows that millions of microbes inhabit the fly larvae gut as decomposers and can degrade antibiotics [35]. With the help of intestinal microbes, fly larvae quickly consumed and metabolized ingested antibiotics [31]. Compared with the traditional method, the gut remediation is an available technology for elimination of pollutants, and it has a low cost [35]. Thus, a large number of bacteria in the intestine could possibly reduce the spread of residual antibiotics into the environment.

It was found that black soldier fly larvae can effectively and rapidly degrade tetracycline and provide an effective strategy to manure treatment [31]. After analysis of degradation pathways of tetracycline by black soldier fly larvae, it was found that nearly 97% of tetracycline was degraded within 12 days in a non-sterile black soldier fly larvae treatment system [31]. The gut microbiota of the black soldier fly larvae largely carried out tetracycline degradation; the tetracycline degradation rate is twice as much that of those achieved in sterile black soldier fly larvae systems [31]. Moreover, intestinal bacterial and fungal communities provided the means to degrade and tolerate tetracycline by detailed microbiome analysis. The researchers further prove that fungi and bacteria significantly helped the black soldier fly larvae to degrade the tetracycline [31]. Moreover, six intestinal microorganisms with tetracycline degradation ability were isolated [31]. The isolates, including two bacteria and four fungi, were identified as *Candida rugosa*, *Galactomyces geotrichum*, *Pichia kudriavzevii*, *Serratia marcescens*, *Serratia sp.*, and *Trichosporon asahii*. Tetracycline degradation reactions included hydrolysis, ring-cleavage, deamination, oxygenation, demethylation, and modification. In conclusion, the degradation of tetracycline antibiotics by black soldier fly larvae is due to the function of intestinal microbes. The study may provide a new idea for promoting antibiotics degradation by regulating the gut microbiota of the larvae.

7.4.2.3 Gut remediation for polycyclic aromatic hydrocarbons (PAHs)

PAHs refers to more than 100 compounds with anthracene, benzo (a) pyrene, naphthalene, and pyrene as the main chemical components [72]. They are produced from oil-, coal-, and coke-fired power plants and petroleum refineries [73]. Besides, asphalt and aluminum also produce PAHs during the production process. It takes a long time for several PAHs photodegradation in air, weeks to months, or more for soil microbes to degrade PAHs [72]. Therefore, PAHs cause severe environmental pollution.

It is reported that earthworms, as ecosystem engineers have a significant effect on the fate of organic contaminants in soil [34]. The positive influence of earthworms on the elimination of organic contaminants has been reported in some studies, and earthworms can promote the elimination of atrazine, polychlorinated biphenyls (PCBs), and PAHs in soil [72, 73]. Millions of microbes as decomposers in the gut of the earthworm play a significant role in degrading PAHs pollutants [35].

It is found that there are millions of microbial decomposers in the earthworm gut [73]. Several studies have found that earthworms can degrade PAHs residues and organochlorine pesticide due to its microbial decomposers of gut [35]. One study showed that the microbiota associated with the intestine of the earthworms was analyzed, and it found that species like *Pseudomonas*, *Azoarcus*, *Paenibacillus*, *Burkholderia*, *Acaligenes*, *Spiroplasma*, and *Acidobacterium* inhabit in the gut of earthworm [74]. Some of these microorganisms, such as *Pseudomonas*, *Acaligenes*, and *Acidobacterium* participate in degrading hydrocarbons. Some fungi such as *Pencillium*, *Aspergillus*, and *Mucor* are found in the gut of earthworm, and they are known to degrade hydrocarbons [35].

7.5 The enhancement of gut remediation

Because of the benefits of gut remediation, the products of live probiotics are increasing applications. Typically, the live probiotics are added to fresh liquid foods, such as yogurt and oral liquid [75]. However, the live bacteria that can reach the effective part of the intestine is negligible, which usually is lower than the recommended concentration of at least 10^7 CFU g^{-1} [76–78]. The reasons are multiple; for example, (i) the resistance ability of the probiotics against the stress is weal because most of them do not generate spores, subsequently, most of them are dying when producing, transporting, and storing the probiotics products. (ii) There are the natural barriers of the high acidity and bile content in the upper gastrointestinal tract (UGT), which can kill most of the live microorganisms that humans consumed [77, 79–84]. Therefore, the development of bioaugmentation technologies protecting live probiotics is urgent.

Microencapsulating techniques, which are a natural or synthetic polymer encapsulating material coating microcapsules with diameters of 3–800 mm, are currently commonly applied for bioaugmentation. Normally, the microcapsules are with a semi-permeable or sealed capsule film. Pieces of solid evidences have shown that the microcapsules effectively enhance the resistance of microorganisms against environmental stress such as high temperature, dryness, stomach acid, and bile content, improving the stability of probiotics [79, 81, 85–88]. Besides, microcapsules can protect probiotics microorganisms from shear stress, provide a better microenvironment for microorganism survival, and facilitate increasing the concentration of the products. In summary, when the microcapsules are formed, the microorganisms are coated with the wall-materials to preserve their remediation activities better. Meanwhile, under appropriate conditions, the probiotics can be released when the wall-material is destroyed. The advantages include: i) microencapsulation changes the shape of microecological preparation products with converting probiotics into stable powder, which facilitates convenience of transportation and storage; ii) It can effectively prevent the inactivation of bacteria, and improve the stability of microecological environment due to the protection of microcapsules; iii) enteric wall-materials can protect the probiotics from low pH to transport probiotics to the target region of intestinal tract; iv) The water-insoluble wall-material can transform the microorganisms dissolving in the water uniformly. Therefore, microencapsulation is expected to improve the stability of probiotics during production, storage, and consumption, and produce microecological preparations that are resistant to the storage, high temperature, high pressure, and acid resistance.

7.5.1 Alginate-based micro-hydrogels

Alginate-based micro-hydrogels have been the most commonly used encapsulating carriers because of its bio-capability, safety, and low cost [81, 88–93]. Studies have shown that the counts of live Lactobacillus at low temperatures and 60 °C were increased when use sodium alginate solution and $CaCl_2$ solution to microencapsulate them. However, the effect of alginate remains controversial. Although some studies have reported that the alginate encapsulation enhanced survival of lactic acid bacteria in simulated gastrointestinal conditions [94–98], there are still some researchers found that encapsulation of probiotic bacteria in alginate beads did not effectively protect the microorganisms from a high acidity [99].

7.5.2 Starch-based micro-hydrogels

Starch, including (micro-)porous starch, was applied in the microencapsulation as well (Patents: CN104388416A and CN101904420A). The research showed that the embedding rate of lactic acid bacteria microcapsules with microporous starch reached more than 90%; meanwhile, it enhanced the storage stability in conventional aqueous solution, distilled water (Patents: CN104388416A).

7.5.3 Combinational wall-materials

There are shortcomings with the single type of wall-materials. For example, sodium alginate gel is porous and sensible to extreme pH values affecting both the release and protection of the compounds [84]. Thus, more studies tend to use the combinational well-materials for encapsulation. The survival rate of bacterial cells in alginate beads containing chitosan, alginate beads containing resistant starch (Hi-maize) and chitosan, or in chitosan-coated alginate capsules, in alginate-coated gelatin microspheres, in a combination of alginate with starch, in carrageenan-alginate beads, in alginate-citric pectin matrixes, were higher than that of pure alginate beads [76, 100–103].

7.5.4 Other water-insoluble wall-materials

With the development of technology, more and more wall-materials were produced to generate high capability microcapsules, e.g. agarose, Pectin-iron, zeolite, synthetic resin capsules, zeolite, exopolysaccharides (EPS), and fats/waxes [75, 86, 104–107]. Besides, more new microencapsulation techniques with bioaugmentation were developed and are developing, which expand the application of probiotics in gut bioremediation.

7.6 The Future of Gut Remediation

The gut remediation to remove pollutants is promising biotechnology, which improves new ideas in the protection of human health from pollution. It is a convenient technology for pollutant elimination in vivo because the functional intestinal microorganisms can be obtained by simple culture, and it colonizes in the gut for several weeks or longer. Besides, compared to traditional methods, gut remediation efficiently reduces the secondary pollutants to natural as well as the biological environment. With the development, such as isolation of functional microbes, genetic modification, and development of new microbial augmentation technology, gut remediation will further reduce pollutant emissions and accumulation and shows a low cost in the process of practical application in the future.

References

1. Chen HM et al (2000) Chemical methods and phytoremediation of soil contaminated with heavy metals. Chemosphere 41(1-2):229–234
2. Antoci A, Galeotti M, Sordi S (2018) Environmental pollution as engine of industrialization. Communications in Nonlinear Science and Numerical Simulation 58:262–273
3. Wang QYD, Cui Y, Liu X (2001) Instances of soil and crop heavy metal contamination in China. Soil Sediment Contam 10(5):497–510
4. Aoshima K (2012) [Itai-itai disease: cadmium-induced renal tubular osteomalacia]. Nihon Eiseigaku Zasshi 67(4):455463.
5. Collen A, Smith JJB (2015) 10% Human: How Your body's Microbes Hold the Key to Health and Happiness. 9(780316):380102
6. Cani PD, Delzenne NM (2011) The gut microbiome as therapeutic target. Pharmacology & Therapeutics 130(2):202–212
7. Woodhouse CA et al (2018) Review article: the gut microbiome as a therapeutic target in the pathogenesis and treatment of chronic liver disease. Alimentary Pharmacology & Therapeutics 47(2):192–202
8. Knight DJW, Girling KJ (2003) Gut flora in health and disease. Lancet 361(9371):1831–1831
9. Guarner F, Malagelada JR (2003) Gut flora in health and disease. Lancet 361(9356):512–519
10. Gordon JI (2012) Honor thy gut symbionts redux. Science 336(6086):1251–1253
11. Frank DN et al (2007) Molecular-phylogenetic characterization of microbial community imbalances in human inflammatory bowel diseases. Proc Natl Acad Sci U S A 104 (34):13780–13785
12. Schloss PD, Handelsman J (2004) Status of the microbial census. Microbiol Mol Biol Rev 68 (4):686–691
13. Gough E, Shaikh H, Manges AR (2011) Systematic review of intestinal microbiota transplantation (fecal bacteriotherapy) for recurrent Clostridium difficile infection. Clin Infect Dis 53 (10):994–1002
14. Hartmann P, Chen WC, Schnabl B (2012) The intestinal microbiome and the leaky gut as therapeutic targets in alcoholic liver disease. Frontiers in Physiology 3:402
15. Nishida AH, Ochman H (2018) Rates of gut microbiome divergence in mammals. Molecular Ecology 27(8):1884–1897
16. Winek K, Dirnagl U, Meisel A (2016) The Gut Microbiome as Therapeutic Target in Central Nervous System Diseases: Implications for Stroke. Neurotherapeutics 13(4):762–774
17. O'Hara AM, Shanahan F (2006) The gut flora as a forgotten organ. Embo Reports 7 (7):688–693
18. Pei Y et al (2018) Microbial community structure and function indicate the severity of chromium contamination of the Yellow River. 9:38
19. Yu Z et al (2014) A mer operon confers mercury reduction in a Staphylococcus epidermidis strain isolated from Lanzhou reach of the Yellow River. 90:57–63
20. Zheng Z et al (2015) A Bacillus subtilis strain can reduce hexavalent chromium to trivalent and an nfrA gene is involved. 97:90–96
21. Chen Y et al (2018) Long-term and high-concentration heavy-metal contamination strongly influences the microbiome and functional genes in Yellow River sediments. 637:1400–1412
22. Jiang Y et al (2015) Pseudomonas sp. LZ-Q continuously degrades phenanthrene under hypersaline and hyperalkaline condition in a membrane bioreactor system. 1(3):156–167

23. Huang H et al (2017) The naphthalene catabolic protein NahG plays a key role in hexavalent chromium reduction in Pseudomonas brassicacearum LZ-4. 7(1):1–11
24. Wu W et al (2016) Genome sequencing reveals mechanisms for heavy metal resistance and polycyclic aromatic hydrocarbon degradation in Delftia lacustris strain LZ-C. 25(1):234–247
25. Yu X et al (2016) Simultaneous aerobic denitrification and Cr (VI) reduction by Pseudomonas brassicacearum LZ-4 in wastewater. 221:121–129
26. Huang H et al (2016) A novel Pseudomonas gessardii strain LZ-E simultaneously degrades naphthalene and reduces hexavalent chromium. 207:370–378
27. Xu R et al (2018) Co-expression of YieF and PhoN in Deinococcus radiodurans R1 improves uranium bioprecipitation by reducing chromium interference. 211:1156–1165
28. Boopathy R (2000) Factors limiting bioremediation technologies. Bioresource Technology 74 (1):63–67
29. Souiri M et al (2009) Escherichia coli-functionalized magnetic nanobeads as an ultrasensitive biosensor for heavy metals. Procedia Chem 1(1):1027–1030
30. Biswas JK et al (2018) Exploring potential applications of a novel extracellular polymeric substance synthesizing bacterium (Bacillus licheniformis) isolated from gut contents of earthworm (Metaphire posthuma) in environmental remediation. Biodegradation 29(4):323–337
31. Cai M et al (2018) Systematic characterization and proposed pathway of tetracycline degradation in solid waste treatment by Hermetia illucens with intestinal microbiota. Environ Pollut 242(Pt A):634–642
32. Halttunen T et al (2008) Combining strains of lactic acid bacteria may reduce their toxin and heavy metal removal efficiency from aqueous solution. Lett Appl Microbiol 46(2):160–165
33. Feng PY et al (2019) A Review on Gut Remediation of Selected Environmental Contaminants: Possible Roles of Probiotics and Gut Microbiota. Nutrients 11(1)
34. Carter LJ et al (2014) Fate and uptake of pharmaceuticals in soil-earthworm systems. Environ Sci Technol 48(10):5955–5963
35. Sinha RK, Bharambe G, Ryan D (2008) Converting wasteland into wonderland by earthworms—a low-cost nature's technology for soil remediation: a case study of vermiremediation of PAHs contaminated soil. The Environmentalist 28(4):466–475
36. Nagaoka M et al (1995) Structural studies on a cell wall polysaccharide from Bifidobacterium longum YIT4028. Carbohydr Res 274:245–249
37. Landersjo C et al (2002) Structural studies of the exopolysaccharide produced by Lactobacillus rhamnosus strain GG (ATCC 53103). Biomacromolecules 3(4):880–884
38. Frece J et al (2005) In vivo Testing of Functional Properties of Three Selected Probiotic Strains. World Journal of Microbiology and Biotechnology 21(8):1401
39. Avall-Jaaskelainen S, Lindholm A, Palva A (2003) Surface display of the receptor-binding region of the Lactobacillus brevis S-layer protein in Lactococcus lactis provides nonadhesive lactococci with the ability to adhere to intestinal epithelial cells. Appl Environ Microbiol 69 (4):2230–2236
40. Jadán-Piedra C et al (2017) The use of lactic acid bacteria to reduce mercury bioaccessibility. Food Chemistry 228:158–166
41. Kelleher SL et al (2002) Supplementation of infant formula with the probiotic lactobacillus reuteri and zinc: impact on enteric infection and nutrition in infant rhesus monkeys. J Pediatr Gastroenterol Nutr 35(2):162–168
42. Scholz-Ahrens KE et al (2007) Prebiotics, probiotics, and synbiotics affect mineral absorption, bone mineral content, and bone structure. J Nutr 137(3 Suppl 2):838S–846S
43. Zhai Q et al (2016) Oral administration of probiotics inhibits absorption of the Heavy Metal Cadmium by Protecting the Intestinal Barrier. Appl Environ Microbiol 82(14):4429–4440
44. Yu L et al (2016) Potential of Lactobacillus plantarum CCFM639 in Protecting against Aluminum Toxicity Mediated by Intestinal Barrier Function and Oxidative Stress. Nutrients 8(12):783
45. Wu G et al (2017) Gut remediation: a potential approach to reducing chromium accumulation using Lactobacillus plantarum TW1-1. Sci Rep 7(1):15000

46. Zmora N et al (2018) Personalized Gut Mucosal Colonization Resistance to Empiric Probiotics Is Associated with Unique Host and Microbiome Features. Cell 174(6):1388–1405. e21
47. Bisanz JE et al (2014) Randomized open-label pilot study of the influence of probiotics and the gut microbiome on toxic metal levels in Tanzanian pregnant women and school children. MBio 5(5):e01580-14
48. Zhai Q et al (2014) Protective effects of Lactobacillus plantarum CCFM8610 against chronic cadmium toxicity in mice indicate routes of protection besides intestinal sequestration. Appl. Environ. Microbiol. 80(13):4063–4071
49. Li B et al (2014) Mercury nano-trap for effective and efficient removal of mercury (II) from aqueous solution. Nature communications 5:5537
50. Liu YR et al (2016) Effects of cellular Sorption on mercury bioavailability and methylmercury production by Desulfovibrio desulfuricans ND132. Environmental Science & Technology 50 (24):13335–13341
51. Chang D-E et al (2004) Carbon nutrition of *Escherichia coli* in the mouse intestine. Proceedings of the National Academy of Sciences of the United States of America 101 (19):7427–7432
52. Liu M et al (2019) Hg^{2+}-binding peptide decreases mercury ion accumulation in fish through a cell surface display system. Science of The Total Environment 659:540–547
53. Cabral L et al (2016) Methylmercury degradation by Pseudomonas putida V1. Ecotoxicology and environmental safety 130:37–42
54. Zhang W et al (2018) Risk assessment of total mercury and methylmercury in aquatic products from offshore farms in China. Journal of Hazardous Materials 354:198–205
55. Cerveny D et al (2016) Young-of-the-year fish as a prospective bioindicator for aquatic environmental contamination monitoring. Water research 103:334–342
56. Hsu-Kim H et al (2013) Mechanisms regulating mercury bioavailability for methylating microorganisms in the aquatic environment: a critical review. Environmental science & technology 47(6):2441–2456
57. Zhou HY, Wong MH (2000) Mercury accumulation in freshwater fish with emphasis on the dietary influence. Water Research 34(17):4234–4242
58. Harada M (1995) Minamata Disease: Methylmercury Poisoning in Japan Caused by Environmental Pollution. Critical Reviews in Toxicology 25(1):24
59. Tuzen M et al (2009) Mercury (II) and methyl mercury determinations in water and fish samples by using solid phase extraction and cold vapour atomic absorption spectrometry combination. Food and Chemical Toxicology 47(7):1648–1652
60. Liu M et al (2019) Reducing methylmercury accumulation in fish using Escherichia coli with surface-displayed methylmercury-binding peptides. Journal of Hazardous Materials 367:35–42
61. Uchikawa T et al (2010) The influence of Parachlorella beyerinckii CK-5 on the absorption and excretion of methylmercury (MeHg) in mice. The Journal of toxicological sciences 35 (1):101–105
62. Nakamori M et al (2016) Oral administration of erythromycin decreases RNA toxicity in myotonic dystrophy. Ann Clin Transl Neurol 3(1):42–54
63. Zhai Q et al (2020) Oral Supplementation of Lead-Intolerant Intestinal Microbes Protects Against Lead (Pb) Toxicity in Mice. Frontiers in Microbiology 10:3161
64. Ezzariai A et al (2018) Human and veterinary antibiotics during composting of sludge or manure: Global perspectives on persistence, degradation, and resistance genes. Journal of Hazardous Materials 359:465–481
65. Liu M et al (2019) Pretreatment of swine manure containing β-lactam antibiotics with whole-cell biocatalyst to improve biogas production. Journal of Cleaner Production 240:118070
66. Kafaei R et al (2018) Occurrence, distribution, and potential sources of antibiotics pollution in the water-sediment of the northern coastline of the Persian Gulf, Iran. Sci Total Environ 627:703–712

67. de Cazes M et al (2016) Erythromycin degradation by esterase (EreB) in enzymatic membrane reactors. Biochemical Engineering Journal 114:70–78
68. Minami T et al (1996) Effects of erythromycin in chronic idiopathic intestinal pseudo-obstruction. Journal of Gastroenterology 31(6):855–859
69. Kohno Y et al (1989) Comparative pharmacokinetics of clarithromycin (TE-031), a new macrolide antibiotic, and erythromycin in rats. Antimicrobial Agents and Chemotherapy 33 (5):751–756
70. Dubreuil JD (2014) Escherichia coli | Enterotoxigenic E. coli (ETEC). In: Batt CA, Tortorello ML (eds) Encyclopedia of Food Microbiology, 2nd edn. Academic Press, Oxford, pp 728–734
71. Liu M et al (2020) Reducing residual antibiotic levels in animal feces using intestinal Escherichia coli with surface-displayed erythromycin esterase. Journal of Hazardous Materials 388:122032
72. Hernández-Castellanos B et al (2013) Removal of benzo (a) pyrene from soil using an endogeic earthworm Pontoscolex corethrurus (). Applied soil ecology 70:62–69
73. Kersanté A et al (2006) Interactions of earthworms with atrazine-degrading bacteria in an agricultural soil. FEMS microbiology ecology 57(2):192–205
74. Singleton DR et al (2003) Identification of uncultured bacteria tightly associated with the intestine of the earthworm Lumbricus rubellus (Lumbricidae; Oligochaeta). Soil Biology and Biochemistry 35(12):1547–1555
75. Weinbreck, F., I. Bodnár, and M.J.I.j.o.f.m. Marco, Can encapsulation lengthen the shelf-life of probiotic bacteria in dry products? 2010. 136(3): p. 364-367.
76. de Araújo Etchepare M et al (2016) Effect of resistant starch and chitosan on survival of Lactobacillus acidophilus microencapsulated with sodium alginate. 65:511–517
77. Broeckx G et al (2016) Drying techniques of probiotic bacteria as an important step towards the development of novel pharmabiotics. 505(1-2):303–318
78. Hotel, A.C.P. aand AA.J.P. Cordoba, Health and nutritional properties of probiotics in food including powder milk with live lactic acid bacteria.. 2001. 5(1): p. 1-10.
79. Sohail A et al (2011) Survivability of probiotics encapsulated in alginate gel microbeads using a novel impinging aerosols method. 145(1):162–168
80. Doherty S et al (2011) Development and characterisation of whey protein micro-beads as potential matrices for probiotic protection. 25(6):1604–1617
81. Doherty S et al (2012) Survival of entrapped Lactobacillus rhamnosus GG in whey protein micro-beads during simulated ex vivo gastro-intestinal transit. 22(1):31–43
82. Dimitrellou D et al (2016) Survival of spray dried microencapsulated Lactobacillus casei ATCC 393 in simulated gastrointestinal conditions and fermented milk. 71:169–174
83. Huang S et al (2016) Double use of highly concentrated sweet whey to improve the biomass production and viability of spray-dried probiotic bacteria. 23:453–463
84. Mortazavian A, Sohrabvandi SJEP, Tehran (2006) Probiotics and food probiotic products; based on dairy probiotic products..
85. Oliveira AC et al (2007) Stability of microencapsulated B. lactis (BI 01) and L. acidophilus (LAC 4) by complex coacervation followed by spray drying. 24(7):685–693
86. Jiménez-Pranteda ML et al (2012) Stability of lactobacilli encapsulated in various microbial polymers. 113(2):179–184
87. Krasaekoopt, W., B. Bhandari, and H.J.I.d.j. Deeth, Evaluation of encapsulation techniques of probiotics for yoghurt. 2003. 13(1): p. 3-13.
88. Su R et al (2011) Encapsulation of probiotic Bifidobacterium longum BIOMA 5920 with alginate–human-like collagen and evaluation of survival in simulated gastrointestinal conditions. 49(5):979–984
89. Shaharuddin, S. and I.I.J.C.p. Muhamad, Microencapsulation of alginate-immobilized bagasse with Lactobacillus rhamnosus NRRL 442: Enhancement of survivability and thermotolerance. 2015. 119: p. 173-181.

90. Sohail A et al (2013) The viability of Lactobacillus rhamnosus GG and Lactobacillus acidophilus NCFM following double encapsulation in alginate and maltodextrin. 6 (10):2763–2769
91. Sousa S et al (2015) Characterization of freezing effect upon stability of, probiotic loaded, calcium-alginate microparticles. 93:90–97
92. Yeung TW et al (2016) Microencapsulation of probiotics in hydrogel particles: enhancing Lactococcus lactis subsp. cremoris LM0230 viability using calcium alginate beads. 7 (4):1797–1804
93. Yeung TW et al (2016) Microencapsulation in alginate and chitosan microgels to enhance viability of Bifidobacterium longum for oral delivery. 7:494
94. Lee, K.-Y. and T.-R.J.A.E.M. Heo, Survival of *Bifidobacterium longum* immobilized in calcium alginate beads in simulated gastric juices and bile salt solution.. 2000. 66(2): p. 869-873.
95. Adhikari K et al (2000) Viability of Microencapsulated Bifidobacteria in Set Yogurt During Refrigerated Storage1. 83(9):1946–1951
96. Abbaszadeh S et al (2014) The effect of alginate and chitosan concentrations on some properties of chitosan-coated alginate beads and survivability of encapsulated Lactobacillus rhamnosus in simulated gastrointestinal conditions and during heat processing. 94 (11):2210–2216
97. Mandal S et al (2014) Enhancement of survival of alginate-encapsulated Lactobacillus casei NCDC 298. 94(10):1994–2001
98. Kim S-J et al (2008) Effect of microencapsulation on viability and other characteristics in Lactobacillus acidophilus ATCC 43121. 41(3):493–500
99. Sultana K et al (2000) Encapsulation of probiotic bacteria with alginate–starch and evaluation of survival in simulated gastrointestinal conditions and in yoghurt. 62(1-2):47–55
100. Chávarri M et al (2010) Microencapsulation of a probiotic and prebiotic in alginate-chitosan capsules improves survival in simulated gastro-intestinal conditions. 142(1-2):185–189
101. Nag, A., K.-S. Han, and H.J.I.D.J. Singh, Microencapsulation of probiotic bacteria using pH-induced gelation of sodium caseinate and gellan gum. 2011. 21(4): p. 247-253.
102. Homayouni A et al (2008) Effect of microencapsulation and resistant starch on the probiotic survival and sensory properties of synbiotic ice cream. 111(1):50–55
103. Coghetto CC et al (2016) Electrospraying microencapsulation of Lactobacillus plantarum enhances cell viability under refrigeration storage and simulated gastric and intestinal fluids. 24:316–326
104. Ghibaudo F et al (2017) Pectin-iron capsules: Novel system to stabilise and deliver lactic acid bacteria. 39:299–305
105. Takei T et al (2017) Air drying on superamphiphobic surfaces can reduce damage by organic solvents to microbial cells immobilized in synthetic resin capsules. 54:28–32
106. Hassanzadeh AM et al (2017) Immobilization and microencapsulation of Lactobacillus caseii and Lactobacillus plantarum using zeolite base and evaluating their viability in gastroesophageal-intestine simulated condition.
107. Alehosseini A et al (2019) Agarose-based freeze-dried capsules prepared by the oil-induced biphasic hydrogel particle formation approach for the protection of sensitive probiotic bacteria. 87:487–496

Chapter 8
Current Policies and Policy Implications for Environmental Pollution

Huawen Han, Haiying Huang, and Xiangkai Li

8.1 Introduction

In the past two decades, the average growth rate of China's GDP is 9.7% [1]. Such economic development and rapid urbanization exacerbates the environmental pollution of China [2]. These environmental problems mainly contain water pollution [3], air pollution (CO_2, SO_2, $PM_{2.5}$) [4], soil pollution [5], etc. According to the statistics from Ministry of Environmental Protection of China, $PM_{2.5}$ concentration in some prefecture-level cities greatly exceeded the average value of 25 μg m^{-3} recommended by the WHO [6]. Severe air deterioration 42% of premature deaths globally [7]. Thus, the Chinese government successively enacted a series of strategies to achieve the sustainable development of environment [8, 9]. With economic growth and increased public awareness, the core principles of China's environmental protection shifted from the previously simple control of the Three Wastes towards the construction of ecological civilization [10]. This evolution in environmental management system was driven by economic progress, with a significant environmental Kuznets inverted U curve between economic growth and environmental pollution [11, 12]. Environmental economists have traditionally argued that market-based tools are more cost-effective than command control policies in all relevant jurisdictions if there is heterogeneity in the cost of controlling air pollutants [13]. The cost-effectiveness of market-based tools essentially relies on setting policy

H. Han · X. Li (✉)
Ministry of Education Key Laboratory of Cell Activities and Stress Adaptations, School of Life Sciences, Lanzhou University, Lanzhou, People's Republic of China
e-mail: xkli@lzu.edu.cn

H. Huang
Institute of Virology (VIRO), Helmholtz Zentrum München, German Research Center for Environmental Health, Neuherberg, Germany

Germany Insitute of Virology, Technical University of Munich, München, Germany

X. Li, P. Liu (eds.), *Gut Remediation of Environmental Pollutants*,
https://doi.org/10.1007/978-981-15-4759-1_8

objectives to cut the total cost of reducing emissions by a given amount of pollutants. As for China's environmental strategy, environmental protection work was gradually incorporated into the national five-year plan since 1996. Emission reduction targets dominate the environmental agenda between 11th and 13th five-year plans (2006–2020) to indicate the potential use of market-based tools. Taking consideration of public complaints about air quality, China's central government took direct measures to control pollutant concentrations in the Environmental Prevention and Control Action Plan [14]. Thus, market-based tools are unsuitable for cost-effective improvements in environmental quality.

Although the Chinese government has carried out many environmental and ecological projects to handle environmental problems, such as carbon reduction emissions and developing renewable resources, new environmental problems are still inevitable with economic development. In this chapter, we provide detailed information on the development phase of environmental policies and analyzed the success or failure in environmental regulations from different countries; this contributes to explore the major challenges and shortcomings in environmental management. This process would be beneficial to formulate flexible environmental policies.

8.2 Policy Evolvement of Environmental Management

Environmental problems trigger various diseases (e.g. pulmonary dysfunction, respiratory disease, etc.) [15]. Taking consideration of the damage resulting from toxic pollutants, the Chinese government initiated to take some effective measures to address environmental problems. Indeed, the central government has put environmental governance on the agenda since Stockholm conference in 1972. With China's reform and opening up for 40 years, many factors, such as economic development, urbanization, population, energy price, third-party monitoring, public participation, have a direct or indirect the on policy formulation and reforms in Chinese environmental governance [16–18]. In spite of the spatial variability in regionalized environmental regulations [16, 19], China's environmental governance mainly contains authoritarian control and market-based mechanisms [17]. In this section, the evolvement of environmental management is discussed in more detail depending on the goals of China's five-year development plan. Additionally, we provide a comparative analysis of environmental policies between China and other countries.

8.2.1 Development Phases of Environmental Managements in China

Extensive studies have been conducted to elucidate the evolution of environmental policies in China, such as rural energy policies [20], urban minerals policies [21], marine policies [22], etc. In the past four decades, China's environmental managements have experienced the following changes shifting end-of-pipe treatment, pollution prevention and process control, regional environmental governance and a legal means and economic instruments-based approach. This section gives the detailed information on the characteristics of these three stages.

8.2.1.1 End-of-pipe and Damage Control (1972–1991)

In order to narrow the economic gap with developed countries, China faces the great leap forward for the movement of "conquer nature," accompanied with the severe deforestation, water pollution, soil erosion, air pollution. In particular, the Stockholm conference in 1972 adopted the declaration on the human environment and proposed that June 5 of each year be designated as "world environment day." Thus, the Chinese government valued the environment issue highly and introduced a series of measures and actions in the spirit of the Stockholm conference. In 1973, the first national environmental protection conference passed first environmental protection document "several provisions on protecting and improving the environment," which marked the beginning of China's strict environmental protection work. With China's reform and opening up in 1978, the first "environmental protection law of the People's Republic of China" (EPL) was passed, and the state environmental protection bureau (NEPA) was established under the ministry of rural and urban construction. The key principles in environmental protection contain "prevention is primary, then control," "polluter of pollution control," and "strengthening environmental management" [23]. Subsequently, the second national environmental protection conference confirmed that environmental protection has been listed as a strategic task of China's modernization construction in 1983. This indicates environmental protection plays an important role in China's economic and social development, and has a far-reaching impact on the implementation of China's environmental protection program in the future. In 1988, the amendment to the China's environmental protection law marked the beginning of comprehensive environmental legislation to combat the deteriorating environment [24].

At this stage, the focus of environmental management is to achieve the comprehensive improvement of the urban environment and control industrial pollution. Different regions and industries should formulate their own environmental targets, and enumerated 51 key cities for environmental protection. Chinese government encourages enterprises to formulate regulations and take the initiative in pollution control and environmental protection. With regard to major industries and densely populated areas, extensive studies have been conducted to explore the pollution status (e.g. air, water, and solid waste pollution) and corresponding environmental control measures [25, 26]. According to the survey results, large industries were a major source of pollution and urged central and local government to adjust environment action plan, such as strict emission standards, licensing system for state-owned enterprises. Additionally, the maximum allowable emission level of waste (e.g. gas, wastewater, etc.) has become kernel tool to maintain environmental quality. During pollution control of chemical and heavy metal industries, the end-management approach has made significant progress via point source pollution control.

Following the current governance model, environmental agencies have chosen a top-down approach, lack of other stakeholders, or local authority involvement. The guiding principle of environmental protection developed in the first stage is "environmental protection is coordinated with economic and social development," but economic development must precede over environmental protection. This policy emphasized that construction and operation of new industrial enterprises must install appropriate waste treatment facilities. Furthermore, internationally renowned criteria, including "the polluter pays principle" and "the priority principle for reducing pollution at source," are successively introduced.

8.2.1.2 Pollution Prevention and Process Control (1992–2001)

Since the Rio Conference in 1992, China enacted Ten Strategic Policies for environment protection to realize sustainable development [27]. In 1994, the Chinese government promulgated Agenda 21; this agenda has identified detailed planning objectives on population, environment, and economic development in the twenty first century [28]. Subsequently, China implemented "two fundamental transformations": (1) shifting from planned economy to market-oriented economy; (2) the transition from extensive to intensive economic growth mode. To avoid the flood disaster in 1998, the Chinese government recognized the urgency of ecological conservation, and "pollution control and ecological conservation" are put in coequal and significant position. In this respect, a number of policies were implemented, such as banning the logging of natural forests in the reaches of Yangtze River and Yellow River, ecological restoration in China's western region. This marked a transformation for China's ecological conservation.

During this stage, the style of policy and decision-making are gradually changing with the emergence of new environmental problems, leading to the shift in environmental management form passive cleanup actions to active pollution prevention policies [14]. Furthermore, the total investment in pollution control has been soaring rapidly and reached 346 billion Yuan during the Ninth Five-Year Plan Period (1996–2000), accounting for about 1% of China's GDP [10]. The public takes an active part in environmental and biodiversity conservation. Compared with environmental governance adopted in the first phase, the second phase has proven to be a broader tool than top-down command control strategy. On account of US cap-and-trade approach in the 1970s, sourcing-oriented pollution prevention, combined with total emission control facilitated the introduction of financial incentives, environmental responsibility, and emissions trading [29]. For example, the establishment of sewage permit system was launched to remove water and air pollutants in Shanghai and 16 other cities. In 1999, the environmental protection fund (EDF) firstly initiated a true emissions trading scheme [30].

Generally, the most of environmental management are implemented via a "campaign or storm" approach, including "pollutant total emission control plan", "cross-century green engineering plan", "332111" plan. These programs can achieve specific goals within given time and reduce emissions from the point pollution source. Nonetheless, sustained economic growth remains a major challenge to existing environmental regulation. Although reforestation has obtained significant progress in the second half of the 1990s, soil erosion and debris flow are still inevitable. These lessons call for establishing a long-term mitigation profile, including the pollution effects of potential future production growth [30]. Especially after China joining WTO, international demand for green products has also urged Chinese companies to pass a higher level of ISO14000 certification, and to emphasize cleaner production, eco-labeling systems [31].

8.2.1.3 Integrating Environment and Economy (2002–The Present)

In nature, the environmental issues are the development issues. Thus, China has put forward a series of new ideas to maintain balance between the economy and environment to guide ecological and environmental protection. Under the premise of strict control of pollutant discharge, the third stage sustained the coordinated relationship between environment and economy at large [32]. After stepping into the twenty first century, many new economic concepts are emerging, including the low-carbon development (2009), the ecological civilization construction (2012), etc. These concepts provide new insight into the relationship between environment and development in China, which in return showed that environmental problems, coupled with the economic development contributed to the concept innovation.

This stage encourages integration of environmental and economic objectives. With the promulgation of the cleaner production promotion law in 2002, it has been considered as a turning point of China's environmental policies ranging from terminal control to pollution prevention. In 2003, National Environmental Safety Strategy Report showed that environmental costs paid for economic development offset the benefits of economic growth in many regions. The decision of the state council on implementing the scientific outlook on development and strengthening environmental protection were issued in 2005. Subsequently, the report of the 17th party congress claimed that "economic development should be coordinated with population, resources and the environment." By the end of 2010, 18 provinces have carried out trials of ecological compensation, compensation schemes contain mineral recovery, watershed environmental protection, and nature reserves [33]. In addition to the energy consumption intensity and carbon emission intensity, some environmental economic instruments (e.g. green credit, green insurance, etc.) have been incorporated into the 12th five-year plan [34]. In short, total quantity control in energy conservation and emission reduction is major objective of the 11th Five-Year Plan and the 12th Five-Year Plan. According to data observed by NASA (USA) satellites, SO_2 emission in China decreased by 70% in 2005–2016. After the 13th Five-Year Plan, the improvement of ecological and environmental quality is core model of environmental governance and promoted the action plans for air, water, and soil pollution. In addition, this planning outline created a set of standards for new vehicle emission, wastewater or solid waste treatment. It is estimated that $600 billion will be used for clean energy, environmental protection, and growing areas, especially booming investment in renewable energy. In 2018, the new Ministry of Ecology and Environment is mainly responsible for pollution control and climate change mitigation, including the country's nascent Emissions Trading Scheme [35]. At the same time, the central government established the overall goal of attaining a completely well-off society by 2020 [36], but there still exists two urgent issues: (1) how to solve contradiction between resource depletion and environmental pollution; (2) how to achieve coordinate development between the ecological civilization construction and moderately well-off society [37, 38].

Overall, the environmental management at all phases belongs to a top-down command control strategy. Looking back to the development of China's policy, many experiences and lessons for the implementation on environment regulation provide warnings for accurate estimation of the future. Indeed, existing policies or approaches are inadequate to meet the challenges facing China in the future. In view of severe environmental situation, Chinese government should implement flexible principles, as well as broad international collaboration and public participation.

8.2.2 Comparison of Environmental Policies between China and Other Countries

8.2.2.1 Policies Adjustment

Global environmental assessments face huge challenges, such as the issue of scale, policy relevance, etc. [39]. An increasing number of institutions and organizations share more information related to "diversity and climate" on a global scale to influence negotiations and decision-making processes [40]. Unfortunately, formulation and implementation of environmental policies in the global level are always questionable, such evidence can be proven form study conducted by Turnhout et.al [39]. In restrictive sense, global forms of environmental knowledge are not very authoritative. Strengthening environmental education offers an advantage on establishing food products with labeled environmental credentials [41].Thus, national governments should timely tailor the related environmental policies to local conditions.

The contradiction between environmental policies and economic competitiveness has become a topic of global concern [42]. A traditional command-and-control environmental regulation has emerged in developed countries since 1970s, such as

USA' Clean Air Act, Swedish industrial pollution control, etc. In the last 20 to 30 years, developed countries gradually adjusted their environmental policies depending on pricing mechanisms [43]. However, this command-and-control environmental regulation are still prevalent in many developing countries, e.g. "Two Control Zone" in China. Tang et al. suggested the implementation of market-based environmental regulation is beneficial to environmental sustainability [44]. Apart from conventional command-and-control or market-based approaches, informational governance of the environment has been adopted as a new mode [45], which provides high-effective strategy to incentivize environmental policy implementation in democratic societies.

8.2.2.2 Pollution Control

Haze pollution was regarded as a necessary consequence of industrial progress and attracted worldwide environmental concern [46]. Despite air pollution research have been launched for 60 years in Europe and the USA, haze still occured in Los Angeles and London [47]. China undergos considerable trajectory previously experienced by developed countries, large-scale haze pollution shrouds the northeast region along with industrialization and urbanization, especially in urban agglomerations [48–50]. On account of "blue sky fabrication in China," related control policies are mainly implemented in international mega-events in China, such as the 2008 summer Olympic Games, the 2014 APEC summit, and the 2016 G20 summit [51]. This process represents an ad hoc top-down *campaign-style* of governance rather typical of Chinese politics, which is distinct from Air Clean Act of other countries. Particularly, air pollution ranking system in China motivates environmental administrations in bottom cities to strengthen air pollution control [52]; this top-down environmental information disclosure only maintains the short-term effect. In 2014, the amended Environmental Protection Law canceled the upper limits on fines for factories that cannot reach emissions standards, promoting the construction of ecological civilization [53]. In contrast, green technological innovation and industrial structure optimization driven by environmental regulation favors the mitigation of haze pollution [35, 54].

With regard to persistent organic pollutants (POP), the United Nations Environment Programme has enumerated 12 POP. In terms of expensive and time-consuming characteristic in POP detection, only few countries, such as the USA, Germany, or the Arctic region, detected the concentrations of POPs in blood, urine, and/or milk samples from human [55]. Most comprehensive studies were conducted to evaluate the effect of human exposure to environmental chemicals and provided a detailed information on their distribution in the US population [56]. In European Union, a new legislation (REACH), involved in registration, evaluation, authorization, and restriction of chemical substances, came into effect in 2007 (EC, 2001), resulting in the registration of 4725 chemical substances. Moreover, European Pollutant Release and Transfer Register database share available information on pollutants discharge. Although China has achieved significant advancements in monitoring the levels of POPs in different habitats via chemical management policy and legislations, e.g. National Implementation Programme of the Stockholm Convention [57], China still lags far behind developed countries because of lacking systematic research methods [56]. In Austria and some European countries (e.g. Netherlands, Norway and Sweden, etc.), the implementation of environmental taxes on chemical compounds has proven their potential importance in reducing fertilizer use [58].

In the waste management sector, it is estimated that the annual total amount of e-waste in China will reach 883,800 and 955,400 tons by 2015 and 2020, respectively [59]. Thus, Life Cycle Assessment (LCA) has become a standard tool for decision- and policy-making in developed countries for seeking more integrated and sustainable waste management systems [60]. In contrast, emerging nations are still in the transitional stage from open dumpsites to controlled landfills owing to high urbanization growth and economic expansion [61]. Similar situation appears to be relatively homogeneous in Latin America [62]. The latter still has some obstacles to the application of new technologies. As for BRICS countries, South Africa has no legislation to address e-waste pollution [63]. Although all the countries in Africa authorized the Basel Convention, the majority of them never enact an E-waste law [64]. The core principle of waste management is designed to attain circular economy; it focuses on boosting reuse and expanding their life span [65]. The European Union and China already have legislation on circular economies, including Circular Economy package [66] and Circular Economy Promotion Law [67].

8.2.2.3 Chinese Politics

The Chinese-style of decentralization has a negative impact on the efficiency of local environmental governance [68], which makes many cadres to emphasize economic development at the expense of the environment [69]. A growing literature provides evidence that cadres' terms and characteristics obviously influenced the local environmental pollution [70, 71], exclusively belonging to China. The effect of cadres' term on environmental pollution shows an "inverted U" curve; highly educated cadres acquaint the equal importance of environmental protection and economic development, contributing to the control of environmental pollution [70]. Furthermore, cadres in central and western regions of China are more likely to interfere in the environment problems. Another study found that a prefecture party secretary's years in office exhibited a U-shaped relationship with average annual $PM_{2.5}$ concentration [71].

8.2.2.4 Urbanization

The close relationship between urbanization and environmental pollution is general consensus worldwide, the unique difference is that the pace of urbanization in western countries precedes the developing countries. Although there has no linear

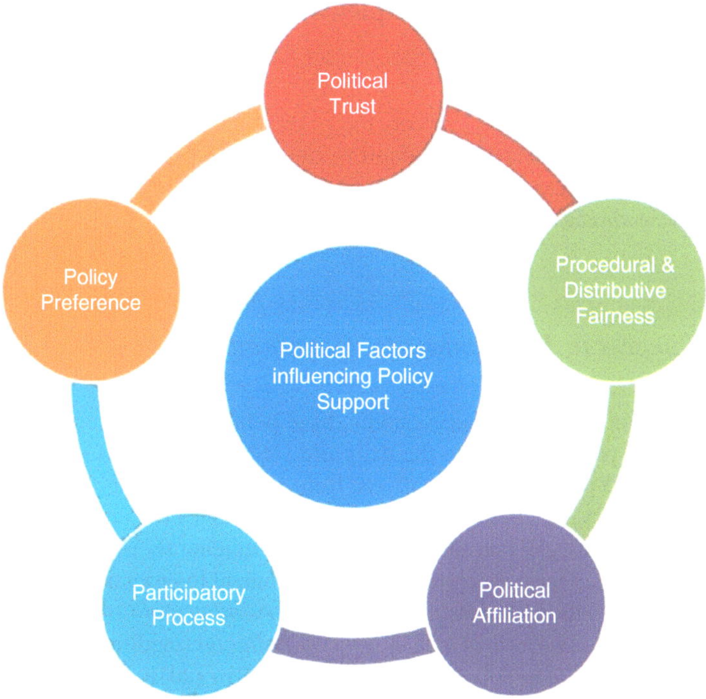

Fig. 8.1 The political factors are associated with public support of environmental policies [77]

relationship between urbanization and environmental pollution [54, 72], urbanization accelerated the process environmental pollution with an increase in the proportion of secondary industries. Another study on the spillover effect of urbanization found the environmental pollution exhibited significant agglomeration characteristics, and such spillover effect is adverse in the eastern region [73].

Overall, the state-oriented development in Asian countries emphasizes the role of government effect in enforcement effectiveness of environmental policies and on the relocation of industry [74], this is unique from environmental policies of western countries [75]. More importantly, environmental regulations with clear objectives and flexible approach contribute to the formation of diversified market-based environmental regulation policies [44, 76]. In addition, public support are crucial to the effective implantation of environmental protection policies [77] (Fig. 8.1).

8.3 Environmental Policy Implementation: Successes and Failures

Environmental regulation (ER) reflects the government's direct and indirect interventions on pollution behaviors to improve the environment. This section briefly provides some case study for environmental performance, including success and failures. These classical examples can be from China, USA, Northern Europe, etc.

8.3.1 *Success in Environmental Governance*

With the blowout trend of China economy, haze pollution dominated by PM_{10} and $PM_{2.5}$ seriously posed a threat to the daily life [16]. According to statistics, about 100 cities from 25 provinces have suffered from one month a year of haze pollution all over the country. A 2015 report by Beijing Morning Post noted that versatile "anti-smog tea" are widely available in Chinese medicine shops, pharmacies, and online sites; this data further urged Chinese government to enact a range of policies to combat haze pollution, such as *Air Pollution Prevention and Control Action Plan* (2013), *Prevention and Control of Atmospheric Pollution* (modified version, 2015), *Three-Year Action Plan to Win the Blue Sky Defense War* (2018) [51]. Energy saving and emission reduction policies carried out by central government and local governments stimulate air pollution control [78]. In fact, extensive studies focus on the implementation of environmental regulation via different methods. In the environmental regulation study from Quebec, Canada, companies must provide more accurate pollutant emission reports, further resulting in reduction of air pollutant emissions [79], similar result of environmental regulation was observed in Zhenjiang polluting enterprises on the basis of GMM estimation analysis from 1993 to 1997 [80]. Shapiro et.al found that environmental regulation contributes to a 75% reduction of air pollution from manufacturing data in the USA from 1990 to 2008 [81]. However, some scholars have achieved contradictory conclusions. For example, the Clean Air Act showed low efficiency on the decreasing sulfur dioxide concentration (1970s–1990s) of USA [82]. The role of strict traffic restrictions on air pollution of Beijing is still controversial [83]. Guo et al. pointed out that fragmented environmental regulatory framework causes underuse of regulatory resources, and jurisdictional density has a negative effect on air quality [84].

In most cases, it is believed that China's environmental regulation plays a positive role in alleviating environmental pollution. Guo et al. reported the current environmental regulations of China can effectively control carbon emissions [85], supported by Tapio decoupling models, differential GMM methods, and peak forecasting models. The positive effects of the green traffic pilot cites in reducing SO_2, NO_2, and PM_{10} are more obvious, and their concentrations decreased by 10.7%, 11.2%, and 9.8%, respectively [86], suggesting the potential of green traffic system in controlling air pollution. By detecting air quality scores of 26 cities in the Yangtze

River Delta region of China [87], the air pollution control policies have shown clear improvements in Shanghai, Jiangsu, and Zhejiang provinces. Among them, Zhejiang province has shown the characteristics of "Campaign-style Governance" right before the G20 Summit. It is worth noting some cities still showed continuous deterioration in air quality scores. Zhang et al. utilized super-slack-based measure (Super-SBM) model to evaluate the environmental efficiency of 283 Chinese cities from 2003 to 2016 [88]. Although overall environmental efficiency performance across the Chinese cities is low, the surveyed period showed an obvious increase in environmental efficiency. This finding revealed a win-win balance between environmental protection and economic development in the Chinese context. Using a threshold regression model and panel data of 30 provincial-level regions administrative regions in China, Zhang et al. found that environmental regulation significantly decreased the amount and intensity of carbon emissions [89]. In terms of performance of Chinese Two Control Zones policy for controlling acid rain and sulfur dioxide, difference-in-differences (DID) analysis showed that environmental regulation exhibited a lower level of polluting industrial activities, accompanied with the transfer to non-targeted regions [90]. Recently, Zhang et al. found current environmental regulation policies obviously restrained the haze pollution and achieved the expected effects [91]. In the assessment of emissions in the USA and the effectiveness of environmental policies, five policies, including Air Pollution Control Act (1965), Air Quality Act (1967), the amendment of Clean Air Act (1970, 1977, and 1990), showed a sharp decrease on per capita NO_x (Nitrogen Oxides) and VOC (Volatile Organic Compounds) emissions by means of using fractional integration techniques [92]. It is controversial that the emission reduction in the EU is inferior to the reduction in USA [93]. The possible reason is associated with the mandate and roles of Environmental Protection Agency (EPA).

The emission trading system (ETS) pilot contributes to the emission reduction process and stimulates environmental and ecological governance [94, 95]. To evaluate collaborative governance effects of emissions trading system (ETS) implementation on air pollution, DID method, and mediating effect model, China's ETS pilots showed an obvious "reduction effect" on haze concentration and SO_2 emissions with time extension [96]; their concentration decreased by 0.933 mcg/m^3 and 0.7452 tons, respectively. This study further confirms the rationality and high effectiveness of ETS by boosting the popularization of green technologies among enterprises. Furthermore, transaction volume of China Certified Emission Reduction and the total penalty amounts incurred are main driving factors to curb haze pollution. In a study of regional green innovation [97], the overall effect of ETS was insignificant. Intriguingly, some provinces that independently implemented ETS achieved better results than those approved by the central government.

8.3.2 *Failure in Environmental Policies*

Despite the implementation of several policies, Tehran was still considered as the most polluted non-Eastern Asian megacity with [98]. According to annual average concentration in Tehran from 2007 to 2018 [99], its PM concentration is 4–5 fold higher than the standard point. Taksibi et al. provided a detailed information on air pollution in the megacity of Tehran [100]; the spatial distribution of energy was a crucial factor of effectiveness of mitigation actions. Because of ignoring source distribution effects and geographical conditions, the actual deviation from pollutants emission levels further misguides environmental impact assessments.

In the migration study of pollution-intensive industries throughout China's Guangdong Province, Non-Pearl River Delta has evolved into a pollution haven [101]. Another study from the water crisis of Taihu Lake Watershed (TLW) in 2007 reported pollution haven hypothesis (PHH) only works efficiently in the short term owing to tightened environmental regulations [102]; this response is markedly different from the usual response caused by standard environmental policy-making and administrative procedures. Actually, the majority of polluting industries has transferred destinations to rural areas away from public eye, which coincides with the finding of Zhu et al. [103]. In these rural areas, the deficiency of environmental protection treatment facilities may trigger more severe pollution incidents and ecological damage in the long run. Although previous studies confirmed enforcement effectiveness of environmental policies caused firm to migrate in China [104], stringent environmental policies were compromised by differences across the watershed. Compared to Wuxi (the site of the crisis), other regions (e.g. Huzhou) are rather insensitive to environmental regulations enforced by pollution incidents [102].

In Brazil, Strategic Environmental Assessment (SEA) has a reputation to handle the climate change issues in the planning process [105] and promote sustainability in decision-making [106]. The quality of 35 SEA reports (1997–2014) merely reached 37% of framework criteria owing to a missing link between climate change policies (e.g. National Policy of Climate Change, NPCC) [107]. The evidence can be reinforced what was concluded in the previous studies [108, 109]. Thus, constructing a legal framework is the prerequisite of to promote the integration of NPCC objectives into plan-making. In this respect, the lessons learned from the Brazilian context would encourage other countries to reinforce the SEA capacity to improve the effectiveness of climate change public policies. The IPCC Summit in 2015 clearly pointed out that effective strategies implementation can relieve climate change, but their effectiveness varied greatly in different countries [107]. On the other hand, systematic reviews and updating meta-analyses facilitate the formulation and regulation of environmental policy [110].

The serious pollution of Ganges River provides a typical model to explore developing-country environmental health and policy [111]. To address water pollution in India, National River Conservation Plan (NRCP, 1985) and Central Pollution Control Board (CPCB, 2009) were successively promulgated, respectively, but little evidence are successful to improve water quality in Indian cities [112]. The failure of these environmental regulations is attributed to poor inter-agency cooperation, funding imbalances across sites, and low sewage treatment capacity [113]. Recently, Indian judiciary has gradually stepped in environmental activism [114]. In the case study of Supreme Court rulings that targeted industrial pollution in the Ganga River [111], DID analysis revealed that this adjudication alleviates river pollution and one-month infant mortality over ten years.

8.4 Challenges of Environmental Protection

Global industrialization and urbanization have been rapidly developed in the late twentieth century and the early twenty first century. Subsequently, with the vigorous development of industrial and agricultural production, human living standards have improved. However, environmental pollution, such as air pollution, global warming, water pollution, soil pollution, and freshwater crisis, has gradually emerged along with boomed development. More seriously, environmental pollution has become a limiting factor that restricted the progress of the modern industry. Thus, human beings are facing significant challenges to balance the industrialization and environmental protection. To this purpose, humans are exploring how to maintain productivity without increasing the ecological crisis and building new ways to retreat the pollution which occurred.

8.4.1 Serious Environment Pollution

In recent years, China's society and economy keep developing with the boomed global industrialization. Likewise, problems of environmental pollution have occurred as well. According to the Global Environmental Performance Index (EPI) report released by the Columbia University, the World Economic Forum shows that China's GDP growth rate ranking rose from fourth place in 2006 to the current second place. However, the environmental performance index dropped from 56 points in 2006 to 43 points in 2014, with the ranking dropped from 94th to 118th. Moreover, the ecological efficiency index has always been in the back row. The environmental problems, e.g. atmospheric, water, and soil pollution has attracted the most public attention.

8.4.1.1 Air Pollution

A large number of studies have shown that various impurities in the air (such as sulfur dioxide, nitrogen dioxide, PM_{10}, and $PM_{2.5}$) may damage multiple systems and organs of the human body in absolute concentrations, which has a major impact on human lethality population [115]. In 2016, the International Energy Agency

(IEA) released a special report on energy and air pollution, stating that air pollution has become the fourth largest threat to human health after secondary hypertension, dietary risks, and smoking. Air pollution led to premature death of about 6.5 million people worldwide. This number far exceeds the number of deaths caused by humans and malaria each year.

According to the "Communique of the State of the Environment of China, 2014," only 16 of the 161 cities whose air quality was monitored reached the average air quality standards, while the rest 145 cities failed. As a result, an estimated 600 million people live in the low quality of air. Additionally, according to the Global Environmental Performance Index (EPI) report released by Yale University in 2014, China's air quality (18.81 points) ranked 176 out of 178 participating countries, with the average $PM_{2.5}$ exposure level (2.44 points), the first to last, and $PM_{2.5}$ exceeding the standard rate (0 points). Most pressing, the haze days in some regions, for example, Beijing-Tianjin-Hebei, the Yangtze River Delta, and the Pearl River Delta, exceeded 200 days per year. Thus, Zhu Chen, an academician of the Chinese Academy of Sciences, and Jinnan Wang, an environmental protection expert of China, claimed that the health effects of pollution are a crucial problem that cannot be ignored. Meanwhile, the Chinese government kept treating air pollution actively [115].

8.4.1.2 Water Pollution

In 2014, 968 surface water sections of 423 major rivers and 62 key lakes were monitored. The rivers and lakes reached Grade I to Inferior Grade V accounted for 3.4%, 30.4%, 29.3%, 20.9%, 6.8%, 9.2%, respectively (Fig. 8.2). Namely, the aquatic environment was less than Grade V accounted for 16% in total [China Environmental Bulletin 2014]. It was estimated that eutrophication occurred in more than 25% of lakes and reservoirs.

The quality of groundwater is even worse. In the 4896 groundwater monitoring points in total, the water quality reached the Grades of Excellent Quality, Good Quality, Normal Quality, Worse Quality, and Extreme Bad Quality accounted for 10.8%, 25.9%, 1.8%, 45.4%, and 16.1%, respectively (Fig. 8.3).

For the coastal water, the monitoring points below Grade IV or Less-Grade-IV accounted for 26.2%, while the monitored points with the acceptable grades (i.e., Grade I–III) were 73.8%.

Moreover, because of the over-exploitation of water sources, the water ecology in China was extremely imbalanced. Water shortage existed extremely in China. Thus, water pollution and scarcity are still the most severe environmental pollution in China.

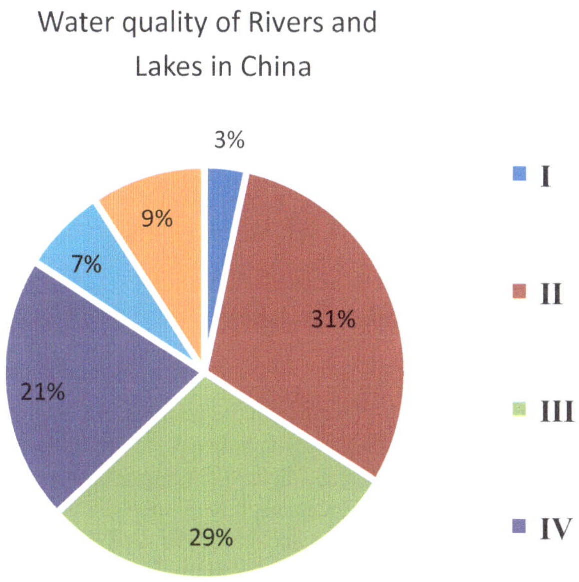

Fig. 8.2 The quality of water in rivers, lakes in China

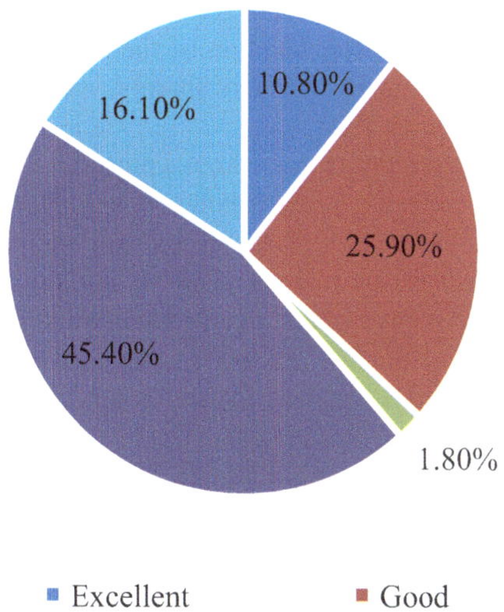

Fig. 8.3 The quality of groundwater in China

8.4.1.3 Soil Pollution

By the end of 2013, there were 641.6684 million hectares of agricultural land across the country. The total area of existing soil erosion in the country is 2.491 million square kilometers, accounting for 31.12% of the total census area. Compared with the survey results of the overall national soil environmental cleanup during the Seventh Five-Year Plan period, China's soil environment has shown a sharp deterioration in the past 40 years. The country has 300 million acres of heavy metal polluted arable land, 32.5 million acres of arable land polluted by sewage irrigation, and 2 million acres of solid waste storage and land destruction. Soil pollution in some areas is serious, and the quality of the farming soil environment is worrying [116]. The soil environment of industrial and mining wasteland is a serious problem [117]. Soil pollution endangers food safety [118, 119]. The stability and function of soil ecosystems are affected [120].

8.4.1.4 Ecological Degradation

In 2012, the "China Ecological Footprint Report 2012" jointly released by the Worldwide Fund for Nature and the Institute of Geographic Sciences and Resources of the Chinese Academy of Sciences showed that 80% of China's provinces currently have "ecological deficits." Qinghai, Inner Mongolia, Xinjiang, Yunnan, and Hainan are "ecological surpluses." In 2013, "excellent" and "good" counties in 2461 counties nationwide accounted for 46.7% of the country's land area, "good" counties were 23.0%, and "poor" and "bad" counties were 30.3%. The quality of natural ecosystems such as forests and grasslands is low, the functions of grassland ecosystems are damaged, and grasslands in traditional pastoral areas are seriously degraded. The function of the wetland ecosystem is degraded, the wetland disappears, the area shrinks sharply, and the natural regulation capacity of the wetland decreases.

In response to the above problems, China has adopted a series of major measures to protect and improve the ecological environment, increased the construction of the ecological environment, and fully protected and improved the ecological environment in some areas of China, mainly reflected in important ecological processes. These environmental protection elements mainly include progress has been made in key ecological processes such as afforestation, soil and water conservation, grassland construction, and land remediation. The key prevention and control measures for soil and water conservation in the upstream and middle reaches of important rivers have been fully implemented. The protection of natural forest resources in key areas; the establishment of different types of nature reserves, scenic spots, and forest parks; the construction of ecological agricultural pilot demonstration areas and ecological demonstration areas has been developing steadily, and the environmental protection legal system has been gradually improved. The state also plans to invest a lot of money in pollution control projects to promote coordinated economic and ecological development. However, we cannot fail to acknowledge that China's ecological environment is still in a difficult position. At present, the deterioration of the ecological environment in some areas has not been effectively contained, and the ecological environment continues to deteriorate. We must fully realize that the results of environmental protection work are still fragile, and the goals to be achieved are preliminary and phased. Therefore, we must face up to the status quo of pollution problems and seek causes and solutions.

8.4.2 Extensive Economic Growth

Extensive economic growth is the root cause of China's increasingly serious environmental pollution and ecological damage [121]. The Chinese government has been exploring an intensive economic growth mode that can both develop rapidly and reduce environmental losses [122]. Therefore, changing the mode of growth, adjusting the industrial structure, and promoting the greening of the national economy are fundamental ways of environmental protection in China [122]. Incorporate environmental protection into the green growth process of the national economy and increase the role and status of environmental protection in the growth of the new economy. In order to realize the greening of the national economy, one is to transform traditional industries with the concepts of green, low-carbon and recycling, and implement the green version of the "Made in China 2025" strategy; the second is to develop and provide good environmental quality and ecological services through policy support. The implementation of "Air Pollution Control Action Plan," "Water Pollution Control Action Plan," and "Soil Pollution Control Action Plan," promotes the construction of major projects, energy conservation, and consumption reduction, low-carbon economy, environmental protection industry. Circular economy has become a new pillar Industrial and economic growth points [123]. The third is to establish a green consumption model. Give full play to the role of new media such as the Internet, carry out education on ecological civilization values, popularize knowledge of green economy, promote environmental protection laws and regulations, and promote green consumption to enter schools, institutions, enterprises, communities, and families. By establishing the public's concept of environmental protection, the whole society is guided to establish a sustainable consumption or green consumption model, and to realize the greening of the entire national economic system [124].

8.4.3 Imperfect Environmental Monitoring Indicators

Industrial production has caused serious environmental pollution problems, and these pollutants will also have a long-term impact on human and social development [125, 126]. For example, humans and various organisms drink contaminated water, causing disease and death [127], and some species may even become extinct [128]. Most pollutants also enter the intestines of humans and certain animals through the food chain, and the accumulation of toxic substances can also cause serious health problems [129–131]. At present, the world has recognized the seriousness of this problem, and formulated specific and strict laws and regulations to control environmental pollution, repair environmental pollution, and realize the harmonious development of human and nature [132]. Industrial development and agricultural production have caused global environmental pollution and risks, climate anomalies, and long-term toxic effects of pollutants, all of which seriously threaten human health. Therefore, environmental managers are facing severe challenges, and all risk assessments are inherently uncertain. This depends on the quality of the input data, so scientific data sets are essential for risk assessment of new chemicals. To ensure the reliability of the risk assessment results, it is necessary to evaluate the important information and data collected. Developed countries: the USA, Japan, the European Union, and other countries and regions have all experienced their own health risk assessment development process and gradually formed a scientific risk assessment method system. Environmental monitoring is essential in world monitoring, such as air quality, water quality, and soil quality. Some newly revised laws and regulations have strengthened the monitoring of pollutant emissions, and these environmental protection regulations have also provided detailed guidance for China to improve pollution in the future. The heavy metal pollution index is taken as an example, its pollution derives from natural processes and human activities. Anthropogenic sources of heavy metal pollution include mining, smelting, fossil fuel combustion, waste disposal, corrosion, and agricultural practices. For example, industrial wastewater irrigation has caused heavy metals to contaminate large areas of arable land, and it has also contaminated millions of tons of grain in China every year. Regulatory standards for heavy metal levels have been developed for agricultural soils, but the scope is wide. Many biogeochemical properties/parameters have been used to detect the soil pollution level, such as chemical indicators, biochemical indicators (e.g. enzymatic activity), microbiological indicators (microbial biomass and microbial community structure), soil animal indicators, and plant indicators. However, the most commonly used indicator of soil heavy metal pollution is still the total/recyclable content, although the amount that can be extracted is often closely related to the uptake or availability of plants.

8.4.4 S & T Investment in Environmental Protection Needs to be Strengthened

It is necessary to strengthen scientific and technological investment in the field of environmental protection and produce excellent new technologies that can serve environmental protection. The development of new scientific research technologies and their application to environmental monitoring will lead to more sensitive and reliable environmental pollution monitoring. In 2017, Anhui Institute of Optics and Fine Mechanics, Hefei Institute of Materials Science, Chinese Academy of Sciences, launched the construction of "National Engineering Laboratory for Advanced Technologies and Equipment for Atmospheric Environmental Pollution Monitoring." With the development of science and technology, the continuous improvement of computer technology and performance, computational fluid dynamics has penetrated into many related disciplines and engineering applications. In order to meet the needs of atmospheric diffusion and environmental protection in China, using computational fluid dynamics theory to study actual engineering problems in pollutant diffusion analysis has important engineering value and social significance. Tianfu Software Co., Ltd. can use the CFD method to predict the concentration distribution and diffusion of pollutants at different heights, thus providing a useful tool for dealing with the diffusion of pollutants.

In addition, many new scientific studies will provide scientific support and guidance for better environmental governance. Harmful algal blooms are a phenomenon in lakes and rivers, recently, the deep learning models has been applied to predict algal blooms in South Korea's rivers; this method has improved advanced warnings [133]. Bacteriophages are viruses that explicitly infect bacteria in nature. Coliphages have acted as indicators of fecal pollution for water quality [134]. In our lab, we have collected the potential bacterial strains for applicants from the contaminated sediment. For example, *Pseudomonas gessardii* LZ-E or *Pseudomonas brassicacearum* LZ-4 can simultaneously degrade naphthalene and reduces hexavalent chromium [135, 136]; *Pseudomonas sp.* LZ-Q can degrade phenanthrene under hypersaline and hyperalkaline condition in a membrane bioreactor system [137]. We have proved that the nano-attapulgite clay compounded hydrophilic urethane foams as biofilm support can enrich efficient degraders, bacteria, and archaea in wastewater treatment [138]. Hg^{2+} is one kind of toxic heavy metals; the surface-engineered *E. coli* can reduce Hg^{2+} accumulation in fish muscle by modified fish gut microbiota [139].

8.4.5 Incomplete Laws

Policy deployment experts and scientists and their expertise are an important part of policy deployment. They are decision makers, and they make scientific decisions based on science and the desire for sustainable development. The environmental protection plan has been a national plan since 1972. The pollution control of the Huaihe River started from the "Ninth Five-Year Plan" and after years of continuous pollution control, certain benefits have been achieved. However, things are still tricky in some environments. It is necessary to continue to refine and formulate more specific policies and regulations in the field of environmental protection. Since 2014, environmental legislation has entered a new stage. The Environmental Protection Law, the Air Pollution Control Law, the Water Pollution Control Law, the Environmental Impact Assessment Law, and other laws have been revised, and the Environmental Protection Law has been passed.

At present, China's industrialization and urbanization process is at a new stage. The formulation of the environmental protection strategy contributes to form a new path of ecological priority and green development. The entire environmental and economic issues are increasingly intertwined, entering a complex and sensitive division of the relationship between ecology, environmental protection, and economics. Comparative studies of different impact mechanisms and action mechanisms in the short and medium term have been implemented to achieve overall economic and environmental protection. Second, Strengthen the coordination and cooperation between environmental and economic policies, social policies, resources, ecology, and the environment, and analyze and demonstrate the integration of technology and economic policies. Solve the problem of overlapping policy effects, achieve synergies, and enhance the predictability of policies. Establish a policy evaluation technology system, focus on target analysis and policy toolbox reserve research, strengthen forward-looking and predictive analysis, improve the new era system of ecological civilization, establish a scientific and efficient ecological environment policy system, and strive to improve the efficiency of environmental governance.

8.5 Future Prospects

In 2019, the United Nations Environment Programme released the sixth edition of the Global Environment Outlook at the United Nations Environment Assembly. The report pointed out that the global environmental pollution situation is very serious, and the harm to human health is increasing. If urgent measures are not taken immediately, by 2050, pollutants will affect human fertility and neurodevelopment, resulting in millions of deaths. Actively responding to environmental changes is beneficial to human health. Some efforts have been made around the world to control environmental pollution. These efforts have achieved significant results. Although environmental policies are the basis for controlling environmental pollution, many factors lead to unsatisfactory performance. Therefore, human effort is still needed.

First, we need to construct the balance between the rapid development of human society and environmental protection, so human society needs to make sustainable development one of its long-term development strategies.

Second, accelerate the research and development of environmental protection technologies and apply these technologies to environmental protection and improvement promptly. With better scientific and technical support, managing the environment and predicting potentially harmful substances is becoming increasingly useful. For example, environmental policy-making is combined with computational fluid dynamics. In addition, the combination of intestinal governance and multi-scale environmental policies also provides new directions.

In a contaminated environment, some environmental pollutants, such as antibiotics, toxic heavy metals, and some organic pollutants, may through food or drinking waters enter the human body, causing damage to human health. Human gut microbiota is a complex and dynamic ecosystem. Microbes are the most widely distributed life form on the earth, with the largest biomass and the largest number of organisms. They contain vibrant species and genetic resources, affecting the entire earth's ecosystem. The clear consensus mechanism will bring revolutionary new ideas to solve environmental problems facing human society, such as health, food, and the environment, and provide unusual solutions. Penicillin is a metabolite of a microorganism. Its discovery has saved countless people's lives and significantly improved human life expectancy. The intestinal bacteria were proved to participate in biotransformation of xenobiotics. And they can metabolize environmental chemicals [140]. Recent research indicates that gut repair can be a novel environmental restoration method that will reduce the toxic effect of the target to a certain extent. For example, scientists show that gut microflora is able to reduce Cr (VI) to Cr (III) in the protection against metal toxicity [141]. Use of antibiotics is common in livestock industries; the raised pigs can reduce the antibiotic content in their bodies by adjusting their intestinal microbial system; by adjusting the intestines of fish, the intestinal microorganisms of fish can reduce or eliminate the toxic effects of pollutants [142, 143]. There is also a crayfish breeding industry in China. We can also consider adjusting the microbial system in the crayfish to reduce the accumulation of harmful substances in the crayfish.

Third, improve scientific data collection and risk assessment of new chemicals. Chemical health risk assessment includes basic information reporting of chemical substances, environmental risk screening, investigation and monitoring of chemical substance occurrence, and risk assessment [144]. The competent ecological environment department shall formulate technical methods, procedures, and specifications for environmental risk assessment of chemical substances to establish and improve the basic database for environmental risk assessment of chemical substances. Competent authorities need to focus on controlling chemicals that are inherently hazardous, persistent, bioaccumulative, or that may exist in the environment for a long period and that pose potential dangers to the ecological environment and human health. Encourage the development and promotion of environmentally friendly alternative chemicals and alternative technologies. Support international cooperation in environmental risk assessment and control of chemical substances. These will actively prevent harmful substances from entering the environment [145].

Fourth, formulate personalized governance programs. In different countries and regions, there are various environmental problems and similar ecological pollution problems. Each country or region should propose a customized solution to their problems.

Fifth, strengthen global cooperation. Environmental protection is not restricted by national borders. The world needs to unite to actively respond to environmental pollution and environmental protection. Countries around the world should cooperate to make contributions and efforts to control environmental pollution.

References

1. Jiang Y, Guo Y, Zhang Y (2017) Forecasting China's GDP growth using dynamic factors and mixed-frequency data. Econ Model 66:132–138
2. Liang L, Wang Z, Li J (2019) The effect of urbanization on environmental pollution in rapidly developing urban agglomerations. J Clean Prod 237:117649
3. Chen B et al (2019) In search of key: protecting human health and the ecosystem from water pollution in China. J Clean Prod 228:101–111. %@ 0959-6526.
4. Sweerts B et al (2019) Estimation of losses in solar energy production from air pollution in China since 1960 using surface radiation data. Nat Energy 4(8):657–663. %@ 2058-7546

5. Huang Y et al (2019) Current status of agricultural soil pollution by heavy metals in China: a meta-analysis. Sci Tot Environ 651:3034–3042. %@ 0048-9697
6. Lelieveld J et al (2015) The contribution of outdoor air pollution sources to premature mortality on a global scale. Nature 525(7569):367–371. %@ 1476-4687
7. Glinianaia SV et al (2004) Does particulate air pollution contribute to infant death? A systematic review. Environ Health Perspect 112(14):1365–1370. %@ 0091-6765
8. Liu J, Diamond J (2008) Revolutionizing China's environmental protection. Science 319 (5859):37–38. %@ 0036-8075
9. Lu Y et al (2015) Addressing China's grand challenge of achieving food security while ensuring environmental sustainability. Sci Adv 1(1):e1400039. %@ 2375-2548.
10. Wang L (2010) The changes of China's environmental policies in the latest 30 years. Procedia Environ Sci 2:1206–1212
11. Liang W, Yang M (2019) Urbanization, economic growth and environmental pollution: evidence from China. Sustain Comput: Infor 21:1–9
12. Yao S, Zhang S, Zhang X (2019) Renewable energy, carbon emission and economic growth: a revised environmental Kuznets Curve perspective. J Clean Prod 235:1338–1352
13. Zhao X, Sun B (2016) The influence of Chinese environmental regulation on corporation innovation and competitiveness. J Clean Prod 112:1528–1536. %@ 0959-6526
14. Zhang K-m, Wen Z-g (2008) Review and challenges of policies of environmental protection and sustainable development in China. J Environ Manag 88(4):1249–1261. %@ 0301-4797
15. Song W et al (2016) Public health in China: an environmental and socio-economic perspective. Atmos Environ 129:9–17. %@ 1352-2310.
16. Ruan Y et al (2019) High doses of copper and mercury changed cecal microbiota in female mice. Biol Trace Elem Res 189(1):134–144
17. Niu X et al (2020) Has third-party monitoring improved environmental data quality? An analysis of air pollution data in China. J Environ Manag 253:109698
18. Zhang G et al (2019) The impact of the policy and behavior of public participation on environmental governance performance: empirical analysis based on provincial panel data in China. Energy Policy 129:1347–1354
19. Wu M et al (2020) Antibiotic-induced dysbiosis of gut microbiota impairs corneal development in postnatal mice by affecting CCR2 negative macrophage distribution. Mucosal Immunol 13(1):47–63
20. Wu S (2019) The evolution of rural energy policies in China: a review. Renew Sust Energ Rev 119:109584. %@ 1364-0321
21. Chang W et al (2018) China's urban minerals policies: evolution, problems and countermeasures—A quantitative research. J Clean Prod 197:114–123
22. Su M, Yang Y (2018) Evolution of district marine policies in China: the case of Shandong Province. Mar Policy 89:124–131
23. Guogang H et al (1991) China's environmental protection objectives by the year 2000. Int J Soc Econ 18:180–192
24. Bachner B (1992) Coming home to roost: pollution, law and economics in the People's Republic of China. Geo Int'l Envtl L Rev 5:635
25. Vermeer EB (1990) Management of environmental pollution in China: problems and abatement policies. China Inform 5(1):34–65
26. Jinhua S (1990) The assessment methodology for eutrophication level of lakes in China [J]. Environ Poll Cont 12(5):2–6
27. Roth D (1994) China's environomic challenge. J Environ Develop 3(2):139–145
28. Bradbury I, Kirkby R (1996) China's agenda 21: a critique. Appl Geogr 16(2):97–107
29. Florig K, Spofford W (1994) Economic incentives in China's environmental policy. Washington, DC.
30. Wang Y-b, Wang S-f, Zhou P-j (2006) Estimating water environmental carrying capacity in Shijiazhuang city using system dynamics [J]. Environ Sci Technol 29(3):26–27

31. Panitchpakdi S, Clifford ML (2002) China and the WTO: changing China, changing world trade. Wiley, Chichester
32. Xie Z (2020) China's historical evolution of environmental protection along with the forty years' reform and opening-up: moving from the control of three wastes to the construction of ecological civilization. Environ Sci Ecotech 1:100001
33. Yang Y, Yao C, Xu D (2020) Ecological compensation standards of national scenic spots in western China: a case study of Taibai Mountain. Tour Manag 76:103950
34. Liao Z (2018) Content analysis of China's environmental policy instruments on promoting firms' environmental innovation. Environ Sci Pol 88:46–51
35. Tilt B (2019) China's air pollution crisis: science and policy perspectives. Environ Sci Pol 92:275–280
36. Jin L, Tam T, Tao L (2019) Well-off but powerless? Status incongruence and psychological well-being in contemporary China. Soc Sci Med 235:112345
37. Zheng D, Shi M (2017) Multiple environmental policies and pollution haven hypothesis: evidence from China's polluting industries. J Clean Prod 141:295–304
38. Grumbine RE, Xu J (2013) Recalibrating China's environmental policy: the next 10 years. Biol Conserv 166:287–292
39. Turnhout E, Dewulf A, Hulme M (2016) What does policy-relevant global environmental knowledge do? The cases of climate and biodiversity. Curr Opin Environ Sustain 18:65–72
40. Mitchell RB et al (2006) Global environmental assessments: information and influence. MIT Press
41. Tong Q et al (2020) The roles of pollution concerns and environmental knowledge in making green food choices: evidence from Chinese consumers. Food Res Int 130:108881
42. Reinstaller A (2008) The technological transition to chlorine free pulp bleaching technologies: lessons for transition policies. J Clean Prod 16(1):S133–S147. %@ 0959-6526
43. Albrizio S, Kozluk T, Zipperer V (2017) Environmental policies and productivity growth: evidence across industries and firms. J Environ Econ Manag 81:209–226. %@ 0095-0696
44. Tang H-l, Liu J-m, Wu J-g (2020) The impact of command-and-control environmental regulation on enterprise total factor productivity: a quasi-natural experiment based on China's "Two Control Zone" policy. J Clean Prod 254:120011
45. Mol APJ (2006) Environmental governance in the information age: the emergence of informational governance. Environ Plann C 24(4):497–514. %@ 0263-774X
46. Fu H, Chen J (2017) Formation, features and controlling strategies of severe haze-fog pollutions in China. Sci Total Environ 578:121–138. %@ 0048-9697
47. Shi H et al (2016) Preventing smog crises in China and globally. J Clean Prod 112:1261–1271
48. Yang T et al (2017) Model elucidating the sources and formation mechanisms of severe haze pollution over Northeast mega-city cluster in China. Environ Pollut 230:692–700
49. Zhai Q et al (2017) Effects of subchronic oral toxic metal exposure on the intestinal microbiota of mice. Sci Bull 62(12):831–840
50. Du Y et al (2018) Direct and spillover effects of urbanization on PM2.5 concentrations in China's top three urban agglomerations. J Clean Prod 190:72–83
51. Shen Y, Ahlers AL (2019) Blue sky fabrication in China: science-policy integration in air pollution regulation campaigns for mega-events. Environ Sci Pol 94:135–142
52. Shi C, Guo F, Shi Q (2019) Ranking effect in air pollution governance: evidence from Chinese cities. J Environ Manag 251:109600
53. Fa ZRGHB (1989) Environmental protection law of the People's Republic of China: promulgated by the standing comm. Nat'l People's Cong 26:2014–2004
54. Zhou Q et al (2019) The non-linear effect of environmental regulation on haze pollution: empirical evidence for 277 Chinese cities during 2002–2010. J Environ Manag 248:109274
55. Porta M et al (2008) Monitoring concentrations of persistent organic pollutants in the general population: the international experience. Environ Int 34(4):546–561. %@ 0160-4120
56. Lau MHY et al (2012) Environmental policy, legislation and management of persistent organic pollutants (POPs) in China. Environ Pollut 165:182–192

57. Die Q et al (2015) Persistent organic pollutant waste in China: a review of past experiences and future challenges. J Mat Cycles Waste Manag 17(3):434–441
58. Söderholm P, Christiernsson A (2008) Policy effectiveness and acceptance in the taxation of environmentally damaging chemical compounds. Environ Sci Pol 11(3):240–252
59. Fang YM et al (2010) Current status and suggestions on electronic waste recycling in Guangdong. Technol Invest China 4:69–70
60. Laurent A et al (2014) Review of LCA studies of solid waste management systems—Part I: lessons learned and perspectives. Waste Manag 34(3):573–588
61. Ferronato N et al (2019) Introduction of the circular economy within developing regions: a comparative analysis of advantages and opportunities for waste valorization. J Environ Manag 230:366–378. %@ 0301-4797
62. Margallo M et al (2019) Enhancing waste management strategies in Latin America under a holistic environmental assessment perspective: a review for policy support. Sci Total Environ 689:1255–1275
63. Borthakur A (2020) Policy approaches on E-waste in the emerging economies: a review of the existing governance with special reference to India and South Africa. J Clean Prod 252:119885
64. Balde CP et al (2017) United Nations University (UNU). International Telecommunication Union (ITU) & International Solid Waste Association (ISWA), Bonn/Geneva/Vienna
65. Cobo S, Dominguez-Ramos A, Irabien A (2018) From linear to circular integrated waste management systems: a review of methodological approaches. Resour Conserv Recycl 135:279–295. %@ 0921-3449
66. E, U.C (2015) Closing the loop-An EU action plan for the circular economy. Communication from the commission to the European parliament, the council, the European economic and social committee and the committee of the regions. COM 614(2):2015
67. Geissdoerfer M et al (2017) The circular economy–A new sustainability paradigm? J Clean Prod 143:757–768. %@ 0959-6526
68. Li K, Lin B (2017) Economic growth model, structural transformation, and green productivity in China. Appl Energy 187:489–500
69. van der Kamp D, Lorentzen P, Mattingly D (2017) Racing to the bottom or to the top? decentralization, revenue pressures, and governance reform in China. World Dev 95:164–176
70. Yu Y, Yang X, Li K (2019) Effects of the terms and characteristics of cadres on environmental pollution: evidence from 230 cities in China. J Environ Manag 232:179–187
71. Cao X, Kostka G, Xu X (2019) Environmental political business cycles: the case of PM2.5 air pollution in Chinese prefectures. Environ Sci Pol 93:92–100
72. Hao Y et al (2020) Reexamining the relationships among urbanization, industrial structure, and environmental pollution in China—New evidence using the dynamic threshold panel model. Energy Rep 6:28–39
73. Liu X, Sun T, Feng Q (2020) Dynamic spatial spillover effect of urbanization on environmental pollution in China considering the inertia characteristics of environmental pollution. Sustain Cities Soc 53:101903
74. Lai Y, Tang B (2016) Institutional barriers to redevelopment of urban villages in China: a transaction cost perspective. Land Use Policy 58:482–490
75. Walder AG (1995) Local governments as industrial firms: an organizational analysis of China's transitional economy. Am J Sociol 101(2):263–301. %@ 0002-9602
76. Pham DDT, Paillé P, Halilem N (2019) Systematic review on environmental innovativeness: a knowledge-based resource view. J Clean Prod 211:1088–1099
77. Wan C, Shen GQ, Choi S (2017) A review on political factors influencing public support for urban environmental policy. Environ Sci Pol 75:70–80
78. Zheng S, Yi H, Li H (2015) The impacts of provincial energy and environmental policies on air pollution control in China. Renew Sust Energ Rev 49:386–394
79. Laplante B, Rilstone P (1996) Environmental inspections and emissions of the pulp and paper industry in Quebec. J Environ Econ Manag 31(1):19–36

80. Dasgupta S, Laplante B, Wang H et al (2002) Confronting the environmental Kuznets curve. J Econ Perspect 16(1):147–168
81. Shapiro JS, Walker R (2015) Why is pollution from US manufacturing declining? The roles of trade, regulation, productivity, and preferences (No. W20879).
82. Greenstone M (2004) Did the Clean Air Act cause the remarkable decline in sulfur dioxide concentrations? J Environ Econ Manag 47(3):585–611
83. Zhong N, Cao J, Wang Y (2017) Traffic congestion, ambient air pollution, and health: evidence from driving restrictions in Beijing. J Assoc Environ Resour Econ 4(3):821–856
84. Guo S, Lu J (2019) Jurisdictional air pollution regulation in China: a tragedy of the regulatory anti-commons. J Clean Prod 212:1054–1061
85. Wenbo G, Yan C (2018) Assessing the efficiency of China's environmental regulation on carbon emissions based on Tapio decoupling models and GMM models. Energy Rep 4:713–723
86. Qiu L-Y, He L-Y (2017) Can green traffic policies affect air quality? Evidence from a difference-in-difference estimation in China. Sustainability 9(6):1067
87. Yang W, Yuan G, Han J (2019) Is China's air pollution control policy effective? Evidence from Yangtze River Delta cities. J Clean Prod 220:110–133
88. Zhang Y et al (2019) How is the environmental efficiency in the process of dramatic economic development in the Chinese cities? Ecol Indic 98:349–362
89. Zhang W et al (2020) Environmental regulation, Foreign investment behavior, and carbon emissions for 30 provinces in China. J Clean Prod 248:119208
90. Chen B, Cheng Y-s (2017) The impacts of environmental regulation on industrial activities: evidence from a quasi-natural experiment in Chinese prefectures. Sustainability 9(4):571
91. Zhang M et al (2019) How does environmental regulation affect haze pollution governance?-An empirical test based on Chinese provincial panel data. Sci Total Environ 695:133905
92. Gil-Alana LA, Solarin SA (2018) Have U.S. environmental policies been effective in the reduction of U.S. emissions? A new approach using fractional integration. Atmos Pollut Res 9 (1):53–60
93. Martin N et al. (2016) Comparative study on the differences between the EU and US legislation on emissions in the automotive sector. European parliament, policy department a: economic and scientific policy, Tech. Rep.
94. Zhang Z (2015) Carbon emissions trading in China: the evolution from pilots to a nationwide scheme. Clim Pol 15(sup1):S104–S126. %@ 1469-3062
95. Pang T, Duan M (2016) Cap setting and allowance allocation in China's emissions trading pilot programmes: special issues and innovative solutions. Clim Pol 16(7):815–835. %@ 1469-3062
96. Yan Y et al (2020) Emissions trading system (ETS) implementation and its collaborative governance effects on air pollution: the China story. Energy Policy 138:111282
97. Shen C et al (2020) The effect of environmental policy tools on regional green innovation: evidence from China. J Clean Prod 254:120122
98. Heger M, Sarraf M (2018) Air pollution in Tehran: health costs, sources, and policies. World Bank
99. Yousefi S, Shahsavani A, Hadei M (2019) Applying EPA's instruction to calculate air quality index (AQI) in Tehran. J Air Pollut Health 4(2):81–86
100. Taksibi F, Khajehpour H, Saboohi Y (2020) On the environmental effectiveness analysis of energy policies: a case study of air pollution in the megacity of Tehran. Sci Total Environ 705:135824
101. Shen J et al (2019) Does migration of pollution-intensive industries impact environmental efficiency? Evidence supporting "Pollution Haven Hypothesis". J Environ Manag 242:142–152
102. Yuan F et al (2019) Water crisis, environmental regulations and location dynamics of pollution-intensive industries in China: a study of the Taihu Lake watershed. J Clean Prod 216:311–322

103. Zhu S, He C, Liu Y (2014) Going green or going away: environmental regulation, economic geography and firms' strategies in China's pollution-intensive industries. Geoforum 55:53–65. %@ 0016-7185
104. Tang S-Y, Lo CW-H, Fryxell GE (2003) Enforcement styles, organizational commitment, and enforcement effectiveness: an empirical study of local environmental protection officials in urban China. Environ Plan A 35(1):75–94. %@ 0308-518X
105. Larsen SV, Kørnøv L (2009) SEA of river basin management plans: incorporating climate change. Impact Assess Proj Apprais 27(4):291–299. %@ 1461-5517
106. Lobos V, Partidario M (2014) Theory versus practice in strategic environmental assessment (SEA). Environ Impact Assess Rev 48:34–46. %@ 0195-9255
107. do Nascimento Nadruz V et al (2018) Identifying the missing link between climate change policies and sectoral/regional planning supported by strategic environmental assessment in emergent economies: lessons from Brazil. Renew Sust Energ Rev 88:46–53
108. Malvestio AC, Montaño M (2013) Effectiveness of strategic environmental assessment applied to renewable energy in Brazil. J Environ Assess Policy Manag 15(02):1340007. % @ 1464-3332
109. Gallardo ALCF, Duarte CG, Dibo APA (2016) Strategic environmental assessment for planning sugarcane expansion: a framework proposal. Ambiente & Sociedade 19(2):67–92. %@ 1414-753X
110. Koricheva J, Kulinskaya E (2019) Temporal instability of evidence base: a threat to policy making? Trends Ecol Evol 34(10):895–902
111. Do Q-T, Joshi S, Stolper S (2018) Can environmental policy reduce infant mortality? Evidence from the Ganga Pollution Cases. J Dev Econ 133:306–325
112. Greenstone M, Hanna R (2014) Environmental regulations, air and water pollution, and infant mortality in India. Am Econ Rev 104(10):3038–3072. %@ 0002-8282
113. Suresh BSV et al (2007) Sewage Canal: how to clean the Yamuna. Centre for Science and Environment, New Delhi
114. Singh A (2014) Judicial activism on environment in India. Available at SSRN 2383144.
115. Guan W-J et al (2016) Impact of air pollution on the burden of chronic respiratory diseases in China: time for urgent action. Lancet 388(10054):1939–1951
116. Yang Q et al (2018) A review of soil heavy metal pollution from industrial and agricultural regions in China: pollution and risk assessment. Sci Total Environ 642:690–700
117. Yang H et al (2014) Soil pollution: urban brownfields. Science 344(6185):691–692
118. Wu L, Zhu D (2014) Food safety in China: a comprehensive review. CRC Press
119. Zhang X et al (2015) Impact of soil heavy metal pollution on food safety in China. PLoS One 10(8):e0135182
120. Chen Y et al (2018) Long-term and high-concentration heavy-metal contamination strongly influences the microbiome and functional genes in Yellow River sediments. Sci Total Environ 637:1400–1412
121. Zhang K-m, Wen Z-g (2008) Review and challenges of policies of environmental protection and sustainable development in China. J Environ Manag 88(4):1249–1261
122. Xinghua W, Weimin H (2007) Choice of economic growth styles in china and transition from the extensive style to the intensive one [j]. Econ Res J 7(42):15–22
123. Qu C et al (2016) China's soil pollution control: choices and challenges. ACS Publications
124. Seyfang G (2005) Shopping for sustainability: can sustainable consumption promote ecological citizenship? Environ Politics 14(2):290–306
125. Kanu I, Achi O (2011) Industrial effluents and their impact on water quality of receiving rivers in Nigeria. J Appl Technol Environ Sanitation 1(1):75–86
126. Lu Z-N et al (2017) The dynamic relationship between environmental pollution, economic development and public health: evidence from China. J Clean Prod 166:134–147
127. Gleick PH (2002) Dirty-water: estimated deaths from water-related diseases 2000–2020. Citeseer

128. Hua AK (2015) An indication of policy study towards water resources in Malacca state: a case study of Malacca river, Malaysia. Int Res J Soc Sci 4(6):15–20
129. Pandey G, Madhuri S (2014) Heavy metals causing toxicity in animals and fishes. Res J Animal Vet Fish Sci 2(2):17–23
130. Jiwan S, Ajah K (2011) Effects of heavy metals on soil, plants, human health and aquatic life. Int J Res Chem Environ 1(2):15–21
131. Mudgal V et al (2010) Effect of toxic metals on human health. Open Nutraceuticals J 3 (1):94–99
132. Kneese AV (1971) Environmental pollution: economics and policy. Am Econ Rev 61 (2):153–166
133. Lee S, Lee D (2018) Improved prediction of harmful algal blooms in four Major South Korea's Rivers using deep learning models. Int J Environ Res Public Health 15(7):1322
134. McMinn BR, Ashbolt NJ, Korajkic A (2017) Bacteriophages as indicators of faecal pollution and enteric virus removal. Lett Appl Microbiol 65(1):11–26
135. Huang H et al (2016) A novel Pseudomonas gessardii strain LZ-E simultaneously degrades naphthalene and reduces hexavalent chromium. Bioresour Technol 207:370–378
136. Huang H et al (2017) The naphthalene catabolic protein NahG plays a key role in hexavalent chromium reduction in Pseudomonas brassicacearum LZ-4. Sci Rep 7(1):1–11
137. Jiang Y et al (2015) Pseudomonas sp. LZ-Q continuously degrades phenanthrene under hypersaline and hyperalkaline condition in a membrane bioreactor system. Biophys Rep 1 (3):156–167
138. Jiang Y et al (2019) Using nano-attapulgite clay compounded hydrophilic urethane foams (AT/HUFs) as biofilm support enhances oil-refinery wastewater treatment in a biofilm membrane bioreactor. Sci Total Environ 646:606–617
139. Liu M et al (2019) Hg2+-binding peptide decreases mercury ion accumulation in fish through a cell surface display system. Sci Total Environ 659:540–547
140. Claus SP, Guillou H, Ellero-Simatos S (2016) The gut microbiota: a major player in the toxicity of environmental pollutants? NPJ Biofilms Microbiomes 2(1):1–11
141. Upreti R, Shrivastava R, Chaturvedi U (2004) Gut microflora & toxic metals: chromium as a model. Indian J Med Res 119:49–59
142. Kelly BC, Gobas FA, McLachlan MS (2004) Intestinal absorption and biomagnification of organic contaminants in fish, wildlife, and humans. Environ Toxicol Chem: Int J 23 (10):2324–2336
143. Zhang Z et al (2019) Ability of prebiotic polysaccharides to activate a HIF1α-antimicrobial peptide axis determines liver injury risk in zebrafish. Commun Biol 2(1):274
144. van Leeuwen CJ, Vermeire TG (2007) Risk assessment of chemicals: an introduction. Springer Science & Business Media
145. Williams ES, Panko J, Paustenbach DJ (2009) The European Union's REACH regulation: a review of its history and requirements. Crit Rev Toxicol 39(7):553–575

The manufacturer's authorised representative in the EU is Springer Nature Customer Service Centre GmbH, Europaplatz 3, 69115 Heidelberg, Germany. If you have any concerns regarding our products, please contact ProductSafety@springernature.com

Printed and bound by CPI Group (UK) Ltd, Croydon, CR0 4YY

27/07/2026

02177360-0002